Geographic Information Systems in Fisheries

Edited by

William L. Fisher

Oklahoma Cooperative Fish and Wildlife Research Unit,
U.S. Geological Survey, and Department of Zoology,
Oklahoma State University

and

Frank J. Rahel

Department of Zoology and Physiology,
University of Wyoming

American Fisheries Society
Bethesda, Maryland, USA
2004

Suggested citation formats are:

Entire book

Fisher, W. L., and F. J. Rahel, editors. 2004. Geographic information systems in fisheries. American Fisheries Society, Bethesda, Maryland.

Chapter in book

Paukert, C. P., and J. M. Long. 2004. Geographic information systems applications in reservoir fisheries. Pages 85–111 *in* W. L. Fisher and F. J. Rahel, editors. Geographic information systems in fisheries. American Fisheries Society, Bethesda, Maryland.

Cover image by Frank J. Rahel and Douglas C. Novinger.

Printed in the United States of America on acid-free paper.

Library of Congress Control Number 2003114531
ISBN 1-888569-57-3

American Fisheries Society website: www.fisheries.org

American Fisheries Society
5410 Grosvenor Lane, Suite 110
Bethesda, Maryland 20814-2199, USA

Contents

Contributors

Carolyn N. Bakelaar (Chapter 5): Fisheries and Oceans Canada, Great Lakes Laboratory for Fisheries and Aquatic Sciences, Bayfield Institute, Post Office Box 5050, 867 Lakeshore Road, Burlington, Ontario L7R 4A6, Canada; BakelaarC@dfo-mpo.gc.ca.

Timothy A. Battista (Chapter 7): National Oceanic and Atmospheric Administration, National Ocean Service Biogeography Program, 1305 East-West Highway, SSMC4, N/SCI1, Silver Spring, Maryland 20910, USA; Tim.Battista@noaa.gov.

Anthony J. Booth (Chapter 8): Department of Ichthyology and Fisheries Science, Rhodes University, Post Office Box 94, Grahamstown 6140, South Africa; t.booth@ru.ac.za.

Peter Brunette (Chapter 5): Fisheries and Oceans Canada, Great Lakes Laboratory for Fisheries and Aquatic Sciences, Bayfield Institute, Post Office Box 5050, 867 Lakeshore Road, Burlington, Ontario L7R 4A6, Canada.

Paul M. Cooley (Chapter 5): Fisheries and Oceans Canada, Freshwater Institute, 501 University Crescent, Winnipeg, Manitoba R3T 2N6, Canada.

Susan E. Doka (Chapter 5): Fisheries and Oceans Canada, Great Lakes Laboratory for Fisheries and Aquatic Sciences, Bayfield Institute, Post Office Box 5050, 867 Lakeshore Road, Burlington, Ontario L7R 4A6, Canada.

William L. Fisher (Chapters 3 and 10): Oklahoma Cooperative Fish and Wildlife Research Unit, U.S. Geological Survey, and Department of Zoology, Oklahoma State University, 404 Life Sciences West, Oklahoma State University, Stillwater, Oklahoma 74078, USA; wfisher@okstate.edu.

James McDaid Kapetsky (Chapter 6): Consultants in Fisheries and Aquaculture Sciences and Technologies (C-FAST, Inc.), 5410 Marina Club Drive, Wilmington, North Carolina 28409-4103, USA; cfast@sigmaxi.org.

James M. Long (Chapter 4): National Park Service, 1978 Island Ford Parkway, Atlanta, Georgia 30350, USA; Jim_Long@nps.gov.

Geoff J. Meaden (Chapter 2): Fisheries GIS Unit, Department of Geography, Canterbury Christ Church University College, North Holmes Road, Canterbury, Kent, CT1 1QU, UK; g.j.meaden@canterbury.ac.uk.

E. Scott Millard (Chapter 5): Fisheries and Oceans Canada, Great Lakes Laboratory for Fisheries and Aquatic Sciences, Bayfield Institute, Post Office Box 5050, 867 Lakeshore Road, Burlington, Ontario L7R 4A6, Canada.

Charles K. Minns (Chapter 5): Fisheries and Oceans Canada, Great Lakes Laboratory for Fisheries and Aquatic Sciences, Bayfield Institute, Post Office Box 5050, 867 Lakeshore Road, Burlington, Ontario L7R 4A6, Canada; minnsk@dfo-mpo.gc.ca.

Mark E. Monaco (Chapter 7): National Oceanic and Atmospheric Administration, National Ocean Service Biogeography Program, 1305 East-West Highway, SSMC4, N/SCI1, Silver Spring, Maryland 20910, USA; Mark.Monaco@noaa.gov.

Heather A. Morrison (Chapter 5): Fisheries and Oceans Canada, Great Lakes Laboratory for Fisheries and Aquatic Sciences, Bayfield Institute, Post Office Box 5050, 867 Lakeshore Road, Burlington, Ontario L7R 4A6, Canada.

Craig P. Paukert (Chapter 4): U.S. Geological Survey, Biological Resources Division, Kansas Cooperative Fish and Wildlife Research Unit, Division of Biology, 205 Leasure Hall, Kansas State University, Manhattan, Kansas 66506, USA; cpaukert@usgs.gov.

Frank J. Rahel (Chapters 1 and 3): University of Wyoming, Department of Zoology and Physiology, Box 3166, Laramie, Wyoming 82071, USA; frahel@uwyo.edu.

Kevin St. Martin (Chapter 9): Department of Geography, Rutgers University, 54 Joyce Kilmer Drive, Piscataway, New Jersey 08854-8045, USA; kstmarti@rci.rutgers.edu.

Brent Wood (Chapter 8): National Institute of Water and Atmosphere (NIWA), Fisheries Research Centre, Box 14901, Kilbirnie, Wellington, New Zealand; b.wood@niwa.cri.nz.

Preface

Geographic Information Systems in Fisheries was developed to summarize a growing body of information on applications of geographic information systems (GIS) in fisheries research and management. It is the first book of its kind that provides examples of GIS applications for all components of a fishery, (i.e., organisms, habitats, and people) in both freshwater and marine environments. The book is intended for use by fisheries students and professionals alike. We assume that readers of the book have a general background in GIS and fisheries science.

This book is written as a reference for fisheries scientists who are interested in using GIS as a tool for solving fisheries management problems. The book introduces ways GIS can be applied to fisheries, identifies challenges associated with using GIS in aquatic environments, reviews fisheries applications of GIS in freshwater (streams, rivers, lakes, and reservoirs) and marine (nearshore and offshore) environments and in aquaculture, examines GIS as a tool for fisheries decision making, and concludes with the future of GIS in fisheries.

Each chapter in this book was externally peer reviewed, and the comments of the reviewers were thoroughly considered by the authors and editors and addressed in the revisions. We thank the following reviewers for their thoughtful comments: J. Aguilar-Manjarrez, W. S. Arnold, J. S. Ault, M. Beveridge, T. D. Chi, K. Cunningham, W. L. Fisher, M. S. Gregory, S. G. Hinch, N. E. Mandrak, B. Neis, N. P. Nibbelink (2 reviews), R. L. Noble, H. Norris, M. J. Parsley, G. J. Pierce, F. J. Rahel, R. F. Reuter, E. Sheppard, S. P. Sowa, C. E. Torgersen, J. Van Sickle, and T. P. Wilber.

Several authors wish to acknowledge additional help they received in creating their chapters. William Fisher and Frank Rahel wish to thank M. J. Wiley and S. P. Sowa for providing figures for Chapter 3.

Craig Paukert and James Long thank K. Rogers from the Colorado Division of Wildlife for providing one of the figures included in Chapter 4.

James Kapetsky appreciates the many colleagues, too numerous to mention individually, that kindly responded to his requests for materials used in Chapter 6.

Anthony Booth and Brent Wood thank Quester Tangent, who granted permission to use their illustrations, and M. Potgieter, who provided valuable comments on various drafts of Chapter 8.

Kevin St. Martin would like to thank the National Marine Fisheries Service, Northeast Fisheries Science Center, for its generous support during his National Research Council associateship appointment to the center while preparing Chapter 9. He also thanks S. Edwards, M. Pavlovskaya, E. Sheppard, and other anonymous reviewers who provided valuable comments on earlier drafts of the chapter.

William Fisher is grateful to M. Gregory for his comments on Chapter 10 and the invaluable GIS technical training, support, and guidance he has provided over the years.

Since the inception of this book in 1997, the publications staff of the American Fisheries Society (AFS) has been both supportive and patient. We are deeply grateful to them for sticking with us on this book project. In particular, we thank former Book Coordinator Beth Stahle who helped get this project off the ground and former Publications Director Bob Kendall for his support of this project. We also thank those who have succeeded Beth, including Eric Wurzbacher, Janet Harry, Bob Rand, Laura Schlegel, and Deborah Lehman, for their assistance in moving the project along. We are truly indebted to Tracy Klein who, as our Book Production Coordinator, provided timely editorial advice and production ideas in the final phases of book production and Aaron Lerner, Publications Manager, for securing funding and encouraging us to complete the book. Lastly, we acknowledge former and current AFS Executive Directors Paul Brouha and Gus Rassam, respectively, for providing outstanding leadership of the Society and supporting a variety of publication outlets, including books, for publishing scientific information about fisheries.

William Fisher would like to thank the staff at the Oklahoma Cooperative Fish and Wildlife Research Unit, J. Gray, S. Lyons, and H. Murray, and Unit Leader D. M. Leslie, Jr. for their tireless assistance and support during the development of the book. He also thanks his graduate students who, over the years, have educated him about the intricacies of doing GIS. Lastly, he thanks his three sons for periodically asking, "When is the book going to be finished?" and his wife, Kim, for her loving support, patience, and encouragement to "Just get the book finished!"

Frank Rahel thanks his family for their patience and understanding during the many hours spent working on the book. A special thanks goes to his wife, Elizabeth Ono Rahel, for her help with some of the figures in Chapters 1 and 3.

Partial funding for this project was provided by a generous grant from the AFS Computer User Section and from the Oklahoma Cooperative Fish and Wildlife Research Unit. The Oklahoma Cooperative Fish and Wildlife Research Unit is jointly sponsored by the U.S. Geological Survey, Biological Resources Division; Oklahoma Department of Wildlife Conservation; Oklahoma State University; and Wildlife Management Institute.

Any product vendors and trade names mentioned in this book are for the convenience of the readers. Specific mention does not imply endorsements by the American Fisheries Society, the authors and editors, or the employers of the book's contributors.

William L. Fisher
Frank J. Rahel

Symbols and Abbreviations

The following symbols and abbreviations may be found in this book without definition. Also undefined are standard mathematical and statistical symbols given in most dictionaries.

A	ampere
AC	alternating current
Bq	becquerel
C	coulomb
°C	degrees Celsius
cal	calorie
cd	candela
cm	centimeter
Co.	Company
Corp.	Corporation
cov	covariance
DC	direct current
D.C.	District of Columbia
D	dextro (as a prefix)
d	day
d	dextrorotatory
df	degrees of freedom
dL	deciliter
E	east
E	expected value
e	base of natural logarithm (2.71828…)
e.g.	(exempli gratia) for example
eq	equivalent
et al.	(et alii) and others
etc.	et cetera
eV	electron volt
F	filial generation; Farad
°F	degrees Fahrenheit
fc	footcandle (0.0929 lx)
ft	foot (30.5 cm)
ft^3/s	cubic feet per second (0.0283 m^3/s)
g	gram
G	giga (10^9, as a prefix)
gal	gallon (3.79 L)
Gy	gray
h	hour
ha	hectare (2.47 acres)
hp	horsepower (746 W)
Hz	hertz
in	inch (2.54 cm)
Inc.	Incorporated
i.e.	(id est) that is
IU	international unit
J	joule
K	Kelvin (degrees above absolute zero)
k	kilo (10^3, as a prefix)
kg	kilogram
km	kilometer
l	levorotatory
L	levo (as a prefix)
L	liter (0.264 gal, 1.06 qt)
lb	pound (0.454 kg, 454g)
lm	lumen
log	logarithm
Ltd.	Limited
M	mega (10^6, as a prefix); molar (as a suffix or by itself)
m	meter (as a suffix or by itself); milli (10^{-3}, as a prefix)
mi	mile (1.61 km)
min	minute
mm	millimeter
mol	mole
N	normal (for chemistry); north (for geography); newton
N	sample size
NS	not significant
n	ploidy; nanno (10^{-9}, as a prefix)
o	ortho (as a chemical prefix)
oz	ounce (28.4 g)
P	probability
p	para (as a chemical prefix)

p pico (10^{-12}, as a prefix)
Pa pascal
pH negative log of hydrogen ion activity
ppm parts per million
ppt parts per thousand
qt quart (0.946 L)
R multiple correlation or regression coefficient
r simple correlation or regression coefficient
rad radian
S siemens (for electrical conductance); south (for geography)
SD standard deviation
SE standard error
s second
T tesla
tris tris(hydroxymethyl)-aminomethane (a buffer)
UK United Kingdom
U.S. United States (adjective)
USA United States of America (noun)
V volt
V, Var variance (population)
var variance (sample)
W watt (for power); west (for geography)
Wb weber
yd yard (0.914 m, 91.4 cm)
α probability of type I error (false rejection of null hypothesis)
β probability of type II error (false acceptance of null hypothesis)
Ω ohm
μ micro (10^{-6}, as a prefix)
′ minute (angular)
″ second (angular)
° degree (temperature as a prefix, angular as a suffix)
% per cent (per hundred)
‰ per mille (per thousand)

Chapter 1

Introduction to Geographic Information Systems in Fisheries

FRANK J. RAHEL

1.1 INTRODUCTION

Many of the questions that fisheries biologists ask have a spatial component such as why fish abundance, growth, survival, or catch rates vary across aquatic systems; or how the location and juxtaposition of landscape features, such as ocean currents, groundwater upwelling, or human-created migration barriers influence fish populations. Geographic information systems (GIS) are a powerful tool for displaying and analyzing such spatial data. The use of GIS in fisheries management has been gaining momentum since the pioneering studies that combined GIS and remote sensing to identify areas suitable for aquaculture (Kapetsky et al. 1988; Meaden and Kapetsky 1991). Other early studies used GIS to address fisheries management issues in marine environments (Meaden and Chi 1996). Marine fisheries biologists continue to be at the forefront of applying GIS technology for addressing fisheries issues (Nishida et al. 2001; Valavanis 2002), although the number of applications in freshwater environments is increasing.

Our goal is to provide an overview of how GIS can be used to address fishery issues in a variety of aquatic habitats. Various chapters discuss how GIS can be used to address management and ecological questions involving fish and other aquatic organisms in rivers, lakes, reservoirs, estuaries, and the open ocean environment. An effort was made to provide an overview of the kinds of questions amenable to spatial analysis by using GIS, to indicate the special considerations for collecting and analyzing GIS data in various aquatic systems, to provide examples where GIS has been used to address fisheries issues, and to suggest future uses of GIS in each system.

To become proficient in using GIS requires both an understanding of the conceptual issues involved in collecting and manipulating spatial data as well as hands-on experience with the software used to accomplish these tasks. We assume readers have a basic understanding of GIS concepts and have focused our attention on how these concepts can be applied to the management and conservation of aquatic resources. Readers desiring more background on GIS and its application in natural resource management are directed to other books or

Web-based resources on this subject (e.g., Johnston 1998; Longley et al. 2001; ESRI 2003).

1.2 WHAT IS GIS?

The acronym GIS stands for geographic information systems, which is defined as a collection of computer hardware, software, data, and personnel designed to collect, store, update, manipulate, analyze, and display geographically referenced information. As this definition indicates, there is no single GIS system. Computer hardware can refer to desktop or notebook computers, personal digital assistants, or even cellular phones, and there are a number of commercial vendors selling software programs (Longley et al. 2001). Data can be collected for a particular project in the field (e.g., fish abundance and habitat characteristics at a sampling location) or can be obtained from commercial or government sources (e.g., elevation maps, drainage networks, or land-use maps). Personnel often involve specialists with academic degrees emphasizing GIS, although natural resource biologists are increasingly being exposed to GIS as part of their academic training (Fisher and Toepfer 1998).

What separates a GIS from other computer programs, such as statistical packages and spreadsheets, is that a GIS permits spatial operations on the data. To illustrate the difference between aspatial and spatial analyses, consider the data in Table 1.1 and Figure 1.1. An aspatial analysis would involve using the data to determine how fish abundance is related to habitat features, without reference to the geographic location of individual sites. To answer this question, we could perform a regression analysis relating fish abundance to the number of pools and amount of undercut bank. Such fish-habitat studies are common in the fisheries literature and are aspatial because they do not consider how the geographic location of sites influences fish abundance. By contrast, a spatial analysis would involve determining if the geographic location of the site also influenced fish abundance. This would occur if unaccounted for factors that vary spatially across the study area also influence fish abundance. For example, surficial geology can determine how stream habitat conditions and elevation affect water temperatures (Nelson et al. 1992; Rahel and Nibbelink 1999). We could use the Universal Transverse Mercator (UTM) coordinates as variables that reflect the north-south and east-west position of a site and redo the regression analysis to see if spatial location influenced fish abundance. Another type of spatial analysis involves asking how the location of each site relative to other habitat features influences fish abundance. We might hypothesize that stream sites close to roads experience higher fishing pressure and thus have fewer fish than sites located far from roads (Bailey and Hubert 2003). By overlaying a map of the study drainage with a road map, we could calculate the distance of each site from the nearest road. This would allow us to add road distance as a predictor variable in our regression analysis. Furthermore, roads contribute sediments and highway

Table 1.1 An example of the spatial aspects of fisheries data. Fisheries and habitat data were collected at 10 sites numbered as in Figure 1.1. At each site, the abundance of trout greater than 15 cm was estimated by depletion-removal electrofishing for a 100-m reach of stream. Habitat data included the number of pools per 100 m of stream and the percentage of stream bank that was undercut. The Universal Transverse Mercator (UTM) coordinates (N = Northing; E = Easting) were recorded at the midpoint of each reach and were used to calculate the distance to the nearest road. The UTM coordinates are for Zone 13 of the northern hemisphere.

Site number	Number trout/m^2	Number pools/100 m	% undercut banks	UTM coordinates N	UTM coordinates E	Road distance (m)
1	0.04	5	20	4,513,410	587,310	200
2	0.07	1	10	4,510,030	587,998	3,544
3	0.05	5	19	4,441,702	595,005	80
4	0.21	7	24	4,402,759	590,020	3,257
5	0.01	0	2	4,446,281	602,095	160
6	0.20	3	8	4,428,362	610,126	3,554
7	0.18	3	9	4,430,389	612,642	3,623
8	0.11	3	25	4,448,636	641,332	4,359
9	0.10	4	20	4,511,036	645,983	5,889
10	0.16	3	15	4,442,751	650,118	6,572

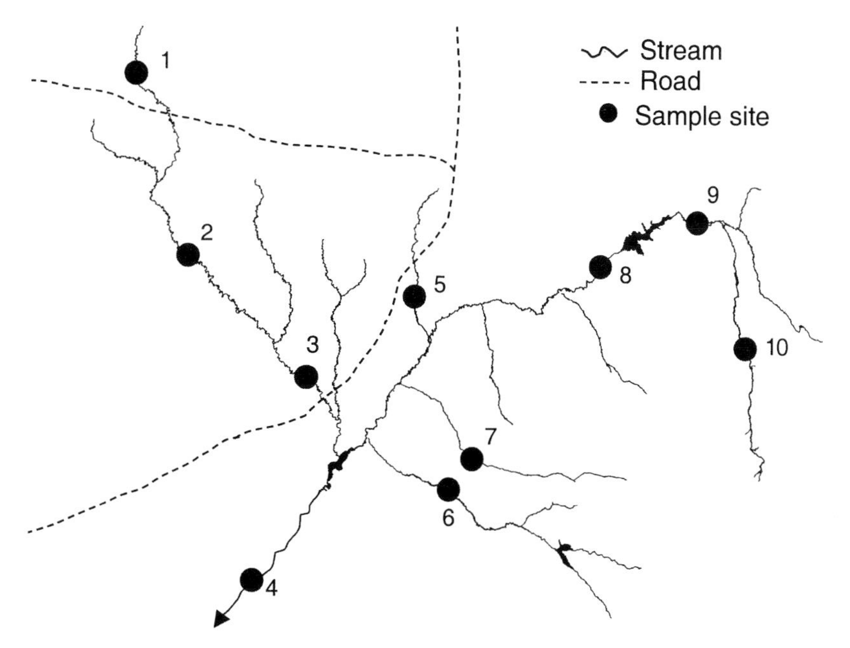

Figure 1.1 The locations of stream reaches where fish sampling was done in a drainage basin. Also shown are roads in the basin. Numbers refer to site identification codes given in Table 1.1.

chemicals to streams that can be detrimental to fish populations, so we also might want to consider whether sites are located upstream or downstream of a road crossing in our analysis. These examples illustrate that geographic location and juxtaposition to other features can be important components of a spatial analysis.

The term GIS as used in this book refers to geographic information systems, that is, the combination of hardware, software, data, and the people that use them to analyze geographically referenced data. A related concept is geographic information science, which is defined as the science dealing with the generic issues that surround the use of GIS technology, impede its successful implementation, or emerge from an understanding of its potential capabilities (Goodchild 1992). Examples of issues addressed by geographic information science include the following: how does a GIS user know that the results are accurate, what principles might help a GIS user design better maps, and how can user interfaces be made more readily understandable by novices (Longley et al. 2001). The formalization of research questions about the use of GIS was an important step because it transformed GIS from being mainly a set of analytical tools into a science concerned with developing better ways to analyze and visualize geographically referenced data.

1.3 FISHERIES DATA IN A GIS FORMAT

Spatially explicit data of interest to fisheries biologists can be classified into two main categories: features and surfaces. Features are discrete objects that can be depicted as points (e.g., fish telemetry locations, gill net locations), lines (e.g., streams, trawl transects), or polygons (e.g., lakes, habitat types in an estuary). Although we may sometimes have to impose arbitrary boundaries in defining features (e.g., habitat types might transition into one another, and we may have to define boundaries between types), once we have agreed on how to define a feature, it can be portrayed as a discrete entity in geographic space. By contrast, surfaces (also called field variables) exist as continuous variables across geographic space. Elevation and rainfall amount are examples of field variables in terrestrial environments. In aquatic systems, water depth would be a field variable. In some cases, we might want to portray a field variable as a feature, that is, by creating a few water-depth categories, such as shallow, intermediate, and deep water, on the basis of particular depth intervals.

Features usually are depicted in a GIS by using vector data (Longley et al. 2001). Data types that can be represented in a vector system include point, line, and polygon features (Figure 1.2). Points are considered zero-dimensional data and are used to represent discrete objects at the particular scale of interest. Examples are fish sampling sites, locations of redds, locations of dams along a drainage network, and locations of individual fish determined by radio telemetry. Lines are considered one-dimensional data, and common examples are stream drainage networks, roads, and trawling

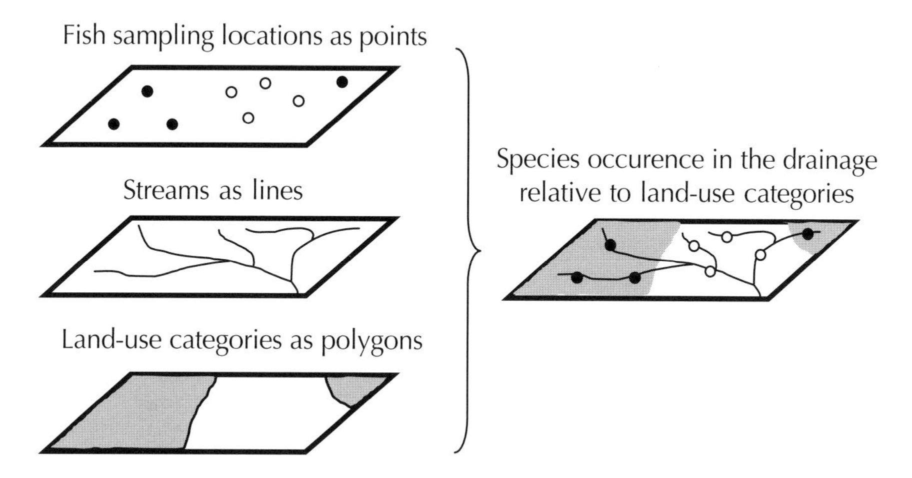

Figure 1.2 Examples of three types of data (points, lines, and polygons) that can be depicted in vector format. Fish sampling locations are represented by points, coded to indicate if the fish species of interest was present (solid circle) or absent (open circle). The stream drainage network is represented by lines. Land-use categories are depicted as polygons, where shaded polygons represent forested areas and unshaded polygons represent crop production areas. Overlays of the coverages can be used to reveal patterns in the data, such as the absence of the fish species in areas where land use involves crop production.

transects. Polygons are two dimensional and enclose an area considered homogenous with respect to some important characteristic. Examples include land-use categories across a watershed, areas of discrete habitat types such as macrophyte beds in a lake or coral reefs in the ocean, or the shape of marine preserves off a coastline. Contour lines are another common type of data that can be connected to form regions having the same value of some feature. Common examples in fisheries data are bathymetry maps or maps showing current velocity categories.

Feature data can change types depending on the nature of the questions being asked. For example, fish sampling sites could be portrayed as points if we are mainly interested in determining if there are any major gaps in our sampling of fish assemblages across a large watershed (Figure 1.3). The same sites could be portrayed as line segments if we are interested in quantifying the proportion of the total stream length sampled in a subwatershed. Finally, if we are interested in quantifying the area of various habitat features within a study reach, we might depict the reach and its habitat patches as polygons.

Features have locations within a GIS; they can be located by using a vector format based on an x,y coordinate system, such as latitude and longitude or the UTM coordinate system (Clarke 1999). The location of a point feature is defined by a pair of x,y coordinates. A straight line (also referred to as an arc) needs two pairs of coordi-

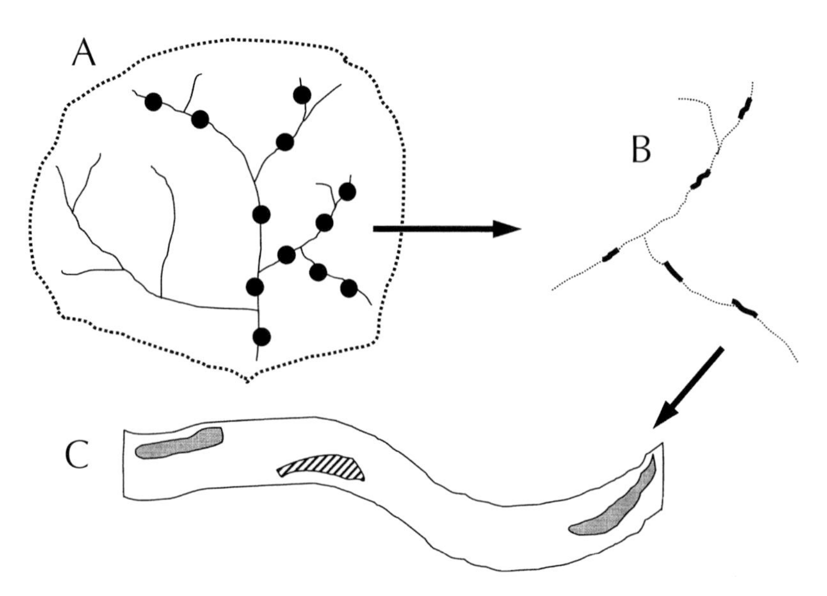

Figure 1.3 The location of sample sites across a basin can be depicted by using points, line segments, or polygons depending on the spatial scale of interest. At the broadest spatial scale, depicting sample reaches as points (A) would indicate areas that have not been sampled. Within a subwatershed, depicting sample sites as line segments (B) would allow calculation of the proportion of the stream network sampled and the proportion of the stream network having various channel features (e.g., in various gradient categories). Depicting a stream reach as a polygon (C) would allow calculation of the proportion of the reach area in various habitat categories, such as macrophyte beds (shaded polygons) or woody debris jams (cross-hatched polygon).

nates, one at the beginning and one at the end. The beginning and ending points are referred to as nodes. If the line bends, such as when depicting a river, then there must be a pair of coordinates at every location where the line changes direction (i.e., at each vertex). A polygon is a line that returns to its starting point and thus encloses an area. To describe the location of a polygon, we need only to specify the points that form its vertices.

Surfaces have no distinct boundaries in the sense of a polygon but rather have measurable values for any particular geographic location. Surfaces usually are depicted in a GIS by using a raster format (Ormsby et al. 2001). A raster format consists of a matrix of square cells (also called pixels), with each cell assigned a value for the field variable of interest. A commonly used raster data type is a digital elevation model, whereby the earth's surface is divided into cells (e.g., 30 m × 30 m), and each cell is assigned an elevation. This allows elevation to be mapped as a surface, that is, the surface of the earth, across the study area. Pixels can vary in size, and, whereas, small pixels capture more fine-scale variation in the surface being described, they also require much more data collection and storage. Most of the data collected by remote sensing is in a raster format, including satellite (e.g., Landsat Thematic Mapper imagery) and aerial photography data (e.g., digital orthophoto quads).

In a GIS, data for a particular characteristic is stored in a layer, also referred to as a coverage or theme. Information about the features in a layer is stored in an attribute table. The table has a record (row) for each feature in the layer and a field (column) for each category of information. Features on a GIS map are linked to information in their attribute table such that if you highlight the feature on a map, you can bring up all the information stored about it in the attribute table. Table 1.1 is an example of an attribute table where the features are sample sites and attributes include information on the location, abundance of fish species, habitat features, and distance to the nearest road for each site

1.4 HOW GIS CAN BE USEFUL TO FISHERIES BIOLOGISTS

A major benefit of using a GIS is the ability to overlay coverages, just as one might overlay map transparencies, to look for emerging patterns. Thus, we could overlay maps showing (a) fish sampling locations in a region, (b) the stream network, and (c) land-use categories in the basin in order to identify patterns in the distribution of a particular species in relation to land use (Figure 1.2). Examples of using map overlays to ask questions about fish-habitat associations include Figure 4.8 in Chapter 4 and Figure 5.6 in Chapter 5. A GIS also can be used to delineate buffers, that is, the space located within a given distance of a feature (Ormsby et al. 2001). Buffers are useful for examining spatial proximity. For example, buffers could be used to identify fishing grounds located within various travel times from a port (Caddy and Carocci 1999) or portions of a reservoir within a certain distance from shore (Chapter 4, Figure 4.4). Buffers also could be used to examine the influence of land use measured at different scales (i.e., various widths of the riparian zone up to the entire watershed) on stream water quality and biotic integrity (Allan et al. 1997).

A GIS especially is useful for asking four types of questions (ESRI 1990; Isaak and Hubert 1997). First, what are the attributes associated with a particular location? This involves linking the location with an attribute table. We could point to a trawl location on a map and obtain information about the date and time the trawling was done, the type of net used, which species were captured, water depth, water temperature, and salinity. We also can overlay coverages to delineate site conditions for a variety of features or surfaces.

A second type of question we can ask by using GIS is which areas or features meet a set of criteria. For example, we might want to identify the locations of trawls that captured a particular species at night during the summer. We then could compare maps showing the locations of trawls that captured the species at various time periods and seasons to gain insights into the diel and seasonal habitat use of the species. Often questions of this nature are answered by overlaying coverages, such as when using GIS to identify sites for aquaculture development. Optimal sites must have a set of abiotic conditions suitable for the organism to be cultured and to be in close proximity to markets (see Chapter 6). Another example is using GIS to identify the best sites for revegetation in a reservoir based on water depth, slope, water clarity, and substrate type (Chapter 4, Figures 4.7, 4.8).

Third, we can ask what spatial patterns exist in the data. Often this involves determining if features of interest are aggregated spatially or how spatial proximity to other features influences the attributes of a feature. We could plot the locations of all trawls to see if certain parts of the study area were not sampled adequately or if trawls capturing a certain species were spatially aggregated near rare habitat features such as underwater reefs, upwelling areas, or river inflows (Chapter 8, Figure 8.5). Zheng et al. (2002) used GIS to determine how the North Atlantic current affected the spatial distribution of an important commercial fish species.

Fourth, we can use GIS data sets in modeling efforts to ask "what if" questions. For example, if we know that a fish species is associated with certain thermal conditions and we can model how thermal conditions will change with climate warming, we can ask how the amount of thermally suitable habitat will change for various increases in water temperatures (e.g., Chapter 3, Figures 3.5, 3.8). Another example involves projecting changes in suitable habitat for fish in reservoirs at various water-level scenarios (Irwin and Noble 1996; Chapter 4, Figure 4.1). A GIS also has great potential for modeling current fish distributions or abundance patterns based on an understanding of habitat relations (Toepfer et al. 2000; Stoner et al. 2001; Filipe et al. 2002).

1.5 APPLICATIONS OF GIS IN THE MANAGEMENT AND CONSERVATION OF FISHES

This book describes the applications of GIS in the management and conservation of fishes. Chapter 2, "Challenges of Using Geographic Information Systems in Aquatic Environments" by G. J. Meaden, provides an overview of the issues involved in collecting and analyzing data from aquatic ecosystems for GIS-based applications. The challenges are discussed in relation to four main categories: intellectual and theoretical, practical and organizational, economic, and social and cultural (Chapter 2, Figure 2.1). A key message is that many of the biotic and abiotic factors important in fisheries are in a dynamic state and operate at multiple spatial and temporal scales. This makes it difficult to link biotic response variables to abiotic driver variables, especially for mobile organisms that are choosing habitat on the basis of a suite of abiotic factors (e.g., temperature, salinity, light levels) and biotic factors (prey availability, predation risk).

Chapters 3, 4, and 5 describe the application of GIS for addressing issues in freshwater environments encompassing streams, reservoirs, and lakes. These chapters describe the techniques used to collect spatially referenced data, examine the problems associated with analyzing such data, and provide examples of using GIS for addressing fisheries issues in each system. Chapter 3, "Geographic Information Systems Applications in Stream and River Fisheries" by W. L. Fisher and F. J. Rahel, discusses systems that are often depicted as linear habitat features in a GIS. A key aspect of GIS applications in riverine environments is the linkage of stream characteristics with watershed characteristics, building on Hynes's (1975) observation that "in every respect, the valley rules the stream." Chapter 4, "Geographic Information Systems Applications in Reservoir Fisheries" by C. P. Pauket and J. M. Long, discusses systems that are hybrids between flowing water habitats (i.e., rivers) and standing water habitats (i.e., lakes). Water-level fluctuations are a key attribute of reservoir habitats, and an important application of GIS is to quantify changes in fish habitat with changes in water level. In Chapter 5, "Geographic Information Systems Applications in Lake Fisheries," authors C. N. Bakelaar, P. Brunette, P. M. Cooley, S. E. Doka, E. S. Millard, C. K. Minns, and H. A. Morrison note that GIS can play a role in addressing four interconnected topics related to the management of lake fish resources: What factors determine the composition and productivity of fish resources? How do habitat features shape the spatial and temporal distribution and dynamics of lake fish stocks? How should fish exploitation be managed to ensure the sustainability and socioeconomic benefits of recreational and commercial fishing? What conservation and protection measures are required in lakes to ensure the continued diversity and productivity of fish resources? Large lakes pose many of the same challenges as the oceans in that we often must rely on remote sensing to collect data on organism abundance and abiotic conditions.

Chapter 6, "Geographic Information Systems Applications in Aquaculture" by J. M. Kapetsky, discusses one of the more applied uses of GIS technology. A key issue in developing aquaculture is to ensure its sustainability, which encompasses the requirements of the organism to be cultured, the responses of the ecosystem in which aquaculture will occur, and the socioeconomic and political climate of the region. All of these factors have spatial elements that make them amenable to GIS analyses for identifying appropriate sites for aquaculture development.

The next two chapters deal with GIS use in marine environments. Chapter 7, "Geographic Information Systems Applications in Coastal Marine Fisheries" by T. A. Battista and M. E. Monaco, focuses on a type of ecosystem that increasingly is being impacted by humans through urbanization, fishing activities, and nonpoint-source water pollution. Among the issues and problems that can be addressed by using GIS are defining important fishery habitats in space and time; simulating impacts to living marine resources from various management alternatives; and defining marine protected areas based on biological, economic, and political considerations. Because of logistical limitations and high cost, there is a limited ability to collect and manage fine-grained data over the large geographic areas of interest to managers in coastal environments and the open ocean. Consequently, spatial interpolation to create surfaces from point observations is an important application of GIS in these environments. The vastness of the open ocean and the lack of identifiable landmarks are among the challenges of working in deepwater marine environments as noted by A. J. Booth and B. Wood in Chapter 8, "Geographic Information Systems Applications in Offshore Marine Fisheries." Another challenge is that many important issues related to fishery resources involve a three-dimensional environment that is difficult to portray in the two-dimensional format of maps. In addition, features of interest to fishery managers, such as abiotic conditions or the location of fish schools, can change so quickly that it is difficult to collect and analyze data rapidly enough to be of use in management decisions. Despite these challenges, GIS has great potential use in identifying areas with appropriate conditions for particular fish stocks and for managing fishing effort in a cost-effective and ecologically sound manner.

Chapter 9, "Geographic Information Systems in Marine Fisheries Science and Decision Making" by K. St. Martin, discusses GIS use from a sociological and political perspective. Implementation of GIS technology in managing fisheries resources has been constrained by several factors, including the characteristics of fisheries resources (e.g., the mobility of fish stocks and the three-dimensional nature of the ocean environment), socioeconomic limitations (such as the level of GIS education and financial resources available to purchase technology), and the fact that spatial domains often are based on statistical and political criteria rather than ecological boundaries. Historically, the dominant mode of fisheries management was based on a homogenous notion of space that is well suited for the calculation of aggregate numbers of fish and fishing effort for large management areas. St. Martin sees management of commercial marine fisheries as moving toward a focus on communities (rather than individual fishers) in allocating resources and a focus on ecosystem well being (rather than single species abundances) in setting harvest limits. This paradigm shift will likely place a higher priority on the spatial aspects of fisheries and thus provide opportunities to use GIS to identify and analyze heterogeneous processes across a diverse landscape.

The final chapter, "Future of Geographic Information Systems in Fisheries" by W. L. Fisher, focuses on major growth areas in the use of GIS in fisheries and aquatic sciences. Undoubtedly, one of biggest growth areas will be in linking GIS systems and remotely sensed data. To date, we have tended to focus our efforts at small spatial scales that are amenable to direct sampling by humans (e.g., measuring habitat conditions across stream transects) or at very large scales where data can be collected remotely by satellites (e.g., Landsat Thematic Mapper imagery). It has proven more difficult to acquire data for intermediate scales, for example, at the level of stream segments encompassing 10^3 to 10^5 m (Fausch et al. 2002). Recent developments in remote-sensing technology such as airborne thermal remote sensing and high spatial resolution hyperspectral imagery will allow us to examine fish habitat relations at a variety of spatial scales. Perhaps the biggest trend in the geospatial industry will be the distribution of GIS data and software applications on the Internet. The continued development of Internet map services will enable widespread access to data on the Web, and this democratization of GIS will facilitate policy decisions based on input from fishers, fisheries scientists, and the public.

1.6 REFERENCES

Allan, J. D., D. L. Erickson, and J. Fay. 1997. The influence of catchment land use on stream integrity across multiple spatial scales. Freshwater Biology 37:149–161.

Bailey, P. E., and W. A. Hubert. 2003. Factors associated with stocked cutthroat trout populations in high-mountain lakes. North American Journal of Fisheries Management 23:611–618.

Caddy, J. F., and F. Carocci. 1999. The spatial allocation of fishing intensity by port-based inshore fleets: a GIS application. ICES Journal of Marine Science 56:388–403.

Clarke, K. C. 1999. Getting started with geographical information systems. Prentice-Hall, Upper Saddle River, New Jersey.

ESRI (Environmental Systems Research Institute). 1990. Understanding GIS, the ARC/INFO method, PC version. Environmental Systems Research Institute (ESRI), Inc., Redlands, California.

ESRI (Environmental Systems Research Institute). 2003. ESRI GIS and mapping software. ESRI. Available: *www.esri.com* (September 2003).

Fausch, K. D., C. E. Torgersen, C. V. Baxter, and H. W. Li. 2002. Landscapes to riverscapes: bridging the gap between research and conservation of stream fishes. BioScience 52:483–498.

Filipe, A. F., I. G. Cowx, and M. J. Collares-Pereira. 2002. Spatial modeling of freshwater fish in semi-arid river systems: a tool for conservation. River Research and Applications 18:123–136.

Fisher, W. L., and C. S. Toepfer. 1998. Recent trends in geographic information systems education and fisheries research applications at U.S. universities. Fisheries 23(5):10–13.

Goodchild, M. F. 1992. Geographical information science. International Journal of Geographical Information Systems 6:31–45.

Hynes, H. B. N. 1975. The stream and its valley. Verhandlungen der Internationalen Vereinigung fur Theoretisch und Angewandte Limnologie 19:1–15.

Irwin, E. R., and R. L. Noble. 1996. Effects of reservoir drawdown on littoral habitat: assessment with on-site measures and geographic information systems. Pages 324–331 in L. E. Miranda and D. R. DeVries, editors. Multidimensional approaches to reservoir fisheries management. American Fisheries Society, Symposium 16, Bethesda, Maryland.

Isaak, D. J., and W. A. Hubert. 1997. Integrating new technologies into fisheries science: the applications of geographic information systems. Fisheries 22(1):6–10.

Johnston, C. A. 1998. Geographic information systems in ecology. Blackwell Scientific Publications, Oxford, UK.

Kapetsky, J. M., J. M. Hill, and L. D. Worth. 1988. A geographical information system for catfish farming development. Aquaculture 68:311–320.

Longley, P. A., M. F. Goodchild, D. J. Maguire, and D. W. Rhind. 2001. Geographic information systems and science. Wiley, Chichester, UK.

Meaden, G. J., and T. D. Chi. 1996. Geographical information systems: applications to marine fisheries. FAO Fisheries Technical Paper 356.

Meaden, G. J., and J. M. Kapetsky. 1991. Geographical information systems and remote sensing in inland fisheries and aquaculture. FAO Fisheries Technical Paper 318.

Nelson, R. L., W. S. Platts, D. P. Larsen, and S. E. Jensen. 1992. Trout distribution and habitat in relation to geology and geomorphology in the North Fork Humbolt River drainage, northeastern Nevada. Transactions of the American Fisheries Society 121:405–426.

Nishida, T., P. J. Kailola, and C. E. Hollingworth, editors. 2001. Proceedings of the first international symposium on GIS in fishery science. Fishery GIS Research Group, Saitama, Japan.

Ormsby, T., E. Napoleon, R. Burke, C. Groess, and L. Feaster. 2001. Getting to know ArcGIS: basics of ArcView, ArcEditor, and ArcInfo. ESRI Press, Redlands, California.

Rahel, F. J., and N. P. Nibbelink. 1999. Spatial patterns in relations among brown trout (*Salmo trutta*) distribution, summer air temperature, and stream size in Rocky Mountain streams. Canadian Journal of Fisheries and Aquatic Sciences 56(Supplement 1):43–51.

Stoner, A. W., J. P. Manderson, and J. P. Pessutti. 2001. Spatially explicit analysis of estuarine habitat for juvenile winter flounder: combining generalized additive models and geographic information systems. Marine Ecology Progress Series 213:253–271.

Toepfer, C. S., W. L. Fisher, and W. D. Warde. 2000. A multi-stage approach to estimate stream fish abundance using geographic information systems. North American Journal of Fisheries Management 20:634–645.

Valavanis, V. D. 2002. Geographic information systems in oceanography and fisheries. Taylor and Francis, New York.

Zheng, X., G. J. Pierce, D. G. Reid, and I. T. Jolliffe. 2002. Does the north Atlantic current affect spatial distribution of whiting? Testing environmental hypotheses using statistical and GIS techniques. ICES Journal of Marine Sciences 59:239–253.

Chapter 2

Challenges of Using Geographic Information Systems in Aquatic Environments

Geoff J. Meaden

2.1 INTRODUCTION

It is pertinent to commence this chapter with some definitions. In keeping with the substance of the book, "aquatic environments" will be confined to the watery milieus associated with fish or fisheries, though these could be both natural environments (marine, lacustrine, or riverine) or artificial environments (for aquaculture or mariculture). Though a "challenge" might seem to be an obvious concept, in view of its wide range of meanings, discussion here is limited to "difficulties that stimulate interest or effort." In the combined case of "fisheries" and "aquatic environments," it is likely that the sum of difficulties will be rather large. It is further clear that there is a sense in which "challenges" have both negative and positive connotations. Though it is the intention to accentuate the latter, if this chapter engenders a catalog of problems, then apologies are advanced! The challenges certainly are being approached in a spirit of giving pointers to achieving the possible, though it also will be made clear that they constitute a wide range of considerations and problems. Finally, in terms of definitions, there are areas of "challenge" that, although vital to the future of the use of geographic information systems (GIS) in fisheries science or management, cannot be considered because they are at the boundary of the overall theme. These include functional and technical areas such as database models and database management, image analysis, hardware and software use and implications, plus copyright factors, as well as human areas such as the retention of highly skilled workers within the research ambience and the acceptance of GIS as a legitimate and useful technology.

The existence of a challenge implies that a goal is being sought. In the case of "fisheries," there exists a wide range of actual and potential goals that vary according to local spatial and temporal manifestations of the fisheries situation. Thus, without going into detail, it will be clear that fish may vary in their abundance from "excellent" or "widespread" to "depleted" or "highly localized," and the causes for this are diverse and complex. However, the goals can be reasonably aggregated into a single overall aim that might be to "create a situation whereby human aspirations for fisheries are maximized and sustained." In order to meet this overall aim, any of a number of management objectives and actions may be necessary. It might be important to set up data collection systems to quantify various facets of the fishing activity; data on local ecosystems may be needed; fishing activities themselves may need to be organized and rationalized; employment opportunities may need to be maximized; and various technical controls may need implementing. But the overall aim in a fishery is to achieve the difficult equilibrium whereby fish numbers are maintained at a level such that some measure of output can be maximized and sustained. For a range of reasons, this overall aim currently is not being achieved in most fisheries. Reasons include factors such as "destruction of the substrate ecology," "setting unrealistic quotas," "netting too much bycatch," or simply "overfishing specific grounds." These are the problems that GIS can help solve. The purpose of this chapter is not to detail these problems but to identify and define the "operational challenges" that presently inhibit the use of GIS in solving them.

The introduction of this book makes it clear that the subject area of "fisheries GIS" incorporates a wide range of topics and that the operation of GIS within the thematic area of "aquatic environments" is indeed a considerable challenge, because many biotic and abiotic factors important in fisheries are in a constantly dynamic state and operate at multiple spatial and temporal scales. This intrinsic difficulty has contributed to the relatively slow development of "fisheries GIS" relative to terrestrial-based GIS. Much of the GIS work presently underway is directed toward comparatively simple mapping tasks of basic cartography, overlaying, buffer generation, and application of some statistical and geostatistical analyses. Most of the answers presently being gathered are relatively minor insights into localized problems. However, many fisheries problems involve complex interlinkages at multiple scales. Interestingly, many of the same complex array of reasons that hinder the growth of fisheries GIS are the same as those that have contributed to the demise of many of the world's fisheries. Here we may consider factors such as data gathering costs, data interpretation, the range of variables and processes involved, and fragmented management. Hence, an important reason for meeting the challenges confronting work in fisheries GIS, is that the output derived will both directly and indirectly go some way toward improving the situation in many inland and marine fisheries.

A further reason for confronting the challenges is that, as fisheries are currently organized, it is often scientists who are made the scapegoats when things go wrong in the complex marine and fisheries biorealms. Since, in GIS, we now have a methodology with the potential to provide answers to a range of fisheries-related problems, and since it is mainly scientists who are running the GIS, it is imperative that we are successful in confronting challenges imposed by this burgeoning technology. It is important that scientists demonstrate that they can supply sensible answers to the panoply of problems and that the burden of responsibility can be shifted to users and politicians where it should rest. By using our systems wisely and imaginatively, the potential to overcoming the spatiotemporal challenges can at least be demonstrated.

There are many ways in which the challenges facing fisheries GIS could be discussed. Herein, headings have been used that reflect GIS implementation procedures, though it also seems appropriate to consider some intellectual and theoretical "abstract realities" contributing to challenges. A schematic diagram (Figure 2.1) portrays the challenges as ordered in this chapter. It is important to make clear

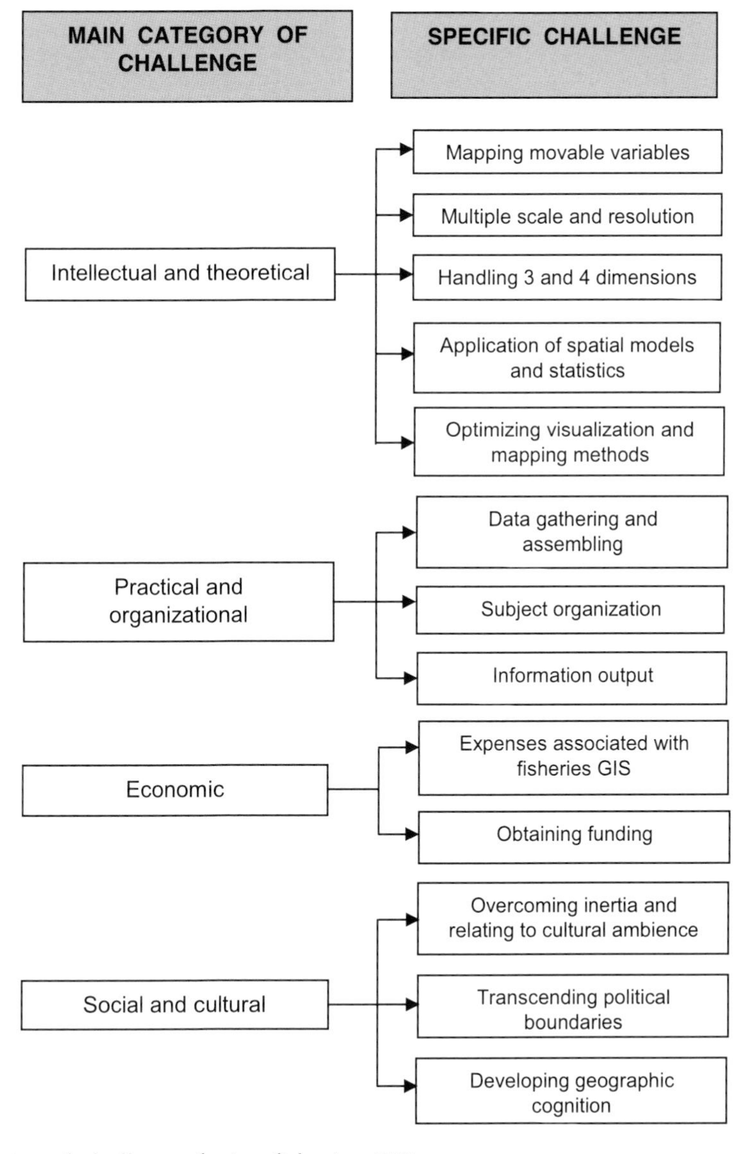

Figure 2.1 Categories of challenge facing fisheries GIS.

that, although challenges have been compartmentalized, there are, in practice, many linkages between them. In most cases, challenges are discussed in terms of both their nature and possible ways of overcoming the inherent problems implied. Most examples used come from the so-called gray literature. This is because relatively little published work appears in the peer-reviewed literature, probably as a result of the applied nature of the subject and because of the integrative nature of GIS-based approaches that may lie at the interface of several scientific and technical domains. Particular use is made of recent conference contributions in the fisheries GIS area, particularly those of the First International Symposium on GIS in Fishery Science (Seattle, Washington, March 1999), the 17th Lowell Wakefield Fisheries Symposium (Anchorage, Alaska, October 1999), and the Second International Symposium on GIS/Spatial Analyses in Fishery and Aquatic Sciences (University of Sussex, Brighton, UK, September 2002).

2.2 INTELLECTUAL AND THEORETICAL CHALLENGES

In the sense that what is being considered here is mostly in the applications domain, there are few theoretical challenges to fisheries GIS per se. In practice, many of the intellectual challenges are concerned with applying and adopting developments that have been worked on in the terrestrial terrain to aquatic situations. However, there are major differences from terrestrial applications in three main senses:

(1) Since most of the aquatic environment is in constant motion, temporal considerations play a far greater role in any GIS activity. This is because analyses must involve change in spatial dispositions through time.

(2) Unlike terrestrial GIS, where considerations of spatial distributions are mostly located in 2-dimensional (D) or 2.5-D space, aquatic distributions are nearly all located in 2.5-D or 3-D space. (2.5-D space refers to considerations where objects are located in both the horizontal and vertical planes but where they are positioned on the land surface, or the sea-bed, i.e., altitude is considered.)

(3) In terrestrial domains, classification (and, therefore, spatial delimitation) is made easier through the existence of many anthropogenic "boundaries," as well as natural demarcations such as river courses, coastlines, ridges, and slope breaks. In the marine domain, the frontier between most classification zones is inevitably "fuzzy," in that precise boundaries are rare or temporary, for example, a boundary current or "the shelf edge."

In a recent study, Meaden (2000) outlined a simple three-stage conceptual model explaining why the adoption of GIS for fisheries-related purposes might be a theoretical challenge. A major consideration identified was that any fisheries GIS project must involve a wide range of task components. Whereas a terrestrial GIS project is concerned with (1) "objects" distributed in (2) a "horizontal plane" at any particular (3) "time," a fisheries GIS project must include these components plus (4) the "vertical plane," plus some (5) "movement processes," and (6) the "dynamics" (speed and direction) of both the "objects" and the "processes." Collectively, the handling and consideration of these components, processes, and dimensions mean that fisheries GIS work is not something that can be approached lightly!

2.2.1 Mapping Movable Variables

The fact that the aquatic environment is constantly and variably mobile and that everything within it also is moving, lies at the core of the GIS applications problem. To put it simply, a conventional terrestrial map is a static object; it shows a situation whereby objects located in space are seen to be permanently fixed. In the marine environment, only major structures associated with the bottom topography may be static, and the fact that everything else moves is a major problem for the data gatherer and, consequently, the GIS worker. This "mobility" factor also is responsible for many related, if more indirect, challenges. It is well documented that a range of secondary energy sources are responsible for movements in the aquatic realm. For instance, water motion may be activated by wind energy that gives rise to waves and advection currents. Geostrophic forces may cause major oceanic circulations, and gravitational forces will cause tidal streams. Hydraulic energy gives rise to shifting water movements in rivers, and to river scour and discharge plumes. While some of the movements have a degree of regularity and can be modeled or predicted with a degree of certainty and replication, many of the movements are highly variable and unpredictable and thus are stochastic. Among the more predictable movements are those that relate to gravitational forces such as tidal currents and streams. Somewhat less predictable are the major oceanic gyral currents associated with the trade winds, geostrophic forcing, and the Coriolis effect, and the major upwellings. Other movements may be relatively regular in direction, but highly variable in extent, for example, the major plumes associated with river discharge. Among the least predictable (or most variable) water movements are the frontal systems, plus advection currents that result from wind forcing. In the riverine or lacustrine situation, water movements again may be more predictable in the directional sense, but the volumes involved are not predictable, and there is a range of factors that lead to minor random chaotic movements.

Why might the fisheries scientist wish to study or model movement? The reasons will be varied but the scientist may wish to fill in knowledge gaps, test theories, interpret existing data, or make predictions, and the fisheries manager is likely to be interested in anticipating movements based upon past research. Thus, because fish species are often highly mobile, fisheries managers need to know movement rates in order to assess interactions between fisheries in different locations and to define the discreteness of stocks (Hilborn 1990). There also are economic benefits to knowing the whereabouts of fish aggregations. Even if individual movements of fauna are random, collectively, the movements may exhibit regularity. So, at an aggregated level, it may be possible to formulate deterministic rules. These may in turn lead to deterministic equations that could be applied within a GIS context.

Some GIS-based work on the study of animal movements within aquatic environments is now appearing. Nams (1999) illustrated how movement patterns associated with various animal activities may be considered under various spatiotemporal domains and may be described by using fractal analysis (Figure 2.2). Schneider et al. (1999)

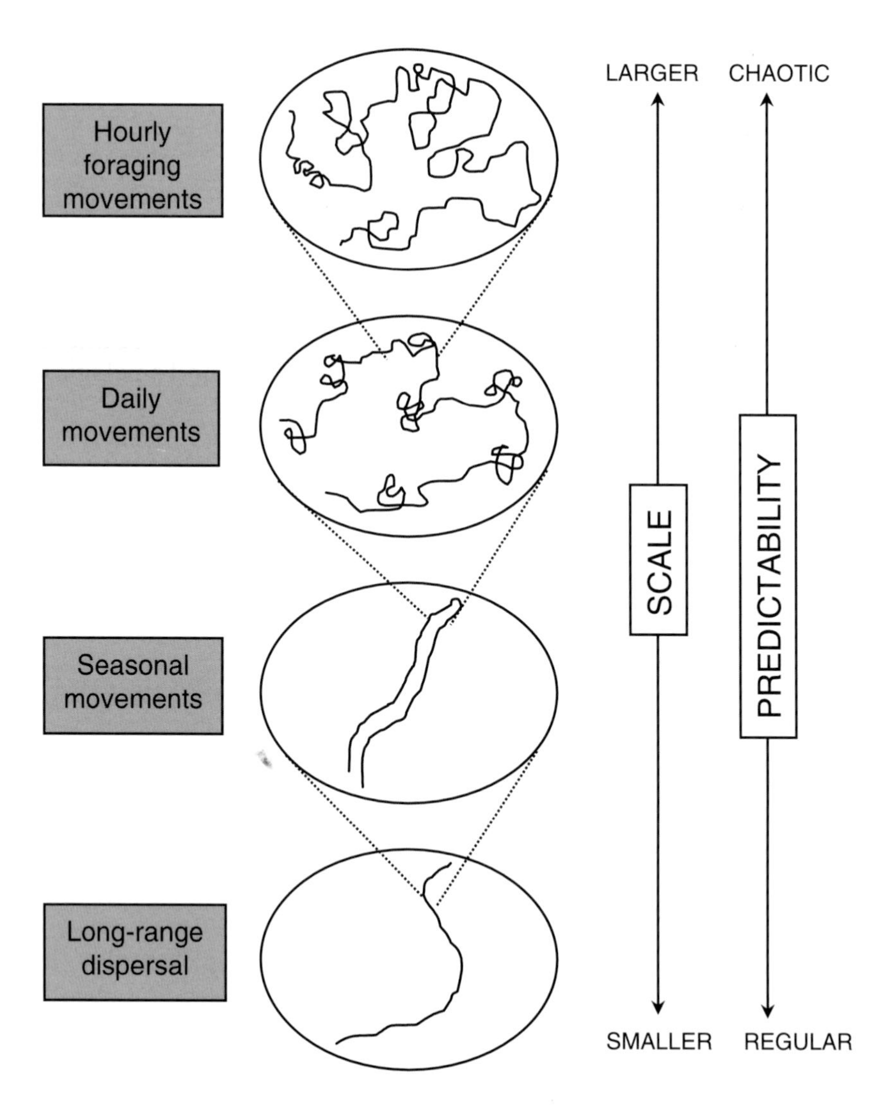

Figure 2.2 A hierarchy of spatiotemporal movement domains in the animal world (adapted from Nams 1999).

attempted to establish the movement domains for Atlantic cod *Gadus morhua* as they occur at particular life stages. However, Hooge et al. (1999) have noted that no previous attempts have been made to integrate movement-analysis tools with GIS for plotting fish movements. In response, his group has developed a software application (The Animal Movement Analyst Extension) that links ArcView GIS with a number of movement-analysis tools. By plotting movements of Pacific halibut *Hippoglossus stenolepis* in Glacier Bay, Alaska, the investigators were able to generate movement hypotheses for testing, and they identified patterns of unexpected movement. Obviously their GIS extension offered the possibility of linking movement paths and types of movement strategies to variables in the aquatic environment. However, there is still much work to be done on integrating general movement models into GIS. For instance, existing transport models could be used as a means of establishing rates and direction of movement (Wroblewski et al. 1989; Hare and Cowen 1996), and Turchin (1998) notes that a variety of diffusion models might be tested within the marine environment. In a more theoretical vein, Fiksen et al. (1995) and Giske et al. (1998) presented movement models that show how the life history and behavior of fish can be modified so as to improve the reproductive rate (fitness) of the individual. Possingham and Roughgarden (1990) presented a modeling framework within which mesoscale features in ocean currents can be integrated with coastal habitat structure to predict the distribution and abundance of a marine organism having a coastal adult phase and a pelagic larval phase. A challenge is to integrate these and other movement models into the standard GIS methodology. For further information on the general applications of movement modeling, the reader might consult Okubo (1980), Brown and Lacey (1990), MacCall (1990), Grindrod (1991), Tyler and Rose (1994), D'Angelo et al. (1995), or Hilborn and Mangel (1997).

2.2.2 Multiple Scale and Resolution

Since the topics of scale and resolution are so diverse, only a brief coverage can be made here. Those wishing to explore additional factors relevant to scale and resolution in the marine environment should consult texts such as Smith (1978), Dickey (1990), Ricklefs (1990), Levin (1992), Mullin (1993), Schneider (1994), and Mann and Lazier (1998). Scale must be considered as having both temporal and spatial dimensions; scale and resolution also must be a consideration when it comes to examining other challenges such as data classification, visualization, and the application of models and statistics. Scale is concerned with the overall size of an area under consideration, whereas resolution is the measurement used for objects within the study area.

It will be obvious that movement of both the aquatic environment and the objects within it takes place across a spectrum of scales or resolutions (Figure 2.3), and the debate over the appropriateness of the scale of study has a long history (Harris 1986). Schneider (1998) has recognized that attention to multiscale spatial and temporal analysis is important because:

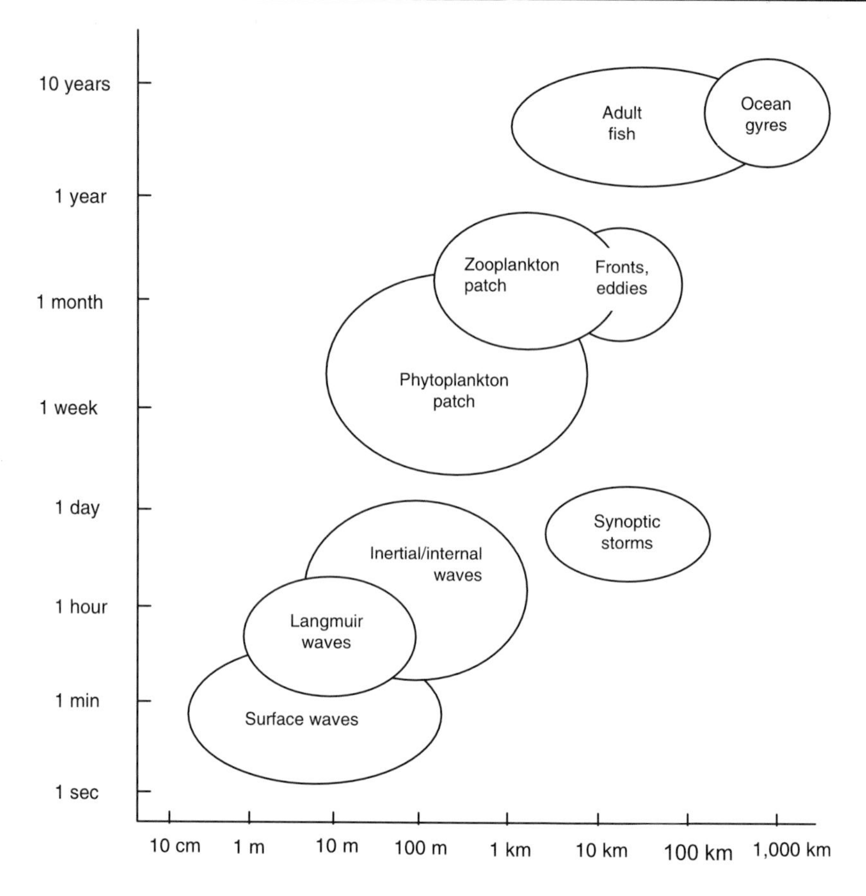

Figure 2.3 Varying time and space scales for selected physical and biological marine variables (adapted from Dickey 1990).

- spatial and temporal patterns depend on scale;
- experimental results cannot be extrapolated to other scales;
- biological interactions with the environment occur at multiple scales;
- population processes often occur at scales that are difficult to investigate;
- environmental problems arise through propagation of effects across scales; and
- there is no single "characteristic" scale for research.

Although we need not delve further into these basic ecological principles, it is important that the fisheries scientist or manager is aware that, depending upon which scale is used for a project, almost any distributional pattern (clustering to random) could be discerned. A major challenge is to select the correct or most appropriate spatial resolution of study, one that is best able to accommodate the relevant moving variables. Although there seems to be an emphasis emerging that it is better to use local patterns or smaller areas as units of study instead of larger, more global units (Anselin 1996; Openshaw and Clarke 1996) in the field of fisheries, it could be argued that the researcher must be flexible enough to make scale changes in either direction.

It also is clear that a temporal resolution for movement must be decided upon. Basically, there will be a relationship between speed of movement of the water itself, or the objects in the water that are being studied, and a relevant periodicity for data gathering. Lucas (2000:57) has remarked that there is, thus, an implication that "at the very least, the user must have an understanding of the relationship between the sampling frequency and the natural variation in the attribute in order to compare multiple data sets." However, the challenge of deciding on the periodicity for any study may be increased greatly since the water body movements (speed and direction) will rarely coincide with movements made by any faunal species under study, that is, except much movement of the plankton or other drifters. Under these circumstances, periodicity of mapping (or the necessary data gathering) may be more difficult to select. And it will be clear that any necessary increase in data periodicity means the data decay rate must be sped up—data will become obsolete more quickly. It can be concluded that, in most cases, resolution depends on (or is limited by) the data available, and it is not easy to consider spatial and temporal resolution separately for either environmental or fisheries variables.

2.2.3 Handling Three and Four Dimensions

Given the challenges invoked through movement in a 3-D or 4-D environment (i.e., two horizontal axes, one vertical axis, and time), it is clear why proprietary marine GIS has failed to be developed. Thus, although much work has been done on 3-D GIS per se, these applications mainly have been for use in terrestrial spheres associated with geology and mineral exploitation (see Raper 2000 for a current review). However, some marine-based work has begun to emerge, though it would be true to say that much of it is at an elemental stage. This work can be conceived as occupying various levels of abstraction and development. Three developments are illustrated below because they exemplify how the challenges faced by working in 4-D are at varying levels of realization according to the conceptualization of the problem.

At the most abstract level is the work on a universal hierarchical 3-D data structure (Meaden 2001). The intention of this structure is to provide an international "working platform" for inputting, analyzing, and displaying marine data. It represents a 3-D or 4-D database concept that is designed specifically to offer universal georeferencing as well as an ability to encompass a wide range of scale variation. Figure 2.4 gives an impression of the data structure. It is composed of a six-level hierarchy of tessellating cuboids, with each level varying in scale by an order of magnitude. The scale varies from 10,000 × 10,000 × 1,000 m to 0.1 × 0.1 × 0.01 m. The tile shape of the cuboid reflects the horizontality of most marine variable distributions. The structure of the cuboids is designed to accommodate latitudinal and longitudinal georeferenced data within the Universal Transverse Mercator projection. The hierarchical nature of the structure allows the user to choose the scale of working according to particular project or research requisites. It is likely that such a data structure will be used in the near future, not only for fisheries and oceanographic GIS-based work, but also as the basis of marine information systems, including those operating in an interactive Internet environment. The challenge will be to realize its implementation in some form or another!

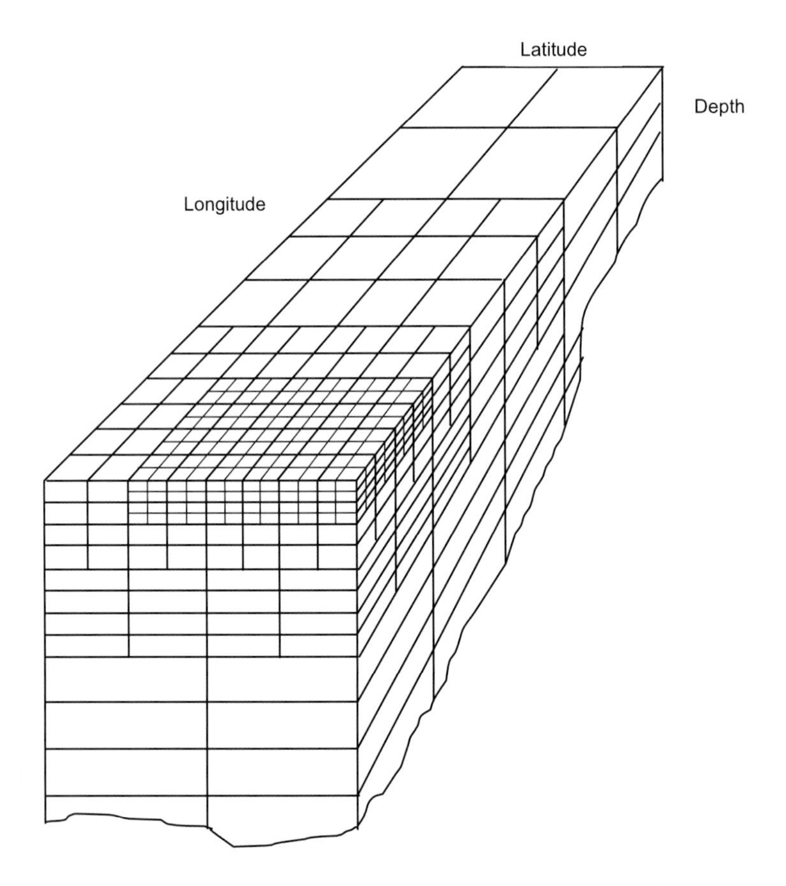

Figure 2.4 An illustrative section of a 3-D, tessellating hierarchical data storage and analysis structure. For illustrative purposes, the model shows only four of the six possible cuboid scales, and cuboids at each scale are only subdivided into two rather than into the 10 necessary to form the next level on the hierarchy.

A significant movement toward creating the necessary structures for data collecting and analyses in a 4-D marine environment is Varma's work based on helical hyperspatial (HH) coding (summarized in Varma 2000). The HH coding allows data to be allocated to positions in 3-D space through recursive subdivision of cuboid space. This is based on Boolean true-or-false analysis of the space in terms of a "yes" or "no" answer to whether a data point lies within the cuboid. Successive iterations may be repeated until a required resolution is reached. A 2-D concept of Varma's idea is illustrated in Figure 2.5, and the 3-D version simply extends this concept by using volumetric octrees (3-D fractals). Once individual units of data have been allocated to 3-D or 4-D space and time, aggregations of this data are used to display whole maps or surfaces for possible use within a GIS environment.

The third example of a 3-D or 4-D marine data structure operating within a GIS is the work demonstrated by Kiefer et al. (University of Southern California, unpublished paper presented at the First International Symposium on GIS in Fishery Science, 1999). The investigators have developed a modular system (Environmental Analysis System) that can input a wide range of spatially referenced oceanographic and fishery data, analyze the data with the use of various algorithms and models, and then produce output in a 3-D format. The data are all managed within commercially available, proprietary software that can be interlinked by using standard data exchange protocols. Their system has been demonstrated by using various data

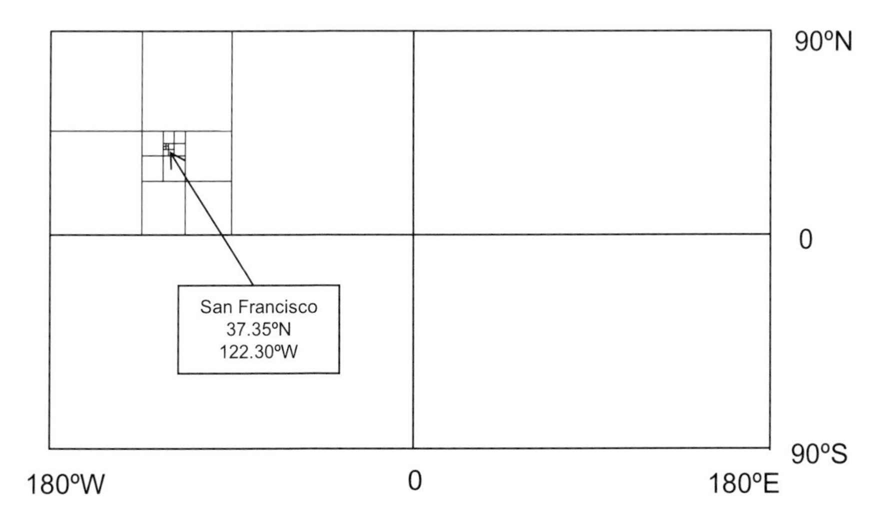

Figure 2.5 A 2-D conception of Varma's (2000) helical hyperspatial (HH) coding method for allocating data to real georeferenced positions. In this example, the HH code has defined the location of San Francisco through a binary process whereby the globe has been successively divided into smaller quadrants until the city's location has been best defined. This is accomplished by converting the latitude and longitude coordinates into two 32-bit binary numbers which are then interleaved to create one 64-bit number giving the full point location.

sets from their local La Jolla (California) fishery, and from the Food and Agricultural Organization (FAO) of the United Nations for some Mediterranean fisheries, and they have looked at the spatial effects of an El Niño event on Peruvian fisheries. The software provides a flexible 4-D representation of data related to latitude, longitude, time, and depth, and the information is visualized through "fly-through" sequencing of space or time.

The purpose of providing these examples is to show three different approaches to the problem of 3-D data structuring, assembly and presentation. None of these examples yet provide the structure behind any proprietary "marine GIS," probably because they each require a great deal of additional work and because they have not necessarily been designed with a commercial application in mind. For each example, the work currently is proceeding, and the challenge will be to direct this work toward meeting the requirements for a full fisheries GIS.

2.2.4 Applications of Spatial Statistics and Models

The world of fisheries science and management is no exception when it comes to having absorbed an abundance of applied conceptual and mathematical models, plus equations and spatial or temporal statistics. Most of these represent an attempt to explain distributions and to make stock or abundance forecasts. While clearly many of these models are not conducive to spatial manipulation or application, given the recent advances made in geostatistical spatial modeling (e.g., Fotheringham and Rogerson 1994; Bailey and Gatrell 1995; Camara et al. 1996; Levine 1996; Zhang and Griffith 1997; Longley et al. 1998; Brady and Whysong 1999; Fotheringham et al. 2000), it is not surprising that there will be a range of challenges invoked in trying to fit various fishery models within the GIS context.

In the context of this chapter, spatial statistics, spatial modeling, and modeling may be considered as one. What is being considered here is the situation whereby GIS forms the software platform or activity surface on which numerical models, usually in the form of equations, are conceived, evaluated, or tested. As an example, we may envisage modeling the spatial diffusion process. Thus, it must be possible to establish an equation that best describes (models) the rate (speed and numbers) of spread of any marine species through a salinity or temperature gradient or, indeed, through any combination of variable gradients. A GIS allows layers representing each variable to be integrated and then incremental iterations of previously recorded movements of a selected species would be simulated by using a raster GIS as the platform. The GIS would reveal the diffusion speed and the direction and would allow for the production of statistical data determining the rate at which each aquatic variable controlled the diffusion. This example is relatively easy to accomplish, and, at the present stage of integrating models to GIS in the fisheries sphere, we are really only using these, what Malczewski (1999) has described as, "basic operations." The challenge for the worker in the fisheries

GIS field is to manage successfully the integration of more complex models involving more advanced GIS functionality.

Spatial analysis and modeling can be achieved either via the use of functions already included within proprietary GIS software or by linking external specialist software to the GIS. So, stock assessment models can be linked with environmental data within a proprietary GIS, though there are a wide range of other models or statistical packages that also could be integrated. Fotheringham et al. (2000) note that although it is not necessary to use GIS for spatial modeling, its use is likely to produce insights that would otherwise be missed. Openshaw and Clarke (1996) suggest the development of a "spatial analysis toolkit," a set of GIS independent spatial analysis tools that could be interfaced with any GIS. The advantages of modeling within a GIS environment include:

• raster data may be conveniently used for various procedures (see above);

• most GIS have built-in formulas that can be used;

• it is easy to add the extra requisite variables that might influence the model;

• it is easy to add weightings;

• it is easy to accomplish temporal iterations for dynamic modeling;

• GIS is designed as a database with visualization properties;

• most GIS can be coupled to external software programs;

• GIS is good for exploratory spatial data analyses; and

• it can be easy to accommodate time, space, or resolution changes.

Since the range of potential modeling is vast and the actual development of models is beyond the interest of this study, all that can be done here is to examine the integration of some fishery models with GIS platforms. This will illustrate how selected challenges have been countered. Most of the brief examples are taken from those given at the recent Seattle or Anchorage Fisheries GIS/Spatial Processes symposia (see Section 2.1).

In an attempt to establish "survivor's habitat" for pelagic fish off California, Logerwell and Smith (2001) used GIS to integrate various aquatic environmental layers to show relationships between pelagic larval recruitment and combinations of environmental conditions. The modeling of this relationship would allow fisheries scientists to make stock predictions or to best target areas for intensive stock assessment. In a similar study, E. D. Brown and B. L. Norcross (Institute of Marine Science, University of Alaska, unpublished paper presented at the 17th Lowell Wakefield Fisheries Symposium, 1999) were able to establish the relationship between environmental factors and the year-class strength of herring in Prince William Sound, Alaska. In the United States especially, there has been much recent interest in the establishment of essential fish habitats. This is in an attempt to identify areas that are in special need of conservation since they are likely to play an important role in the life cycle of various fish species. Some significant work has been done by Rubec et al. (2001) in Florida. This group has established a "habitat suitability model" that identifies optimum areas within southern estuaries for certain species survival. They have shown how the model, based originally on one major estuary, can be used within other estuaries so as to predict likely fish species distributions. In a study based on riverine environments in Oregon, Kelly (2001) has attempted to build a habitat model showing how forest harvest practices might alter stream environments and impact salmon spawning habitats. A final stock prediction model for illustration is that of Fisher et al. (2001) who devised several GIS-based methods for predicting fish population abundance in streams, and the models derived have been shown to be effective in identifying suitable habitat in nonsurveyed areas. From the more general GIS perspective, the building of such predictive models can be a satisfying challenge to take on, not only because of the intellectual work involved, but also because of the importance that suitability models will have in future species conservation.

Not all attempts at fisheries modeling with GIS have been related to developing indicators of stock health through habitat optimizing. Walden et al. (2001), working on the principal groundfish species in the New England fisheries management area, have successfully linked a GIS to a general algebraic modeling system, with the idea of working out a fishing strategy that allows for stock protection in terms of area and time period closures, as well as in maximizing economic returns to the fishers. Two models have been developed that are similar in their attempts to show the distribution of fishing effort through space. In a theoretical mode, Caddy and Carocci (1999) have used an empirical "Gaussian effort allocation" modeling approach to show how fishing effort will gradually vary through time with distance from a port (Figure 2.6). Here fishing effort is shown by arbitrary units representing increases in fishing effort and mortality, and the curves plot progressive shifts in fishing effort through time away from the home port. This model, which is fully active by using IDRISI GIS, has practical utility by aiding fishery managers and coastal area planners to "analyze the likely interactions of ports, inshore fleets, and local nonmigratory inshore stocks, and in providing a flexible modeling framework for decision making on fishery development and zoning issues."(Caddy and Carocci 1999:388). Corsi et al. (2001) have developed a means whereby the relationship is established between total catch, fishing effort, water depth, and distance from the port. They perceive that it should be possible to partition the whole of a fishing area into "homogeneous ecological" zones, each having specific and possibly predictable out-

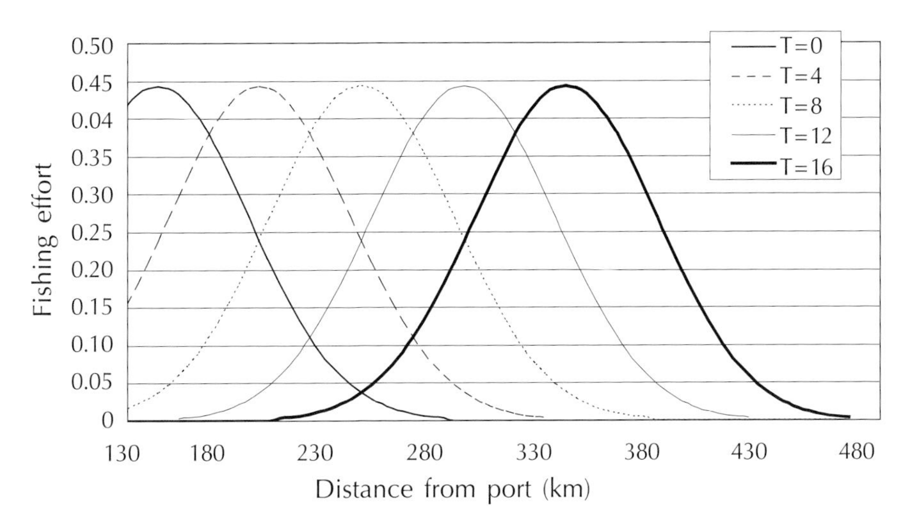

Figure 2.6 Gaussian curves showing how fishing effort may theoretically vary with distance from a port over time (T) (adapted from Caddy and Carocci 1999).

comes. The models illustrated here give an indication of the possibilities that GIS offers to fisheries science and management. Their further development, as one means of improving the plight of fisheries, is likely to be a challenge that will be confronted by a wide range of participants in the fisheries GIS field.

At least two major mathematical challenges face the worker indulging in model applications to fisheries GIS. A major problem for the spatial analyst of fisheries data is spatial autocorrelation, that is, what is an independent unit of fishery or environmental data? Geostatistical analysis can provide answers and may well indicate that, at the finest resolution available for fishery or environmental data, adjacent measurements cannot be considered as independent. A second, mathematically related problem in the field of spatial statistics is that of securing statistical significance from the data being used. It will be appreciated that data gathering and sampling in the marine environment is generally an expensive proposition. This frequently results in small sample sizes and, thus, the lack of statistical significance in a range of testing procedures. In fact, many statisticians find it difficult to compute confidence levels of significance for data gathered in the marine domain, and it will be a major future challenge to ensure that data, and the maps generated from this, can be viewed with a specific degree of confidence. Notwithstanding the challenges faced in fitting mathematical models in a fisheries GIS context, some examples have recently been presented (D. Chen and R. Leickly, International Pacific Halibut Commission, unpublished paper presented at the Second International Symposium on GIS/Spatial Analyses in Fishery and Aquatic Sciences, 2002; R. Grabowski et al., School of Marine Sciences, University of Maine, unpublished paper presented at the Second International Symposium on GIS/Spatial

Analyses in Fishery and Aquatic Sciences, 2002; O. Jensen et al., University of Maryland, unpublished paper presented at the Second International Symposium on GIS/Spatial Analyses in Fishery and Aquatic Sciences, 2002).

2.2.5 Optimizing Visualization and Mapping Methodologies

The basic output from a GIS includes maps, tabular data, and a range of graphics. Since these are the bases upon which decision making is made, it is imperative that they convey meaning in a manner that is both accurate and perceptually easy to synthesize. Thus, at almost any level of cartographic representation, there is a range of rules or conventions that should be adhered to in order to optimize comprehension, though there is license within the rules to accommodate individualistic mapping styles. Challenges occur in this area both with the creation of basic input mapping (modeling of the real world) and in obtaining satisfactory output visualizations.

Data being input to the fisheries GIS will be represented by a mapped model of the real world or by a sample of information concerning some real-world distribution. The main challenges regarding these data inputs revolve around the issue of classification and data representation. Classification is concerned with class allocation in terms of numerical ranges and the number of classes used. While this might not cause problems, and indeed many proprietary GIS have a default range of class options, it still takes experience (or experimentation) before ideal class ranges are obtained, and it should be mentioned that several investigators have shown how it is easy to convey "selective truths," depending on classifications chosen (Pickles 1995; Monmonier 1996). These comments apply equally to output representation. Here the concern may be more with aesthetics, such as appropriate symbol size and placement, pleasing color arrays, and more, but it also is essential that the reader is able to grasp quickly the essentials of the data being presented. In this investigator's experience, the mapping of fisheries or aquatic data invariably involves conveying a wide range of associated data via a minimal number of graphical objects placed on the map. This invariably is only achieved through experimentation.

There is the problem of fuzzy boundaries associated with classification. Thus, while most feature or variable distributions may be relatively discrete, we should be aware of various kinds of spatial and temporal fuzziness. Fuzziness may involve specific nomenclature such as "the Southern Ocean" or general concepts such as "the continental shelf." Regarding the latter as an example, although there may be relatively precise definitions, there is a certain elasticity of definition when it comes to, for instance, data gathering for a specific shelf-based project. Should the shelf be classified as being only waters of less than 200 m? What angle of drop-off should define the shelf's outer limit? Can we legitimately sample on only a proportion of the shelf? Other types of fuzzy boundaries occur as a result of the mobility of the aquatic environment. For instance, the tidal cycles mean that littoral boundaries are constantly in flux, and, for some purposes, it is necessary to define an "intertidal zone." Even where movement occurs slowly, boundaries may be fuzzy. A transect along a longitudinal stream profile reveals that it is often difficult to classify streams into easily identifiable reaches based on physical characteristics. Unfortunately, the number of "fuzzy objects" within the aquatic environment is large indeed. The challenge for the fisheries GIS worker is to define a set of rules whereby some ordering regime is implemented, so that entities can be assigned to classes based on logical and consistent rules and goals. The work of Altman (1994) or Burrough and Frank (1996) is useful for explaining "fuzzy sets" and in showing how they can be used to represent geographical entities with imprecisely determined boundaries.

In Section 2.2.1, it was stated that a fundamental challenge for those working in fisheries GIS was the inconstancy of movement. From the viewpoint of analyses and mapping, this creates a need for sequential mapping to perceive spatiotemporal changes. What time interval (periodicity) should output be presented at? Clearly, this may be a function of speed of movement, but compromise may be needed to avoid excessive output demands. Where annual cycles are involved, sufficient information can often be gained from monthly sequences (e.g., see H. Demarcq, Laboratoire Halieutique et Ecosystems Aquatiques, unpublished poster presentation from the First International Symposium on GIS in Fishery Science, 1999), and some investigators have shown interannual variations (e.g., Frank and Simon 1998). Other investigators have elected to show movement sequences in an almost film animation (or "fly-through") sequence. Thus, Ault and Luo (1998) illustrated patterns of shrimp recruitment and migration in Florida's Biscayne Bay by so-called dynamic sequencing. Although long-term movement patterns over large spatial domains may be easily visualized, the challenge for the GIS worker will be in depicting short-term movements that are of a more unpredictable or chaotic nature.

Probably a greater problem in this visualization area is that of the meaningful mapping of complex surfaces and distributions. Examples here include the mapping of benthic communities or coral reefs. It will be appreciated that the benthos may represent an extremely complex assemblage of numerous species, all within a matrix of varied inorganic sediments. Kohn (1997:205) has described coral reefs as "a heterogeneous mosaic or landscape of substrate types including calcareous sands, smooth reef rock, living and dead coral, and calcareous and fleshy algae, as well as intermediates and combinations of these…" How best to map them? Even in terrestrial space, this is the type of challenge that mostly is avoided. Thus, maps are drawn that largely represent relatively sorted entities. A topographic map will, for instance, separately identify point features such as towns, churches, factories, and linear features such as rivers, roads, and political boundaries. These features will be set within the background matrix of, perhaps, farms or woodland. At a different scale, the town may be mapped in discrete entities such as housing, retail, factories, recreational land, and more. When confronted with tasks such as species mapping in the tropical rain forest, this is an almost impossible challenge. A related example is the challenge of mapping discrete variables that may form either almost continuous distributions or variably patchy distributions, depending on the scale factor used. For example Weber et al. (1986) have shown that the patchiness of krill distributions in Antarctic waters is much greater than the patchiness of the plankton upon which they graze or upon the prevailing water temperatures. Unfortunately for fisheries managers and scientists, many of the important distributions that require spatial analyses, for example, benthos, bottom sediment types, fish assemblages, or even the dispersion pattern of individual animals, are those that play an important part in the decision-making process. Probably the best advances that are occurring in the field of mapping complex entity assortments lies in identifying assemblage groups for mapping or in showing relative densities of target species of specific interest (e.g., maps of fish species assemblages or terrestrial biomes, ecoregions, or agrosystems regions). An interesting study here has focused on the approaches to quantifying the spatial heterogeneity that exists within the water column of marine and large lake systems (L. Kracker, National Oceanic and Atmospheric Administration, National Ocean Service, unpublished paper presented at the Second International Symposium on GIS/Spatial Analyses in Fishery and Aquatic Sciences, 2002).

This section on intellectual and theoretical challenges should conclude by mentioning that everything discussed above has been developing at the forefront of scientific endeavor across a range of technological and conceptual research fronts. It is a major challenge for workers in the field of fisheries GIS simply to keep abreast of current "scientific" developments. Thus, GIS applications here are progressing in parallel with advances in, for instance, data acquisition modes and their integration, computing hardware and software, database systems, geostatistics, Internet progression, visualization, oceanography, and environmental modeling, as well as in fisheries science and management. The integration of advances in these technologies is essential if fisheries GIS is to be fully used.

2.3 PRACTICAL AND ORGANIZATIONAL CHALLENGES

Practical and organizational challenges are those encountered in the day-to-day fisheries GIS operational environment and in the assembling of inputs to this environment. For our purposes, we will restrict attention to matters concerning data gathering and assembly (data inputs), data standardization, the working organizational milieu for fisheries GIS work, and information output considerations.

2.3.1 Data Gathering and Assembling

Although data gathering and assembling considerations lie at the core of all GIS operations, and they usually constitute a considerable challenge, there are reasons why this challenge is exacerbated for those working in fisheries or aquatic environments. Apart from data being difficult and expensive to collect (see below), there exists a situation whereby data needs may be unknown. For instance, as noted in Section 2.2.4, a current concern in fisheries science and management is the establishment of "essential fish habitats." While the major factors in determining areas that might need this protective status may be obvious, how can we be certain that all variables have been considered? In addition, how detailed must the data collection be? Unfortunately, as many working in the field of fisheries GIS are aware, much of the data collected in the past was not qualitatively or quantitatively adequate in terms of present needs, and, in many cases, it simply cannot be used. No doubt data being collected today will be seen in the future to be equally deficient, and there is a fundamental challenge of how to minimize this likelihood.

Data may be from either primary or secondary sources, and these represent convenient headings for discussion:

(1) *Assembling primary data.* The degree of difficulty in collecting this may vary as a function of whether a GIS project is concerned with freshwater or marine operations and with distance from the coast. Clearly, data gathering for land- or coastal-based aquaculture projects usually will be no more of a challenge than it would be for other terrestrial GIS programs. The same applies for most riverine fisheries GIS. Primary data-gathering problems are magnified when there is the need to use a vessel or sophisticated data-gathering devices. Nearly all marine GIS projects encounter these challenges. The problems are invariably more than just financial. Use of a vessel for data gathering involves considerations relating to equipment assembling and use, to team assembling, to tight scheduling of vessels, to reliance on weather, to excessive time inputs relative to the data gathered, and to difficulties in sampling strategies and research replication. These considerations often mean that a major effort is necessary to obtain data that may be very limited in both its spatial or temporal extent.

(2) *Assembling secondary data.* This may be regarded as processed primary data. The challenge here is to obtain exactly the data needed, since most GIS projects will have specific objectives that rarely relate to existing data. The amounts and potential sources of secondary data available are very large indeed. However, the challenge is in locating exactly what is needed. In this investigator's experience, this factor has been little helped by the growth of potentially useful data dissemination methods such as the Internet, electronic data searching, data mining, specialist data archiving groups and mechanisms, and setting up of metadatabases. The reasons for the difficulty in obtaining appropriate secondary data can be ascribed to factors such as:

- the resolution of the data needed;
- the specific nature of most data-collecting events;

• locating data that covers all of the spatial or time period needed;

• finding data that is classified into requisite categories;

• finding data having the correct formatting and structure;

• knowledge on the quality of the data; and

• actually finding an appropriate source since much data are not cataloged.

So, despite the proliferation in Internet-based searching, this investigator has yet to download a single piece of useful data via this mechanism, although some clues have been found to data sources! And these secondary data-sourcing challenges are unlikely to be reduced in the future, mainly because marine research projects are likely to become increasingly specialized with requirements for evermore precise data inputs.

Despite this gloomy prognosis, there are data sources for use in fisheries GIS that may be valuable. Here only limited examples can be given. Remotely sensed imagery, mostly from satellite sensing systems, provides timely access to data on, for instance, water temperature, wave height, wind direction, chlorophyll abundance, turbidity, and shallow water bathymetry. Existing topographic maps, hydrographic maps, and various specialist CD-ROMs provide data on coastline configurations, bathymetry, bottom sediment types, and underwater obstacles. Government fishery offices or research institutions may have data on species distributions, though many of these will be project or time specific. For instance, many governments in Europe collect spatially referenced commercial catch or stock assessment data. Once standardized, this data may yield spatially referenced abundance indices for a range of species. Similar data are available for many North American and Australasian waters. In certain geographic areas, attempts have been made to establish marine information systems, for example, the eastern Canadian seaboard through their metadatabase managed by the Atlantic Coastal Zone Information Steering Committee and accessible through their Internet site (ACZISC 2003). A comprehensive review of secondary fisheries data sources is given in Meaden and Do Chi (1996), and Table 2.1 provides a listing of major fishery data sources. Readers interested in Internet sites providing additional access to marine information should consult Lalwani and Stojanovic (1999). Despite these advances, it is hoped that Internet-based technologies may soon emerge having the efficiency and capacity to be far more effective in satisfying user needs (see Section 2.3.3).

There are a number of plans to combat the marine data gathering problems. The growth in autonomous underwater vehicles (and various other robotic data collectors is likely to be exponential. Westwood (2000) has forecasted that the world growth in total un-

Table 2.1 Selected sources of fishery data.

Data source	Source acronym	Data coverage	Internet address
British Columbia	BC	Western Canada	*www.bcfisheries.gov.bc.ca*
Department of Fisheries and Oceans	DFO	Canada	*www.dfo-mpo.gc.ca/home-accueil_e.htm*
European Union	EU	Europe	*http://europa.eu.int/comm/eurostat/*
Food and Agriculture Organization of the United Nations	FAO	World	*www.fao.org/fi/statist/statist.asp* *http://apps.fao.org/fishery/fprod1-e.htm* *www.fao.org/waicent/faoinfo/fishery/struct/fidif.asp*
International Commission for the Conservation of Atlantic Tuna	ICCAT	Atlantic	*www.iccat.es/*
International Council for the Exploration of the Seas	ICES	Northwestern Europe	*www.ices.dk/fish/statlant.htm*
National Marine Fisheries Service	NMFS	USA	*www.st.nmfs.gov/st1/index.html*
Northwest Atlantic Fisheries Organization	NAFO	Northwestern Atlantic	*www.nafo.ca/*
Pacific States Marine Fisheries Commission	PSMFC	Northeastern Pacific	*www.psmfc.org/efin/*
WorldFish Center	ICLARM	World	*www.cgiar.org/iclarm/*
World Resources Institute	WRI	World	*www.wri.org/wri/facts/data-tables-oceans.html*

manned underwater vehicles will be from US$100 million in 2000 to more than $330 million in 2004. Although these vehicles or their sensing systems will provide huge potential, they usually are aimed at gathering only a limited range of data, for example, typically water quality data, and this is often project specific. In addition, the data may not enter the public domain immediately, and it might have a high price tag. Other data-gathering plans involve the positioning of automatic buoys. They usually are tethered and are able to sense their surroundings as a means of logging longer-term temporal changes, usually in abiotic factors. The French marine research institutes IFREMER (French Research Institute for Exploitation of the Sea) and IRD (Institut de Recherche Pour le Developpment) are already developing these. Some buoys are designed to float in order to get current- or wind-related information. Lohrmann (2000) describes the development of an acoustic wave and current sensor that measures both surface waves and ocean currents from a small device mounted on the seabed (Figure 2.7). Four acoustic beams (V_1 to V_4) sample the wave orbital velocity near to the sea surface by measuring their Doppler shift; "P" is a high-resolution pressure sensor. Another recent technological advance is that underwater wireless communications

Figure 2.7 A seabed-mounted sensor for measuring wave height and current speed (adapted from Lohrmann 2000). P = high-resolution pressure sensor; V_{1-4} = acoustic beams 1–4.

have proved possible (Benthos, Inc. 2003), and this could have a great impact on data-gathering speeds and locational accuracy. For collecting data on marine fish distributions, refinements in acoustic technology are making fish "counting" a more exact science, though there still is some progress to be made toward identifying individual fish species by using sonar techniques. A development that will have major implications in the next decade is the incorporation of electronic log books into commercial fisheries. There are already a number of systems that have been introduced locally to specific fisheries (Kemp and Meaden 1998), but it is almost certain that fishery management authorities in all the major fisheries will soon introduce them as a vital means of determining the spatiotemporal aspects of catch and fishing effort. The data obtained will be matched to environmental parameters within a GIS environment to greatly advance our knowledge of fish distributions. At the present time, there is a degree of caution with regard to the digital capture of data on catches, both from the fishers who see this as intrusive and from management who see the data as "selective" (nonrandom). However, the rapid deterioration in the plight of most fish stocks has acted to concentrate minds on the need for effective and comprehensive data inputs from a source that has the potential for great utility.

A final challenge concerning data gathering is that of standardization. Data collected from various sources inevitably will need to be standardized in terms of its structure, projection, classifications used, format, and import–export potential. Though this is sometimes accomplished with relative ease, it can prove to be a significant challenge. Thus, for instance, new formats keep being introduced along with other systems procedures. Simply keeping up with these changes needs to be part of the standardization process. In the terrestrial sphere organizations such as the Association for Geographic Information, the European Umbrella Organization for Geographical Information, or the Federal Geographic Data Committee have worked to set international standards for data collection, and some considerable advances have been made. Unfortunately, little headway has been made in the marine sphere. Although work in this area is underway within some of the national hydrological services; in other areas, there does not appear to be a lead organization willing to take a standards initiative. This is unfortunate given the fact that much aquatic or fisheries environmental work takes place in an international context, with domestic and foreign vessels interacting for the same resources but in an exclusive economic zone that is under national jurisdiction or across international boundaries (in the case of straddling or shared stocks). In addition, an "explosion" of data gathering is currently underway. An extra reason for the urgent implementation of data standards is the rapidly increasing mobility of the data itself, and Bartlett and Wright (2000:311) have pointed to the increasing "risk of data quality loss and creeping error or uncertainty...," if standards are not internationally agreed upon.

2.3.2 Subject Organization

The thematic area of "GIS in fisheries science or management" does not in itself constitute an area of pure research. It is better described as an applications area (see Section 2.2). Since "fisheries" is an area involving the integration of many disciplines (environmental and fishery science, biology, oceanography, various areas of technology), then it is likely that a range of institutions will be involved in its promulgation. So, the subject is being pursued, usually on a small scale, in a disparate array of institutions that will be dispersed widely, often in peripheral seaboard locations. It is also true that the nature of information technology (IT) strategies will vary greatly among institutions, and this investigator heard recently (at a marine GIS conference) the remark, "We have a chaotic approach to IT as an organization" and that "whenever new managers come in they demand a new set of IT systems!" From a functioning viewpoint, this often means that progress is slow because projects are enacted in comparative isolation, the monetary support may be low, and there may be numerous projects in progress with no cohesive development. These factors may prove a great challenge to workers in this field, that is, in terms of support and the exchange of ideas, data, and information. The situation is made worse by the fact that there is no core of literature on the subject. Much of what is published appears in the gray literature, and there are many cases where the use of GIS is subsumed within the general methodology. Investigators often do not consider that the use of GIS is worthy of note. While in some senses this is an expected development, it does not help those who may be seeking methodological ideas or assistance.

There have been recent attempts to draw together the subject area in an organizational sense, although it is fair to say that advances have been slow. This slowness probably arises from the same forces that operate as agents of subject dispersion and fragmentation. There now have been three major conferences concerned explicitly with spatial aspects of fisheries, the First International Symposium on GIS in Fishery Science, the Spatial Processes and Management of Fish Populations Symposium, and the Second International Symposium on GIS/Spatial Analyses in Fishery and Aquatic Sciences. Each of these has had a significant GIS input. It is clear from the interest in these events that there is a great desire to find out what is happening and thus to at least informally organize in an attempt to share developments. The need to ardently promote cohesion within the community of fisheries GIS workers is a challenge that should be pursued with some vigor, since a failure here will result in much reinventing of the wheel. Other advantages of achieving a level of cohesion and organization include the increased possibilities of data sharing or trading, the regular scheduling of meetings (conferences, symposia, workshops), identifying cores of expertise, and putting governments or other authorities in touch with progress in this field. In the age of the Internet and electronic mailing or discussion lists, better organization should be possible.

2.3.3 Information Output

In this fragmented and diverse operational world of fisheries GIS, it has been difficult to promote the findings of the activities. As was mentioned above, conferences in this subject have only just begun, and, earlier in the chapter, it was indicated that little of the fisheries GIS work was appearing in the established scientific literature. These considerations relate to exhibiting the range of potential that GIS has to offer fisheries, and, to some extent, these considerations are now being dealt with. However, there are other output considerations that still pose a challenge. The first of these is to provide for the immediate and easy access to marine information and make interactive fisheries analyses possible within Internet-based systems. These subjects will not be fully discussed here because they are not confined to the fisheries GIS sphere, but readers wishing to know more could consult Plewe (1997) or Pienaar and van Brakel (1999). However, some important interactive fisheries-based Internet sites recently have come online. For instance, the FAO in Rome, in conjunction with the Support Unit for International Fisheries and Aquatic Research, is offering a unifying aid to fisheries research and GIS development through access to its "oneFish" research site (SIFAR 2003). In addition, fish stock assessment tools are now being offered provisionally over the Internet by the FAO and International Center for Living Aquatic Resources Management backed FiSAT2000 Internet site (FAO and ICLARM 2003). The challenge in using Internet-based technologies to obtain relevant output lies mainly in keeping up with developments in a rapidly changing set of technologies!

Perhaps a more important challenge here is that of instigating GIS-based systems that can provide for near- or real-time information outputs as a means of promoting efficient and timely management decision making. Some considerable efforts are being put into this. Demarcq (unpublished poster presentation from the First International Symposium on GIS in Fishery Science, 1999) shows how satellite imagery can be a source of sea surface temperatures, ocean color, and surface wind data. These can be viewed within a specifically developed software (Coastal Upwelling Structures from Satellite Images, CUSSI) to show relationships between these factors and catch distributions off the coast of Namibia. Moving to a real-time event, Kiyofuji et al. (2001) showed how satellite-derived data were being used to derive the relationship between migration routes of common squid *Todarodes pacificus* and Japanese fleet deployment in the southern Japan Sea. Similar GIS-based work on the coincidence of thermal fronts and optimum fishing locations has now been so successful that commercial companies are selling the GIS-derived data to interested fishers. It is not difficult to envisage other real-time fisheries output information that GIS might provide. For instance, maps could be created that represented fishing costs surfaces or more detailed "profit maximizing surfaces." Questions could be asked such as "given the variable catch potential in different areas and different prices operating at different ports, where ought my fishing effort be best used, and to what extent might I cover x fishing grounds for y species?" Though apparently complex, these are the types of questions that future shipboard GIS might be resolving. Finally, it should be mentioned that output from fisheries applications of GIS are likely to increase in their diversity as users increasingly appreciate the utility of GIS in providing additional output formats (textual, graphical, tabular), and as systems are adopted by a greater range of users in the fisheries industry itself and in associated management, business, scientific, or academic arenas.

2.4 ECONOMIC CHALLENGES

In this brief examination of economic challenges, it is taken for granted that costs for computing-based activities are likely to be relatively high, certainly in terms of establishing the basic working and systems environment. The main themes that concern us here include (1) why fisheries-based GIS operations are relatively expensive, and (2) why funds might be difficult to obtain.

2.4.1 Expenses Associated with Fisheries GIS

It is common knowledge that the costs involved in setting up and running a GIS activity are such that data purchases have risen as a proportion of all costs. This is because labor costs are relatively static over time, hardware and software costs have fallen dramatically, but improvements in GIS functional capability have required greatly improved data inputs in terms of both quality and quantity. Current data costs as a percentage of full GIS operating costs are around 80%, and this situation is likely to continue. For fisheries GIS, data may be especially expensive. It already has been intimated (Section 2.3.1) that data are very difficult to obtain, and these difficulties might only theoretically be overcome with large "cash injections." However, since the use of GIS methods rarely is institutionalized, it is doubtful that cash injections per se could improve the existing situation, certainly not in developing countries where fisheries management issues are critical. The highest data costs are those that incur vessel charter charges for data collection, but the purchase of satellite imagery can still be costly, as can the purchase of many off-the-shelf digital sets. In terms of labor inputs, digitizing can be expensive for individual projects, because, frequently, the necessary data are otherwise unavailable. A further factor applying in the fisheries GIS field, is that many GIS operations do not take place within a larger computing environment, that is, because of the small and fragmented nature of the institutions or groups working in this field. This often necessitates the purchase of a complete suite of equipment and data that may be dedicated to one purpose only.

In many developing countries, the costs involved in establishing a GIS for fishery purposes, may be seen as "a cost too far." Thus, GIS facilities can seldom be seen as a priority. The necessary equipment may need to be imported, and this often incurs high taxation mark-ups. Backup for a system is often minimal. Suitable data for use in a data-hungry environment invariably does not exist. Suitably qualified personnel may be hard to obtain, and the costs of training will be high. Additional support in terms of literature, journals, courses, and conferences can only be purchased at high rates, which are usually synchronized with developed world prices. The general situation, therefore, regarding costs involved for the implementation and management of fisheries-based GIS is that they are varied and often difficult to justify. In areas where fisheries are vital to the economy, there are frequently donor-supported projects, for example, in South Africa, Namibia, and some southeast Asian countries. Most of these projects are managed externally, though there may be some cooperation with local research institutions or universities. Some developing countries have a more secure source of local expertise, for example, Sri Lanka, Indonesia, and some of the Persian Gulf states, and, despite the necessity for initial funding (for equipment and training), programs are subsequently self-sustainable. Given that the plight of fisheries may be just as chronic in developing countries as in many European and North American fisheries, there is a challenge to get successful and affordable geotemporal monitoring and management systems in place and made sustainable.

2.4.2 Obtaining Funding for Fisheries GIS Activities

Unfortunately for those working in any area of GIS, it is frequently a challenge to prove the utility of the system. Any cost:benefit analyses are difficult to perform and must rely on subjective evaluations. The results of any analyses are likely to be highly variable, and the qualitative output from any fisheries GIS also will be variable. Although output can be achieved in an array of formats and the functional analyses performed are often powerful, it is still difficult to predict whether the output is sufficient to enhance decision making to any specific degree. Given this rather nebulous outlook, there is a major challenge for potential users of the systems to persuade management or decision makers to invest in them. If a positive decision is made, then available funding is unlikely to be on a large scale. This is a reflection of the often fragmented nature of the fisheries science and management administrative environment. Despite the fact that the fishing industry is probably the world's second largest economic activity in terms of the numbers of people directly or indirectly employed, the activity mainly takes place within a subsistence or semisubsistence economy, a fact that may even be true in many developed world fisheries. Fisheries, therefore, do not generate sufficient funds to attract what might be seen as a "luxury, peripheral add-on." A GIS activity is often funded from research grants, but invariably most of the grant will go toward some major research activity, with GIS being only a means of carrying out some of the analytical or presentational aspects of the project. Thus, GIS seldom attracts funding in its own right. This lack of status and recognition is a major challenge that needs to be overcome if adequate funding is to be obtained. The major factor that might bring about change is the dire plight that many of the world's fisheries are now in, and the fact that this plight is basically a result of spatiotemporal disequilibrium in the array of factors that control fish harvest or production. Clearly, GIS will eventually be shown to be the best tool to synthesize analyses across the biological, physical, social, and economic factors involved.

2.5 SOCIAL AND CULTURAL CHALLENGES

The final area of challenge to the adoption of GIS for fisheries-related purposes is that concerned with cultural and institutional norms and practices. As with economic considerations, this is an area that, to a large extent, conforms with levels of development, though it is important to say that this is not always the case. Campbell and Masser (1993) have shown that, for GIS to be successfully adopted by any organization or institution, it is important that certain favorable conditions exist. These are:

- there should be a high level of computing skills within the organization;
- the organization should have an innovative environment;
- the initial GIS applications should be simple, and some valued output should quickly be produced;
- projects should be managed carefully, with all GIS team members making valid contributions;
- there should be a stable organizational context; and
- there should be a high level of commitment to the goals of the GIS task.

Given the huge diversity within fisheries management and scientific institutions, it is easy to believe that the working ambience is not always conducive to GIS adoption. Though there is little specific evidence of outdated or entrenched attitudes prevailing in fisheries institutions, it is reasonable to suppose that many "champions" of GIS find it hard to promote the system's utility. After all, evidence for GIS success in the fisheries field is not exactly prolific, or, at least, it has not been adequately disseminated. Given this uncertainty, then it may be problematic to put forward a suggestion that an institution should invest x amount for a GIS and that it should be willing to support it over y period of time. In addition, there is an undoubted lack of appreciation that problems within the fishery domain are rooted in spatial differentiation—fisheries managers (and many others) often are not appreciative of the importance of the geographical perspective. From a different perspective, this investigator has found that there may be reticence on the part of commercial fishermen to adopt electronic catch-per-effort log books (mentioned in Section 2.3.1). It is recognized that the data obtained from log books may easily be linked to a GIS for spatial analyses, but the fishermen do not wish their favored (or successful) fishing grounds to be revealed, that is even though the authorities are bound to anonymity, and they already may be collecting this data in pencil and sheet form. One interesting way of meeting the challenge of adopting GIS to fisheries situations in developing countries comes with the production of the first GIS fisheries training manual (G. de Graaf et al., NEFISCO Foundation, unpublished paper presented at the Second International Symposium on GIS/Spatial Analyses in Fishery and Aquatic Sciences, 2002). This publication sets out a range of exercises, from very basic to quite advanced, and it has specifically shown the exact processes necessary to allow for GIS adoption to be achieved realistically within any fishery research or management area.

Introductions of GIS into certain cultural ambiences may have numerous challenges. Although the established literature has little to say on this regarding the fishery situation, a major FAO report (Campbell and Salagrama 1999) has looked into how indigenous knowledge on fisheries should be integrated into participatory activities among developed and developing peoples, and how a two-way interplay of ideas can best proceed in order to promote research. This process will be important in the fostering of GIS. The present investigator has had varied experience involving cultural aspects of GIS implementations. The challenges are largely centered upon working practices, acceptable standards, and spatial appreciation, and GIS has frequently been heard of as a salvation to existing problems. But when it comes to examining the intricacies of getting the system established and running, it is found that there has been little perception of the requisite inputs, especially that of the necessary data and the levels of expertise, plus the interest and support needed to sustain the system. This is often a fault of the instigating authority because they have tried to implement a technology-driven system into a cultural situation to which the technology has little relevance. Taylor (1991:71) noted, "If it [the GIS] is to be used in the context of development, then it must be introduced, developed, modified and controlled by indigenous people who understand the social, economic and political context of the situation as well as the technical capabilities of GIS." So the challenge here is to make certain that all aspects of GIS adoption can be realistically met within the confines of a particular cultural setting.

Many marine areas, including natural and managed fisheries ecosystems, occupy marine locations that straddle country territorial waters. The management of these areas may be pursued individually or in cooperation with the neighboring territory. Although cooperation is to be desired, there frequently are situations where the arrangements are far from satisfactory and where there is either a conflict in management strategies or where suitable management agreements cannot be reached. Under these circumstances, it may be difficult to implement desirable management systems. In some instances the fisheries of marine areas that straddle two or more territories may be in a critical state. Here it is imperative that cooperative management systems are used, but the difficulties of activating this can be almost insurmountable. For instance, Richards and Bohnsack (1990:51) have outlined the critical state of the Caribbean Sea fisheries ecosystem but have noted that "coordination is difficult because the Caribbean is divided geographically and politically into at least 38 countries or territories with different cultures and varying levels of public education and economic development." In this instance, the introduction of a complex technological system that transcends the 38 borders would be a major challenge indeed.

There is often a lack of the innate understanding concerned with geographical cognition. This means that users of a GIS frequently have little idea on the possibilities of data inputs, on the range of suitable analyses, and on requisite interpretations of the output information. The veracity of the mapped distributions may not be recognized, and important subtle interpretations are missed. As Meaden (2001:19) has noted, "Geographic expertise involves recognition of spatial relationships in terms of adjacency, ubiquity, heterogeneity, contiguity, etc. Visual discrimination is a vital tool (or ability) in seeking out these subtle relationships. Likely flows and interactions must be recognized and understood, as must spatial patterns, surface trends and zonal forms." The existence and appreciation of sound geographical cognition is an absolute imperative for success in most GIS work. So the challenge lies in imparting this. To an extent, this cognition is an innate skill, though undoubtedly more needs to be done in promoting certain geographic basics within geography and GIS courses.

2.6 CONCLUSIONS

In outlining the challenges that confront those working in fisheries GIS, it has not been possible to demonstrate the numerous interlinkages that exist between facets of each challenge. The reader must, therefore, not see challenges as necessarily being separately identifiable. The overcoming of one obstacle to progress might in fact alleviate many problems or reduce the significance of a range of challenges, though progress will always throw up new challenges. What is of paramount importance is that a basic means of attacking some of the challenges are in place. The expansion of GIS in fisheries will probably mean the adoption or achievement of the following:

- a reduction in data costs;

- a proliferation of data-gathering technologies, in conjunction with data-dissemination means;

- better organization among practitioners at an international level;

- more cooperative working (networking) between institutions;

- a range of conferences at a more local (regional) level;

- more fisheries GIS material being exemplified in "recognized" publishing outlets;

- some well-funded GIS-based projects that can serve to exemplify a range of analytical and presentational features;

- a stronger move toward the international standardization of data gathering formats;

- progress in 3-D and 4-D GIS and data storage and modeling structures; and

- more progress in easily accessible marine information sources.

This is a lengthy list, and it illustrates that there may be some way to go before present challenges are alleviated. However, if we know what does need doing, then present challenges are probably on their way to becoming mastered—though only perhaps before the next ones occur.

Perhaps the major challenge that workers in the field of fisheries GIS face is one that is beyond their control. We are presently witnessing enormous changes in the conditions of fish stocks. There are two main (and well known) causes to this—that of lack of ownership of the fish resources, which usually leads to the "race to fish" and, thus, overfishing, and that of environmental change. Overfishing is a function of the lack of, or an inadequate, management system, which itself often results from the inability of scientists, fishers, and politicians to agree on the requisite solutions to individual fishery scenarios in a situation of rapid change. Environmental change is a function of both anthropogenic interference with natural aquatic systems and with enhanced global warming and its effect on fisheries ecosystems through increasing climate variability plus water-level changes. The ultimate challenge faced by workers using GIS will be keeping up with the rate of both sets of change. As soon as new models are developed, they will need to incorporate adjustment mechanisms that may be difficult to calculate. And, although the present rate of fish distributions is changing rapidly, in the future, this change is likely to proceed at an exponential rate. These aspects of change mean that data decay rates are accelerated, usually in an unpredictable fashion. Under these circumstances, it is likely to be extremely challenging to keep abreast with change itself. Despite these difficulties, it will be up to us who are working in this field to prove that GIS has the capacity to meet this challenge and, hence, to bring about a reversal in the fortunes of fisheries.

2.7 REFERENCES

ACZISC (Atlantic Coastal Zone Information Steering Committee). 2003. A guide to coastal information in Atlantic Canada. ACZISC. Available: *www.dal.ca/aczisc/* (September 2003).

Altman, D. 1994. Fuzzy set theoretic approaches for handling imprecision in spatial analysis. International Journal of Geographical Information Systems 8(3):271–289.

Anselin, L. 1996. The Moran scatterplot as an ESDA tool to assess local instability in spatial association. Pages 111–125 *in* I. Masser and F. Salge, editors. Spatial analytical perspectives on GIS: Gisdata4. Taylor and Francis, London.

Ault, J. S., and J. Luo. 1998. Coastal bays to coral reefs: visualization of a spatial multi-stock production model. ICES C.M. 1998/S:1, Lisbon, Portugal.

Bailey, T. C., and A. C. Gatrell. 1995. Interactive spatial data analysis. Longman Scientific and Technical, Harlow, Essex, UK.

Bartlett, D. J., and D. J. Wright. 2000. Epilogue. Pages 309–315 *in* Wright and Bartlett (2000).

Benthos, Inc. 2003. Benthos, Inc. offshore equipment. Benthos, Inc. Available: *www. benthos.com* (September 2003).

Brady, W. W., and G. L. Whysong. 1999. Modeling. Pages 293–324 *in* S. Morain, editor. GIS solutions in natural resource management: balancing the technical-political equation. OnWord Press, Santa Fe, New Mexico.

Brown, K. J., and A. A. Lacey, editors. 1990. Reaction–diffusion equations. Clarendon Press, Oxford, UK.

Burrough, P. A., and A. U. Frank. 1996. Geographic objects with indeterminate boundaries. Taylor and Francis, London.

Caddy, J. F., and F. Carocci. 1999. The spatial allocation of fishing intensity by port-based inshore fleets: a GIS application. ICES Journal of Marine Science 56:388–403.

Camara, A. S., F. Ferreira, and P. Castro. 1996. Spatial simulation modelling. Pages 201–212 *in* I. Masser and F. Salge, editors. Spatial analytical perspectives on GIS: Gisdata4. Taylor and Francis, London.

Campbell, H., and I. Masser. 1993. Implementing GIS: the organizational dimension. Association for Geographic Information, Conference papers, the world of GIS, AGI93, Birmingham, UK.

Campbell, J., and V. Salagrama. 1999. New approaches to participatory research in fisheries. Food and Agriculture Organization of the United Nations and Support Unit for International Fisheries and Aquatic Research, Rome.

Corsi, F., S. Agnesi, and G. Ardizzone. 2001. Integrating GIS and surplus production models: a new approach for spatial assessment of demersal resources? Pages 143–156 *in* Nishida et al. (2001).

D'Angelo, D. J., L. M. Howard, J. L. Meyer, S. V. Gregory, and L. R. Ashkenas. 1995. Ecological uses for genetic algorithms: predicting fish distributions in complex physical habitats. Canadian Journal of Fisheries and Aquatic Sciences 52:1893–1908.

Dickey, T. D. 1990. Physical–optical–biological scales relevant to recruitment in large marine ecosystems. Pages 82–98 *in* K. Sherman, L. M. Alexander, and B. D. Gold, editors. Large marine ecosystems: patterns, processes and yields. American Association for the Advancement of Science, Washington, D.C.

FAO (Food and Agriculture Organization of the United Nations) and ICLARM (International Center for Living Aquatic Resources Management). 2003. FiSAT II. FAO-ICLARM stock assessment tools. FAO and ICLARM. Available: *www.fao.org/fi/statist/fisoft/fisat/index.htm* (September 2003).

Fiksen, O., J. Giske, and D. Slagstad. 1995. A spatially explicit fitness-based model of capelin migrations: the Barents Sea. Fisheries Oceanography 4(3):193–208.

Fisher, W. L., P. E. Balkenbush, and C. S. Toepfer. 2001. Development of stream fish population sampling surveys and abundance estimates using GIS. Pages 253–263 *in* Nishida et al. (2001).

Fotheringham, A. S., C. Brunsdon, and M. Charlton. 2000. Quantitative geography: perspectives on spatial data analysis. Sage Publications, London.

Fotheringham, A. S., and P. Rogerson. 1994. Spatial analysis and GIS. Taylor and Francis, London.

Frank, K. T., and J. E. Simon. 1998. An evaluation of the Emerald/Western bank juvenile haddock closed area. ICES C.M. 1998/U:1, Lisbon, Portugal.

Giske, J., G. Huse, and O. Fiksen. 1998. Modelling spatial dynamics of fish. Reviews in Fish Biology and Fisheries 8:57–91.

Grindrod, P. 1991. Patterns and waves: the theory and applications of reaction–diffusion equations. Clarendon Press, Oxford, UK.

Hare, J. A., and R. K. Cowen. 1996. Transport mechanisms of larval and pelagic juvenile bluefish (*Pomatomus saltatrix*) from South Atlantic Bight spawning grounds to Middle Atlantic Bight nursery habitats. Limnological Oceanography 41:1264–1280.

Harris, G. P. 1986. Phytoplankton ecology. Chapman and Hall, New York.

Hilborn, R. 1990. Determination of fish movement patterns from tag recoveries using maximum likelihood estimators. Canadian Journal of Fisheries and Aquatic Sciences 47:635–643.

Hilborn, R., and M. Mangel. 1997. The ecological detective: confronting models with data. Monographs in Population Biology 28.

Hooge, P. N., W. M. Eichenlaub, and E. K. Soloman. 1999. Using GIS to analyze animal movements in the marine environment. Pages 37–51 *in* G. H. Kruse, N. Bez, A. Booth, M. W. Dorn, S. Hill, R. N. Lipcius, D. Pelletier, C. Roy, S. J. Smith, and D. Witherell, editors. Spatial processes and mangement of marine populations. University of Alaska Sea Grant Report AK-SG-01-02.

Kelly, N. M. 2001. Spatial pattern of forest clearing and potential sediment delivery in a northwestern Oregon watershed. Pages 281–294 *in* Nishida et al. (2001).

Kemp, Z., and G. J. Meaden. 1998. Towards a comprehensive fisheries management information system. Pages 522–531 *in* A. Eide and T. Vassdal, editors. Ninth biennial conference of the International Institute of Fisheries Economics and Trade, volume 2. University of Tromso, Norwegian College of Fisheries Science, Tromso, Norway.

Kiyofuji, H., S. Saitoh, Y. Sakurai, T. Hokimoto, and K. Yoneta. 2001. Spatial and temporal analysis of fishing fleet distribution in the southern Japan Sea in October 1996 using DMSP/OLS visible data. Pages 178–185 *in* Nishida et al. (2001).

Kohn, A. J. 1997. Why are coral reef communities so diverse? Pages 201–215 *in* R. F. G. Ormond, J. D. Gage, and M. V. Angel, editors. Marine biodiversity: patterns and processes. Cambridge University Press, Cambridge, UK.

Lalwani, C. S., and T. Stojanovic. 1999. The development of marine information systems in the UK. Marine Policy 23(4–5):427–438.
Levin, S. A. 1992. The problem of pattern and scale in ecology. Ecology 73(6):1943–1967.
Levine, N. 1996. Spatial statistics and GIS. Journal of the American Planning Association 62(3):381–391.
Logerwell, E. A., and P. E. Smith. 2001. GIS mapping of survivors' habitat of pelagic fish off California. Pages 51–64 *in* Nishida et al. (2001).
Lohrmann, A. 2000. AWAC—acoustic wave and current measurements. Pages 55–56 *in* A. Pink, editor. Integrated coastal zone management: strategies and technologies for ICZM. ICG Publishing, London.
Longley, P. A., S. M. Brooks, R. McDonnell, and W. MacMillan, editors. 1998. Geocomputation: a primer. Wiley, Chichester, UK.
Lucas, A. 2000. Representation of variability in marine environmental data. Pages 53–74 *in* Wright and Bartlett (2000).
MacCall, A. D. 1990. Dynamic geography of marine fish populations. Washington Sea Grant Program, University of Washington Press, Seattle.
Malczewski, J. 1999. GIS and multicriteria decision analysis. Wiley, New York.
Mann, K. H., and J. R. N. Lazier. 1998. Dynamics of marine ecosystems. Blackwell Scientific Publications, Oxford, UK.
Meaden, G. J. 2000. Applications of GIS to fisheries management. Pages 205–226 *in* Wright and Bartlett (2000).
Meaden, G. J. 2001. GIS in fisheries science: foundations for the new millennium. Pages 3–29 *in* Nishida et al. (2001).
Meaden, G. J., and T. Do Chi. 1996. Geographical information systems: applications to marine fisheries. FAO Fisheries Technical Paper 356.
Monmonier, M. 1996. How to lie with maps. University of Chicago Press, Chicago.
Mullin, M. M. 1993. Webs and scales: physical and ecological processes in marine fish recruitment. Washington Sea Grant Program, Seattle.
Nams, V. O. 1999. Animal movement paths and responses to spatial scale. Pages 61–65 *in* F. Huettmann and J. Bowman, editors. Investigation of animal movements, workshop Proceedings. Sir James Dunn Wildlife Research Centre, University of New Brunswick, Fredericton.
Nishida, T., P. J. Kailola, and C. E. Hollingworth, editors. 2001. Proceedings of the first international symposium on GIS in fishery science. Fishery GIS Research Group, Saitama, Japan.
Okubo, A. 1980. Diffusion and ecological problems: mathematical models. Springer-Verlag, New York.
Openshaw, S., and G. Clarke. 1996. Developing spatial analysis functions relevant to GIS environments. Pages 21–37 *in* I. Masser and F. Salge, editors. Spatial analytical perspectives on GIS: Gisdata4. Taylor and Francis, London.
Pickles, J., editor. 1995. Ground truth: the social implications of geographic information systems. The Guilford Press, New York.
Pienaar, M., and P. van Brakel. 1999. The changing face of geographic information on the web: a breakthrough in spatial data sharing. The Electronic Library 17(6):365–371.
Plewe, B. 1997. GIS online: information retrieval, mapping, and the internet. OnWord Press, Santa Fe, New Mexico.
Possingham, H. P., and J. Roughgarden. 1990. Spatial population dynamics of a marine organism with a complex life cycle. Ecology 71(3):973–985.
Raper, J. F. 2000. Multidimensional geographic information science. Taylor and Francis, London.

Richards, W. J., and J. A. Bohnsack. 1990. The Caribbean Sea: a large marine ecosystem in crisis. Pages 44–53 *in* K. Sherman, L. M. Alexander, and B. D. Gold, editors. Large marine ecosystems: patterns, processes and yields. American Association for the Advancement of Science, Washington, D.C.

Ricklefs, R. E. 1990. Scaling pattern and process in marine ecosystems. Pages 169–178 *in* K. Sherman, L. M. Alexander, and B. D. Gold, editors. Large marine ecosystems: patterns, processes and yields. American Association for the Advancement of Science, Washington, D.C.

Rubec, P. J., S. G. Smith, M. S. Coyne, M. White, A. Sullivan, T. C. MacDonald, R. H. McMichael, D. T. Wilder, M. E. Monaco, and J. S. Ault. 2001. Spatial modeling of fish habitat suitability in Florida estuaries. University of Alaska Sea Grant Report AK-SG-01-02:1–18.

Schneider, D. C. 1994. Quantitative ecology: spatial and temporal scaling. Academic Press, San Diego, California.

Schneider, D. C. 1998. Applied scaling theory. Pages 253–269 *in* D. L. Peterson and V. T. Parker, editors. Ecological scale: theory and applications. Columbia University Press, New York.

Schneider, D. C., T. Bult, R. S. Gregory, D. A. Methven, D W. Ings, and V. Gotceitas. 1999. Mortality, movement and body size: critical scales for Atlantic cod (*Gadus morhua*) in the Northwest Atlantic. Canadian Journal of Fisheries and Aquatic Sciences 56(Supplement S1):180–187.

SIFAR (Support Unit for International Fisheries and Aquatic Research). 2003. oneFish. One Internet portal, all fisheries research. SIFAR. Available: *www.onefish.org* (September 2003).

Smith, P. E. 1978. Biological effects of ocean variability: time and space scales of biological response. Journal du Conseil International pour l'Exploration de la Mer 173:117–127.

Taylor, D. R. F. 1991. GIS and developing nations. Pages 71–83 *in* D. J. Maguire, M. F. Goodchild, and D. W. Rhind, editors. Geographical information systems: principles and applications. Longmans, London.

Turchin, P. 1998. Quantitative analysis of movement: measuring and modeling population redistribution in animals and plants. Sinauer Associates Inc., Sunderland, Massachusetts.

Tyler, J. A., and K. A. Rose. 1994. Individual variability and spatial heterogeneity in fish population models. Reviews in Fish Biology and Fisheries 4:91–123.

Varma, H. 2000. Applying spatio-temporal concepts to correlative data analysis. Pages 75–93 *in* Wright and Bartlett (2000).

Walden, J. B., D. D. Sheehan, B. P. Rountree, and S. F. Edwards. 2001. Integrating geographic information system with mathematical programming models to assist in fishery management decisions. Pages 167–177 *in* Nishida et al. (2001).

Weber, L. H., S. Z. El-Sayed, and I. Hampton. 1986. The variance spectra of phytoplankton, krill and water temperature in the Antarctic Ocean south of Africa. Deep-Sea Research 33:1329–1343.

Westwood, J. 2000. The world UUV report. Douglas-Westwood Ltd., Canterbury, UK.

Wright, D., and D. Bartlett, editors. 2000. Marine and coastal geographic information systems. Taylor and Francis, London.

Wroblewski, J. S., J. G. Richman, and G. L. Mellot. 1989. Optimal wind conditions for the survival of larval northern anchovy, *Engraulis mordax*: a modeling investigation. U.S. National Marine Fisheries Service Fishery Bulletin 87:387–398.

Zhang, Z., and D. A. Griffith. 1997. Developing user-friendly spatial statistical analysis modules for GIS: an example using ArcView. Computers, Environment and Urban Systems 21(1):5–29.

Chapter 3

Geographic Information Systems Applications in Stream and River Fisheries

William L. Fisher and Frank J. Rahel

3.1 INTRODUCTION

Streams and rivers are linear habitat features on the landscape that are inherently difficult to study (Fausch et al. 2002). However, streams drain watersheds, and watersheds provide a convenient geographic unit for monitoring and managing the impacts of land-use activities on stream fish habitat and populations. This characterization of stream systems as lines (streams) within polygons (watersheds) illustrates their nested character, the spatial interconnectedness of the aquatic and terrestrial systems, and the potential utility of geographic information systems (GIS) for studying and managing stream ecosystems.

Geographic information applications in stream and river fisheries are increasing in number and complexity. The most common applications are mapping and modeling habitats and fish distributions, mapping watershed land uses and modeling their impacts on habitats and fish populations, and predicting the effects of environmental changes on the temporal and spatial distributions of fishes (Giles and Nielsen 1992; Isaak and Hubert 1997; Fisher and Toepfer 1998). In this chapter, we first describe GIS techniques and the challenges of using them in streams and rivers. We then summarize recent applications of GIS in fisheries science in streams and rivers. We conclude by discussing the current state-of-the-art and potential applications of GIS for riverine fish and fisheries management.

3.2 TECHNIQUES AND CHALLENGES OF USING GIS

Riverine systems present different challenges compared with other freshwater environments such as lakes and reservoirs for mapping and modeling fish distributions and habitats with GIS. This is because lakes and reservoirs are areal features on the landscape and, as such, they are amenable to many of the GIS techniques used for terrestrial ecosystems. Streams are linear features on the landscape, and perhaps the greatest challenge is in acquiring, analyzing, and displaying spatial data for stream networks. This also presents a great opportunity, because streams and rivers are much like transportation systems (roads, highways) for which there have been considerable advances in the application of GIS (Goodchild 2000). In this section, we discuss sources of data and GIS data structures for streams and rivers, analysis techniques for riverine features, and issues related to mapping riverine fish distributions.

3.2.1 Data for Streams and Rivers

3.2.1.1 *Data Sources.* Stream characteristics can be described at multiple spatial scales. Frissell et al. (1986) provided a hierarchical framework for classifying stream habitat at multiple spatial scales in the watershed context. This widely accepted framework classifies streams into habitat systems based on spatial scales that range from fine (microhabitat and channel units) to intermediate (reaches and segments) to coarse (drainage basins). Depending on the scale of choice, acquiring data can be either a small or large challenge. Geospatial data are either obtained from field measurements (primary data) or existing databases (secondary data). Several GIS databases are available at global or continental scales that could be used for applications such as mapping fish species distributions or monitoring aquatic species invasions. Global databases contain a variety of geographic and environmental features such as river drainages, ocean coastlines, elevation, land cover, temperature, and precipitation. For example, the Digital Chart of the World (DCW) is a vector-based map of the world at a scale of 1:1,000,000, which was developed by Environmental Systems Research Institute (ESRI, Redlands, California) for ARC/INFO. The DCW database consists of over 200 attributes in 17 thematic layers with text annotations for cities, mountains, lakes, and other geographic features. Geographic databases for various countries and continents also are available from the United Nations Environmental Program, Global Resource Information Database; U.S. Geological Survey, Data Center; and the Food and Agriculture Organization of the United Nations.

Although global scale data can provide information for continental fisheries issues, these data are not particularly useful to most fisheries researchers and managers who work at landscape or watershed-level scales (e.g., 1:100,000–1:24,000). The GIS databases at these finer scales describe watershed characteristics such as bedrock geology, soils, land use and land cover, elevation, hydrography, and transportation networks (Table 3.1). These databases can be used to develop base maps for stream fisheries projects as well as assess land-water relationships. In North America, several governmental agencies in the United States and Canada have developed GIS databases (Table 3.1). Private vendors such as ESRI and GeoCommunity also provide GIS databases and coverages for a fee.

A limited amount of GIS data is available at even finer scales (e.g., less than 1:24,000), such as individual streams and stream reaches. Extensive stream-reach surveys and inventories of habitat types and fishes have been developed by a few North American federal, state, and provincial agencies (e.g., U.S. Forest Service, Oregon Department

Table 3.1 Watershed characteristics and some sources of digital GIS databases in North America. DEM = digital elevation model; DTEM = digital terrain elevation model; USDA = U.S. Department of Agriculture; DOD = U.S. Department of Defense; DRG = digital raster graphic; DOQ = digital orthophoto quad.

Characteristic	Database	Source and Internet address	Scale	Comments
Geology (bedrock and quaternary)	Geological surveys	Association of American State Geologists *www.kgs.ukans.edu/AASG/AASG.html* Geological Survey of Canada *www.nrcan.gc.ca/gsc*	1:250,000	Small scale allows for landscape-level analyses
Soils	Soil surveys	USDA Natural Resources Conservation Service *http://nasis.nrcs.usda.gov*, *www.ftw.nrcs.usda.gov/stat_data.html* Agriculture and Agri-Food Canada, National Soils Database *http://sis.agr.gc.ca/cansis/nsdb/index.html*	1:20,000–1:250,000	Large-scale data is useful for riparian characterizations
Elevation and watershed boundaries	DEM; DTEM data	U.S. Geological Survey *http://edc.usgs.gov/doc/edchome/ndcdb/ndcdb.html* DOD, National Imagery and Mapping Agency *www.nima.mil* Natural Resources Canada, Canadian Digital Elevation Data *www.geod.nrcan.gc.ca*	1:20,000–1:250,000	Currently, most DEMs are small scale
Hydrography (stream networks)	Digital line graph; TIGER/line files	U.S. Geological Survey U.S. Census Bureau *www.census.gov/geo/www/tiger/index.html* Natural Resources Canada, Geometrics Canada *http://geonames.nrcan.gc.ca*	1:24,000–1:2,000,000	Provide base maps for planning, mapping, and analysis
Dam locations	National Inventory of Dams	U.S. Army Corps of Engineers *http://crunch.tec.army.mil/nid/webpages/nid.html*	Not available	A georeferenced database of U.S. dam locations

Table 3.1 Continued.

Characteristic	Database	Source and Internet address	Scale	Comments
Transportation (road systems)	Digital line graph; TIGER/line files	U.S. Geological Survey U.S. Census Bureau Natural Resources Canada, Geometrics Canada	1:24,000–1:2,000,000	Intersections with streams identify access and impact sites; compute road density within watersheds
Land cover/land use	Land use and land cover; National Wetlands Inventory; DRGs; DOQs	U.S. Geological Survey U.S. Fish and Wildlife Service *www.nwi.fws.gov/* State or provincial agencies	1:100,000–1:250,000; 1:24,000	Useful for watershed and riparian analyses; DRGs are digitized topographic maps; DOQs are rectified aerial images matched to DRGs
Place names	Geographic names information system	U.S. Geological Survey *http://geonames.usgs.gov* Natural Resources Canada, Geomatics Canada	1:24,000	Names of physical and cultural places, features, and areas

of Fish and Wildlife, and British Columbia Fisheries). For example, the U.S. Forest Service has used basin area stream survey methods to inventory physical, chemical, and biological characteristics of streams (Clingenpeel and Cochran 1992) and analyze the cumulative effects of best forest management practices in the Ouachita National Forest in Arkansas (USFS 1994). These surveys map geomorphic channel units (e.g., riffles, runs, pools) in streams, as well as fish distributions, spawning sites, and they also may include biological and water quality data. O'Connor and Kennedy (2002) compared three survey techniques to assess salmonid habitat in three river systems in Northern Ireland. They found that a semiquantitative method, which employed one surveyor who measured only the width of habitat units along the river and incorporated this information into a GIS, was more efficient than a fully quantitative survey of the habitat units, which required two surveyors, and was more accurate than a truncated survey, which was similar to the quantitative survey but measured shorter representative reaches of the study sections.

Base maps for habitat surveys usually include channel boundaries and some instream features (e.g., islands, boulders) digitized from aerial photographs, digital orthophotoquads, or digital raster graphs. However, detection of boundaries and features is dependent on the scale of the image and the size of the stream. Ground resolution is the smallest measurable detail on a remotely sensed image. Ground resolutions range from 1 m on aerial photographs (1:16,000) to 4 m on satellite imagery (1:250,000; Aldrich 1979). Care should be taken to match the scale of the imagery to the size of the stream channel being mapped. Johnson and Gage (1997) describe types of satellite imagery and aerial photography and their applications in the analysis of aquatic ecosystems.

A recent advance in mapping river systems at fine scales is the use of high spatial resolution imagery. Hyperspectral instruments, such as AVIRIS, Probel, and HyMap, and thermal infrared remote sensors mounted on aircraft (airplanes or helicopters) flying at low elevations (300–700 m above the ground surface) allow for acquisition of imagery ranging from 0.2 to 5.0 m, which is useful for detecting habitat features and conditions in streams and rivers. These instruments have been used to identify and classify instream channel units and features (e.g., riffles, runs, pools, woody debris) in 5th order and smaller streams at varying degrees of accuracy (Wright et al. 2000; Marcus et al. 2002). Marcus (2002) used 1-m hyperspectral imagery collected with a Probel sensor to map channel units in the Lamar River, Wyoming. Maximum likelihood supervised classification of the imagery yielded an overall accuracy of 75% and accuracies ranging from 91% for pools, 87% for glides, 76% for riffles, and 85% for eddy drop zones. Torgersen et al. (2001) used forward-looking infrared remote sensing systems at low altitudes to generate continuous data on stream temperatures with ground resolutions of 0.2–0.4 m. This type of thermal remote sensing proved highly effective for studying spatial patterns of stream temperature at a resolution and extent previously unattainable with instream data recorders.

Although the accessibility of spatial data for watershed and landscape features is increasing, the availability of georeferenced fish databases is limited. Such data currently resides in project reports, agency files, museum collection records, and various other publications, but has not been compiled into digital databases for many states and provinces in North America. States in the United States with fish databases that are georeferenced include Illinois, Maryland, Missouri, Virginia, and Oregon. Information on the general distribution of fish (e.g., county, province) and life history characteristics is available on the Internet for California, Colorado, Kentucky, Oregon, Utah, Missouri, Virginia, and Maryland in the United States; British Columbia in Canada; and Belize in Central America. Standardization of data among agencies and entities still remains a need for spatial databases, although there has been considerable progress on this front in the United States through the Federal Geographic Data Committee (FGDC 2003).

3.2.1.2 ***Data Structures.*** Both raster and vector GIS data structures are used to depict characteristics of stream ecosystems, and the choice of a data structure influences the types of analyses that can be performed and the quality of the image display. The raster data structure portrays features as a two-dimensional array of equal-area grid cells that usually are square. Each grid cell is assigned an explicit attribute value for the measurement. Raster data structures are oriented toward analytical operations of data, not display. The vector data structure portrays features as points, lines, or polygons. Objects in a vector data structure are located by coordinate measurements in a spatial reference system. Unlike raster data structures, attributes of each object are stored in a separate database file and linked to that object. A vector data structure is preferred if the depicted objects are discrete and precisely represented (e.g., fish sampling locations, roads, stream or land ownership boundaries) or if the intended analysis involves measurement or calculation using linear features (e.g., stream networks). A raster data structure is preferred when the depicted objects are spatially continuous (e.g., land cover, soil types, elevation, remotely sensed images) or when the intended analysis involves spatially distributed modeling (e.g., cartographic and statistical modeling; Johnston 1998).

3.2.2 Spatial Analysis of Streams and Rivers

A GIS provides a variety of techniques for analyzing riverine ecosystems. These techniques can be used to define stream drainage systems, to describe watershed characteristics, and to characterize fish populations and communities. The GIS techniques can be grouped according to the dimensional space of the feature: zero-dimensional (0-D) or point features (e.g., sample sites), one-dimensional (1-D) or linear features (e.g., stream networks), two-dimensional (2-D) or areal features (e.g., channel units, watershed boundaries), and three-dimensional (3-D) or topographic features (e.g., elevation, bathymetry, volume). Below, we briefly describe the GIS operations associated with these features (except 0-D features, which are incorporated into the other features), provide example applications and discuss challenges associated with their application in riverine ecosystems.

3.2.2.1 ***Linear Features.*** Streams and rivers are linear, 1-D features on the landscape that are relatively stable over time. In a vector GIS, 1-D linear objects are defined geometrically as line segments, strings, and arcs, and topologically as links and chains bounded by nodes (Johnston 1998). In a raster GIS, linear objects are grid cells arrayed in either a straight line or a diagonal stair step. Operations on linear systems are used to classify and analyze linear features (e.g., stream segments) that possess various attributes (e.g., stream habitat types or fish abundances). Such features may be fixed (e.g., roads) or ephemeral (e.g., fish movements), and real (e.g., stream channels) or contrived (political boundaries; Johnston 1998). With a linear system, it is possible to specify a location, such as a stream segment, with a single number in 1-D space. Therefore, 1-D references are much simpler than 2-D references because only one measurement is needed, and the distance between to two linear measurements is much easier to compute than the distance between two points in 2-D or 3-D reference systems (Goodchild 2000).

Measurements of linear features such as stream channels, tend to be less complicated and more accurate with vector than raster data structures in GIS (DeMers 1997; Johnston 1998). This is because streams usually are sinuous, and it is possible for an entire meander to occur within a single grid cell, in which case the length of the stream is underrepresented by the measurement of the grid cell (Figure 3.1). Johnston and Naiman (1990) found that each pixel of a rasterized hydrologic database represented 7.61 m of a stream channel, which was 9% longer than the 7-m pixel dimension obtained for a straight stream channel that paralleled the raster grid. Thus, it is better to use a vector data structure when the analysis relies on the measurement of linear objects such as streams and rivers.

Linear operations are useful in characterizing hydrological aspects of stream ecosystems that have relevance to fisheries. ArcHydro (ESRI) is a recently developed GIS data model that provides a data structure and related operations for a variety of water resources features, including hydrography, drainage features, channels, and networks and for time-series analysis (Maidment 2002). It can be used to create stream networks and watershed boundaries by using GIS operations on raster digital elevation models (DEM; see Section 3.2.2.3). A problem with the stream network derived from a DEM, however, is that it generally does not fit the vector river network (Figure 3.1) because of the vertical resolution of the topographic data and the artificial nature of the network. A. Corley (National Salmon and Trout Fisheries Centre, Environment Agency, UK, unpublished abstract from the Second International Symposium on GIS/Spatial Analyses in Fishery and Aquatic Sciences, 2002) used ArcHydro to preprocess DEMs with a vector (digital line graphs, DLG) river network by using the "agree" method to improve the fit of the DEM-derived river network.

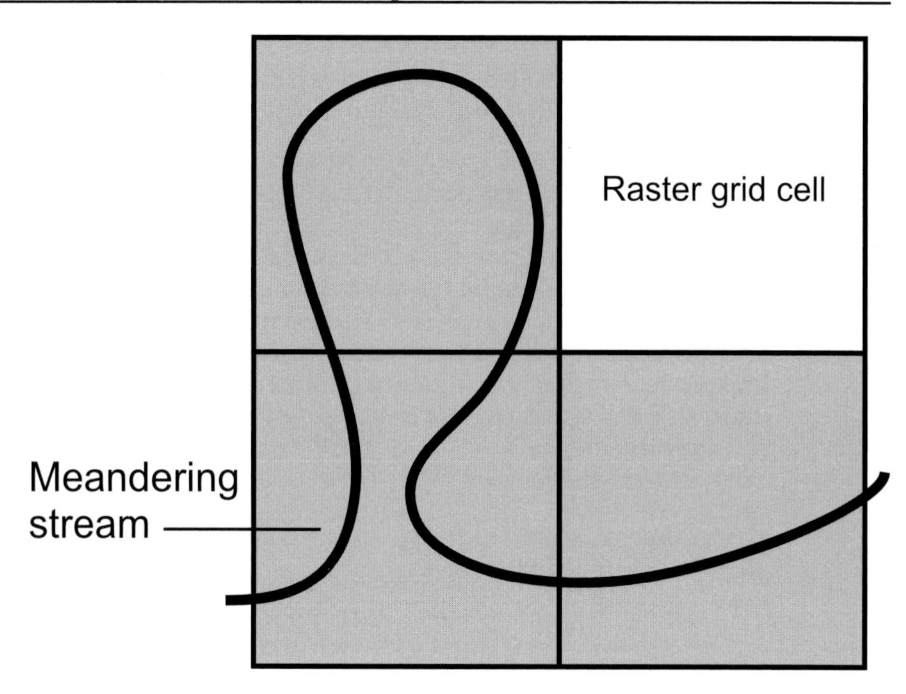

Figure 3.1 The length of meandering stream, where an entire meander is contained within a single grid cell, is underrepresented by measurements based on raster grid cells (from DeMers 1997).

Flow path analysis is a linear operation that can be used with raster or vector data to trace the path of a pollutant from a point source or movement of a fish through a stream drainage network by using ArcGIS Spatial Analyst (ESRI 2001a). Rectilinear (straight line segment) networks are composed of physical pathways (e.g., roads, paths) that guide either unidirectional or bidirectional movements. Although rectilinear networks are important to human-oriented GIS applications such as road transportation analysis (Goodchild 2000), their application to animal movements, particularly fish, are virtually unexplored.

Dynamic segmentation is a powerful tool for manipulating and analyzing linear systems. It allows multiple sets of attributes to be associated with any portion of a linear feature by dividing a linear feature into segments, such that each segment can be identified, manipulated, and assigned attributes independently of the others in the overall linear feature. Attributes of the linear feature can be stored, displayed, queried, and analyzed without affecting the underlying *x,y* coordinates of the linear data (ESRI 2001b). Radko (1997) used dynamic segmentation in ARC/INFO to link fish habitat inventory data sets and other basinwide stream surveys to line segments representing stream reaches and then examined the influence of fine sediments on fish habitat conditions (Figure 3.2). Torgersen et al. (1999) georeferenced stream survey data on distribution, stream temperature, and aquatic habitat of chinook salmon *Oncorhynchus tshawytscha* to digital hydrography layers by using dynamic segmentation tech-

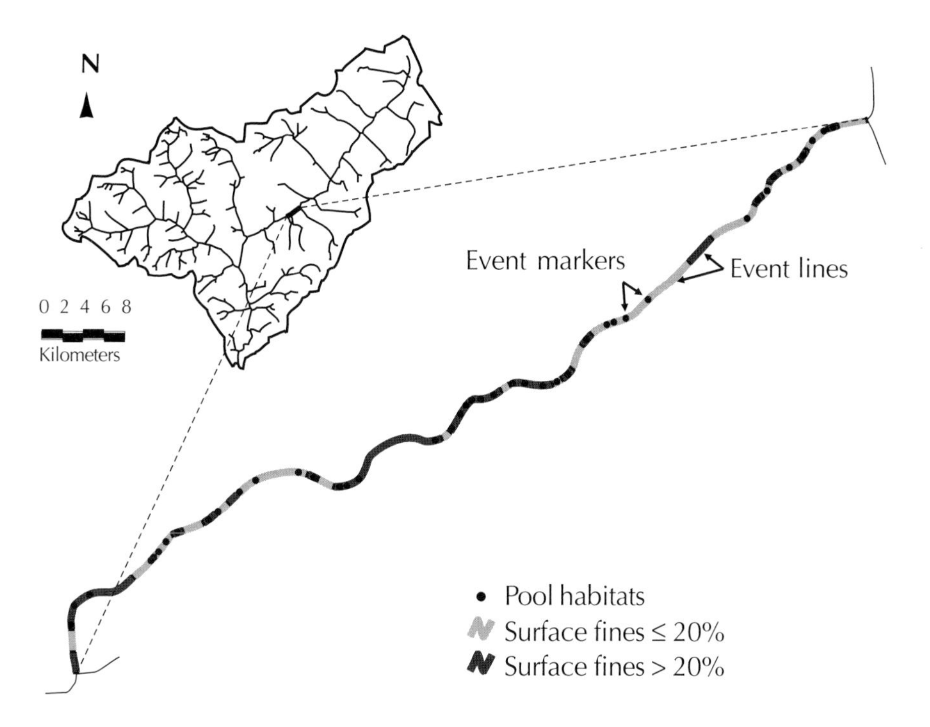

Figure 3.2 An example of using dynamic segmentation and the "eventsource" command in ARC/INFO to query and display the location of pool habitats (event markers) in relation to areas with surface fine sediments (event lines) in a reach of the Yankee Fork of the Salmon River, Idaho (adapted from Radko 1997).

niques. Point data, such as salmon locations and stream temperature sample points obtained from remote sensing, were assigned route-measure coordinates in the GIS data layers. In a similar study on the influence of thermal regime on trout distribution at the watershed level, Gardner et al. (2003) used geostatistical modeling techniques to predict instream temperature from temperature loggers placed throughout the watershed. They created target prediction points by making a systematic set of points every 1,000 m along the stream network by using dynamic segmentation. Distance between the target points and the temperature-logger locations were calculated for three metrics: Euclidean distance, networked distance, and weighted network distance. They derived predicted values for the target points and mapped them by using the "line" option of dynamic segmentation in ArcView (ESRI 2001b).

3.2.2.2 *Areal Features.* Riverine ecosystems consist of many 2-D, areal features such as stream channel units, riparian zones, and land cover types. In a vector GIS, 2-D areal objects are defined geometrically as G-rings and G-polygons, and topologically as GT-rings and GT-polygons, and, in a raster GIS, they consist of grid-cell polygons (Johnston 1998). Generating buffer zones around objects with GIS has several applications in riverine ecosystems. Narumalani et al. (1997) identified critical areas for establishing riparian vegetation by using buffers to delineate existing riparian zones around stream channels that were represented as DLGs. Hunsaker and Levine (1995) defined hydrologically active areas in watersheds by creating equidistant buffers (200 and 400 m) around the stream network with a GIS. Toepfer (1997) assessed habitat complementation (i.e., proximity of two required habitat types) for the leopard darter *Percina pantherina* by creating concentric 3-m buffer zones outward from riffles, where the fish spawns, and measuring the distance of every nonriffle zone (pools and runs), where the fish lives, to the nearest riffle zone.

Overlays can be used to intersect 0-D, 1-D, and 2-D features based on graphical, logical, or arithmetic approaches. In graphical approaches, two or more data layers are superimposed (e.g., DLGs of streams and roads overlain on a land-cover coverage). Logical approaches use Boolean operators (i.e., AND, intersection; OR, union; NOT, negation; and XOR, exclusionary or) to analyze the spatial coincidence of input data layers (e.g., intersecting stream habitat variables such as depth, velocity, substrate, and cover with fish distributions to identify habitat preferences). Arithmetic approaches involve mathematical operations across multiple data layers (e.g., combining information on slope, soil composition, and vegetation to estimate soil loss in a watershed with the Universal Soil Loss equation; Johnston 1998).

3.2.2.3 ***Topographic Features.*** Operations on 3-D, topographic features are used to analyze the change in an attribute, such as surface elevation, bathymetry, or other continuous data surfaces, over space. Elevation data usually are derived from DEMs by using raster GIS or triangular irregular networks by using vector GIS. Surface water hydrologic modeling is a growing field in GIS science that includes various procedures for generating stream networks from DEMs. One of the most comprehensive procedures uses three data layers produced from a DEM to (1) generate a depression-less DEM, (2) define flow directions for each cell, and (3) create a flow accumulation data layer where each cell is assigned a value equal to the number of cells that drain into it (Figure 3.3; Jensen and Dominque 1988; ESRI 1992). The resulting stream drainage network can be used to determine stream lengths, linkages between stream segments, and stream order, which is widely used to relate characteristics and processes of stream ecosystems (Vannote et al. 1980). Watershed delineation by using GIS is performed with two data layers: one that indicates flow direction and the other that contains cells or groups of cells that represent the outflow or pour points of the watershed (e.g., tributaries, monitoring sites, dams, pothole depressions) (Jensen and Dominque 1988). Along with watershed boundaries and stream drainage networks, surface water models (e.g., agricultural nonpoint source, AGNPS, Engle et al. 1993; and soil and water assessment tool, SWAT, Srinivasan and Arnold 1994) have been integrated with a GIS to evaluate the effects of watershed land-use activities on stream water quality, fish populations, and fish habitat (see Section 3.3).

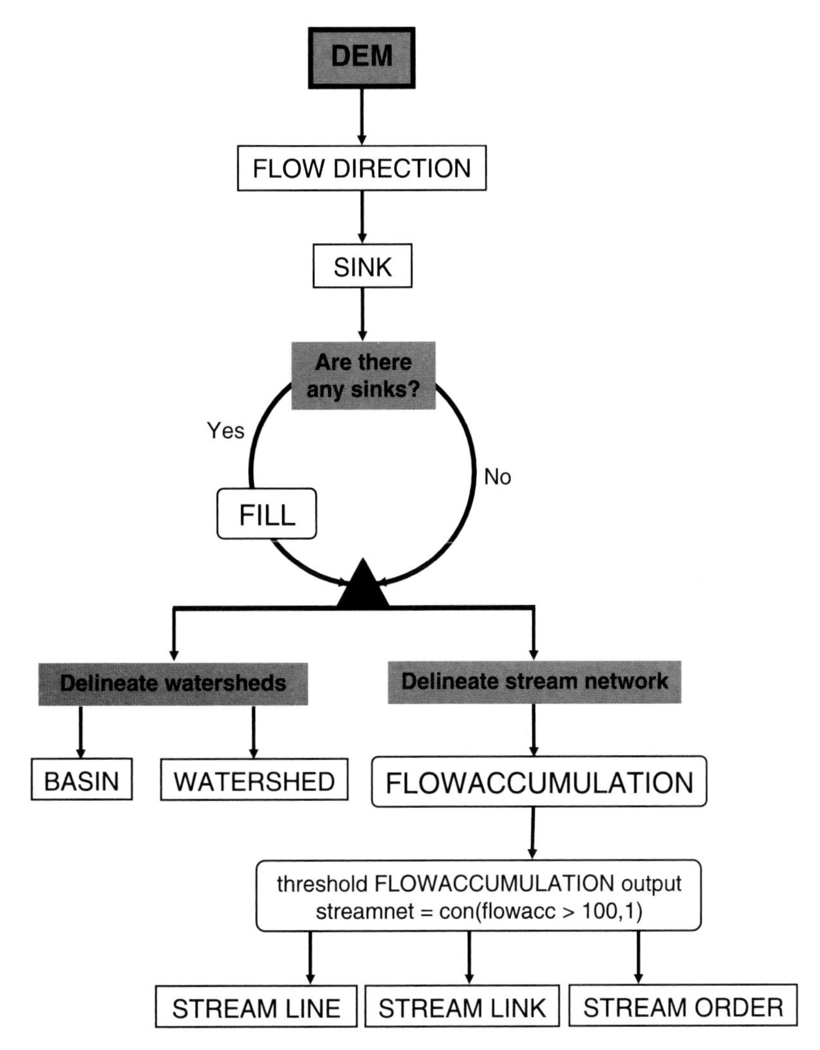

Figure 3.3 Flow diagram of steps used to derive surface characteristics of streams networks and watersheds from a digital elevation model (redrawn from ESRI 1992).

3.2.2.4 ***Spatial Interpolation.*** In streams and rivers, geomorphic channel units such as riffles, runs, and pools can be identified either by subsampling portions of the channel at longitudinal intervals (Toepfer et al. 2000) or by censusing the entire channel (Torgersen et al. 1999). Interpolation to determine boundaries between adjacent habitat types involves two steps: (1) classifying or sampling an area and (2) placing a boundary to distinguish the area from dissimilar areas (Johnston 1998). Typically, boundaries between stream channel units are interpreted by human visualization in the field or on aerial photographs. However, observer variability (Roper and Scarnecchia 1995) and seasonal changes in discharge (Hilderbrand et al. 1999) can influence classification and boundary placement of geomorphic units. Quantitative methods for habitat boundary determination commonly used by terrestrial ecologists are applicable to stream habitat typing. Sample points measured along a transect oriented parallel to the stream channel boundaries can be analyzed in GIS with (1) a moving split-window, in which statistical comparisons between groups of 3–4 points are made as the window moves sequentially along the transect; (2) Thiessen polygons, which use a three-step process to define polygon boundaries that are equidistant from neighboring points; or (3) fuzzy sets, where elements (points) may be partial members of a set, rather than being either members or nonmembers (Johnston 1998).

A variety of methods are available for interpolating continuous data such as stream microhabitat characteristics (e.g., water depth and velocity) measured at sample points along transects perpendicular to the stream channel. Whole area interpolation methods, such as trend surface analysis and Fourier series, use all the points in a study area to interpolate a surface, whereas local interpolation methods, such as splines, moving averages, and kriging, use only neighboring points to estimate values. Kriging is a geostatistical method that uses regression (i.e., semivariogram modeling) to generate a trend surface map based on the spatial correlation among sample points (Johnston 1998). This analysis yields a semivariogram, which is a graph of the average similarity between two successive data points as a function of the distance between the points (Rossi et al. 1992). Gardner et al. (2003) used kriging with three different metrics (i.e., Euclidean distance, instream distance along the stream network, and instream distance along the network weighted by stream order) to model stream temperature at target locations based on data from temperature loggers placed throughout a stream system in New York. All three metrics showed some level of spatial autocorrelation in the data. The two network metrics produced better predictions than the Euclidean metric, and the weighted network metric reduced the influence of data from the tributaries on the downstream main-stem predictions.

3.2.3 Mapping Streams and Rivers

Many users of GIS-generated maps assume they are accurate portrayals of reality. However, errors associated with remote-sensing data acquisition, processing, analysis, and presentation can significantly affect the confidence associated with GIS products and impact their use in decision making. Metadata for GIS coverages and associated databases provides descriptive information about a data set that is important for assessing data quality (Johnston 1998). When GIS maps are used as input to a GIS operation, the errors in the input, along with errors associated with any computational models, will propagate to the output of the operation (Heuvelink 1998). Such error needs to be quantified and documented for subsequent users of the maps. The software tool ADAM traces error propagation in quantitative spatial modeling with GIS (Heuvelink 1998). A study by Lunetta et al. (1997) provides an example of assessing map accuracy. They used GIS to predict the locations of response reaches (low-gradient reaches with less than 4% slope) that typically provide habitat for anadromous salmon in western Washington streams. They assessed the accuracy of their predictions of stream reach classifications with actual measurements of stream slope and found that their identification of response reaches was 96% accurate. They cautioned users of the data that although their model correctly identified most response reaches, the ultimate limiting factor for their predictions was the resolution and quality of the DEM data used to calculate slopes.

In a GIS, maps form the basis for decision making, and an ineffective map design restricts communication of information and may communicate false ideas about the displayed data (Weibel and Buttenfield 1992). The process of map design can be divided into three interrelated components: (1) generalization, which encompasses the abstraction of detail and transformation of cartographic data into a representation usually at reduced scales; (2) symbolization, which creates an image of the mapped phenomena by devising a set of graphical marks (map symbols) that represent the phenomena; and (3) production, or the actual construction of the map display (Weibel and Buttenfield 1992). Streams and rivers are linear features on the landscape, and, as such, depicting them and their attributes in GIS maps becomes problematic when they are viewed at small scales (i.e., large drainage basins or political areas). Viewing fine scale data in streams, such as locations of fishes (e.g., at samples sites or telemetry fixes) or critical fish habitats (e.g., feeding stations or spawning redds) requires zooming into a particular site or stream reach in GIS, which may not always be possible or practical. For example, when distributions of species are represented in a drainage basin, point locations in streams may be too sparse or dense, depending on sampling intensity, to effectively view the distribution at the drainage basin scale (Figure 3.4A). One solution when making GIS maps is to generalize a species' distribution by subdividing the entire drainage basin into sub-basins and representing distributions by sub-basins rather than stream segments. Choropleth maps use ordinal classes of symbols (e.g., shades) or proportional symbols (e.g., symbols that display a quantitative attribute by varying the size of a symbol) to depict patterns in a feature across a large spatial extent (Chrisman 1997). Both proportional (Figure 3.4B) and choropleth (Figure 3.4C) maps can be used to represent quantitative and qualitative properties (abundance, diversity, and health) of fish populations across a drainage basin.

3.3 APPLICATIONS OF GIS TO FISHES AND FISHERIES IN STREAMS AND RIVERS

Habitat conditions that influence fish populations can range from microhabitat features such as localized current speeds to large-scale features such as geology or land-use practices (Bozek and Rahel 1991; Wang et al. 1997; Wiley et al. 1997; Dunham and Rieman 1999). Early studies of fish habitat often measured drainage basin features directly from maps by using rulers or map wheels (Parsons et al. 1982;

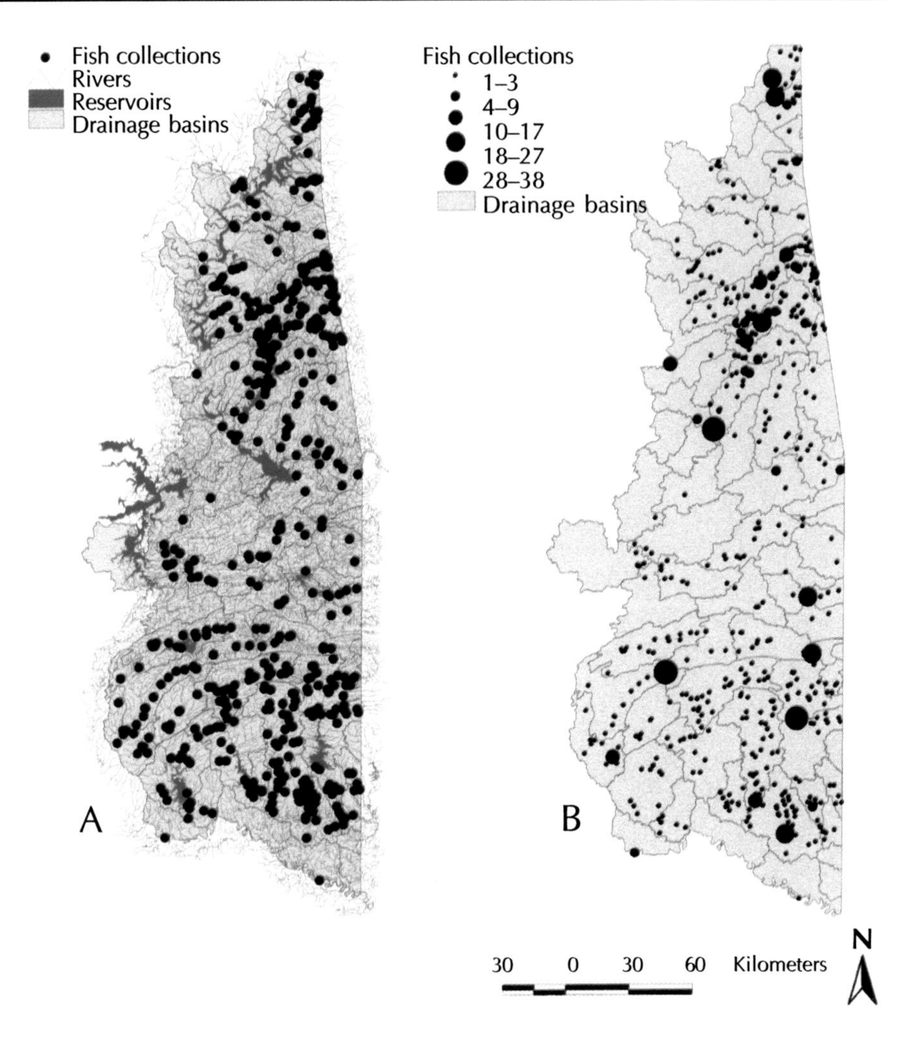

Figure 3.4 The distribution of collections of the central stoneroller *Campostoma anomalum* in rivers from a 30,922-km^2 region of eastern Oklahoma (A). Density of the collections is depicted with proportional symbols (B) and with a choropleth map by using drainage basins (C).

Lanka et al. 1987). Now, however, GIS has become the main tool for quantifying habitat features at the drainage basin scale and for relating such large-scale habitat conditions to fish abundance. Once relationships between fish abundance and large-scale habitat features are understood, we can begin to ask how fish abundance might change with human alteration of the watershed. Geographic information systems have proven useful in modeling how changes in agricultural practices, urbanization, or regional climate might affect fish distributions and the integrity of fish assemblages. Geographic information systems also have many applications in conservation biology, such as identifying areas worthy of protection because they have many rare species or identifying areas with degraded fish assemblages in need of rehabilitation.

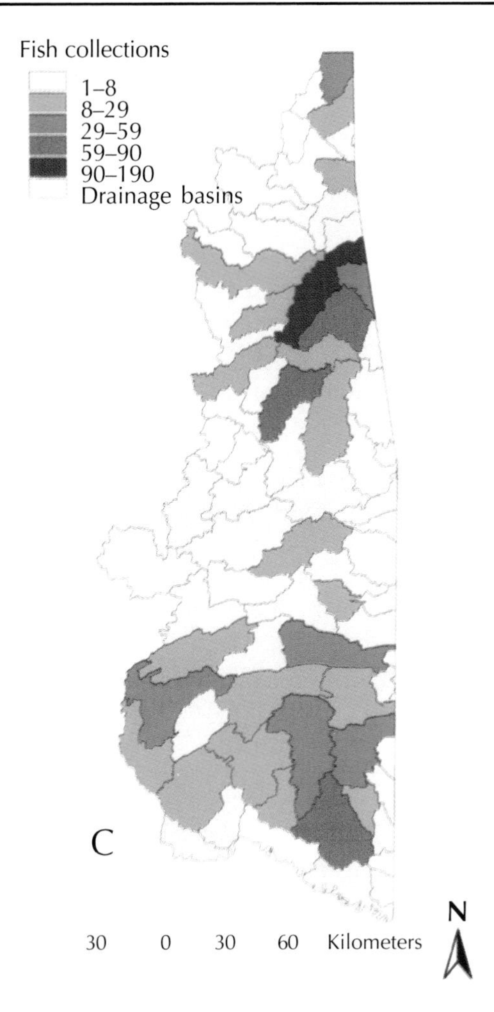

Figure 3.4 Continued.

3.3.1 Mapping and Modeling Fish Habitat and Fish Distributions

A variety of habitat features amenable to GIS analysis have been examined for their relationship to the distribution or abundance of various fish species (Table 3.2). Drainage basin area is easy to measure in a GIS and often is highly correlated with a variety of other basin characteristics. Island biogeography theory suggests there should be a strong linkage between drainage basin area (or its more refined alternative—patch size) and the probability of occurrence for many species. Larger islands are more likely to contain all the required habitats for a species and to support populations large enough to avoid extinction through demographic or environmental stochasticity. Likewise, larger drainage basins or habitat patches should contain more species and have a lower probability of species extinction (Dunham and Rieman 1999). Rieman and McIntyre (1995) used information on thermal limits for bull trout *Salvelinus confluentus* in Idaho to define potential habitat patches as drainage networks above 1,600 m

Table 3.2 Stream or watershed attributes that have provided insight into fish-habitat relations and which can be calculated using GIS.

Attribute	Definition or method of estimation	Examples
Drainage basin area	The area enclosed by a drainage divide or the area from which water drains into the stream upstream of the study site	Gresswell et al. 1997; Porter et al. 2000; Isaak and Hubert 2001
Basin perimeter	Length of the boundary line along a topographic ridge that separates two adjacent drainage basins	Lanka et al. 1987
Basin relief	Highest elevation on the headwater divide minus the elevation at a specified stream location	Lanka et al. 1987; Nibbelink 2002
Watershed channel gradient	Drop in elevation from headwaters to site of sampling divided by length of stream	Porter et al. 2000
Watershed mean slope	Weighted mean estimated by multiplying the midpoint of each degree slope class by the area for that class and summing for all classes; the sum is divided by the total area of the drainage	Gresswell et al. 1997; Isaak and Hubert 2001
Reach slope	Drop in elevation for a given reach of stream divided by the length of the reach	Lunetta et al. 1997
Drainage density	Total stream length divided by drainage basin area	Maret et al. 1997
Mean basin elevation	Sum of the products of the areas between contour lines and the average elevation between contour lines divided by the drainage basin area	Gresswell et al. 1997; Isaak and Hubert 2001
Sinuosity	Ratio of stream length to valley linear distance	Porter et al. 2000
Water yield from a drainage basin	Total precipitation for a drainage basin calculated by multiplying the midpoint of each precipitation class by the area in the class and summing	Gresswell et al. 1997
Habitat patch size	Size of habitat patch that contains sufficient resources to support a population of a species over the long term	Rieman and McIntyre 1995
Isolation of habitat patches	Stream distance between habitat patch and nearest occupied patch; distance of tributary sites from main stem	Dunham and Rieman 1999; Filipe et al. 2002

Table 3.2 Continued.

Attribute	Definition or method of estimation	Examples
Road density	Total length of roads divided by drainage area; proportion of pixels containing a road	Moyle and Randall 1998; Baxter et al. 1999; Dunham and Rieman 1999;
Livestock grazing intensity	Number of livestock per grazing allotment divided by allotment area	Isaak and Hubert 2001
Watershed land use	Proportion of drainage basin in various types of land-use categories	Roth et al. 1996; Wang et al. 1997
Riparian zone condition	Types of vegetation or land use in riparian zone or in various buffer zones away from the stream	Wang et al. 1997; Isaak and Hubert 2001
Thermal zones	Thermal zones across a watershed can be developed from regressions between air or water temperature and elevation; these relationships are used with a digital elevation model to interpolate climate conditions throughout the drainage; thermal remote sensing can be used to develop basin-wide water temperature maps	Keleher and Rahel 1996; Rahel et al. 1996; Rahel and Nibbelink 1999; Torgersen et al. 1999; Isaak and Hubert 2001; Torgersen et al. 2001
Geological characteristics	Geological conditions that determine habitat features important to fish (e.g., groundwater discharge zones) can be identified from geological maps	Nelson et al. 1992; Wiley et al. 1997; Baxter et al. 1999
Area and spatial juxtaposition of channel habitat types	Location and area of riffles, runs, or pools are mapped and then used to calclate stream area in each habitat type or proximity of certain habitat types to others	Tsao et al. 1996; Toepfer 1997; Toepfer et al. 2000; Fisher et al. 2001

in elevation. They found a strong relationship between patch size (drainage area above 1,600 m) and the probability that the patch would contain bull trout. Porter et al. (2000) also found that drainage area was a useful predictor of presence or absence for many fish species in British Columbia streams.

Thermal conditions are another habitat feature amenable to GIS analysis. A common approach is to develop a regression relationship between air or water temperatures and site features such as elevation, aspect, or latitude. The regression model can be combined with a DEM and other data layers to generate a thermal map of the drainage. Species abundance or distribution limits then can be related to thermal conditions for any sites within the study area. Keleher and Rahel (1996) noted that salmonids were absent where mean July air temperatures exceeded 22°C and used a GIS to project areas likely to be suitable for salmonids throughout the Rocky Mountain region of the United States (Figure 3.5). At a finer level of spatial resolution, Rahel and Nibbelink (1999) used a GIS and a DEM to demonstrate that the geographic range of brown trout *Salmo trutta* in southeastern Wyoming was characterized by a thermal window of 19–22°C mean July air temperature. Sites with warmer temperatures lacked any salmonids and sites with cooler temperatures were dominated by brook trout *Salvelinus fontinalis*.

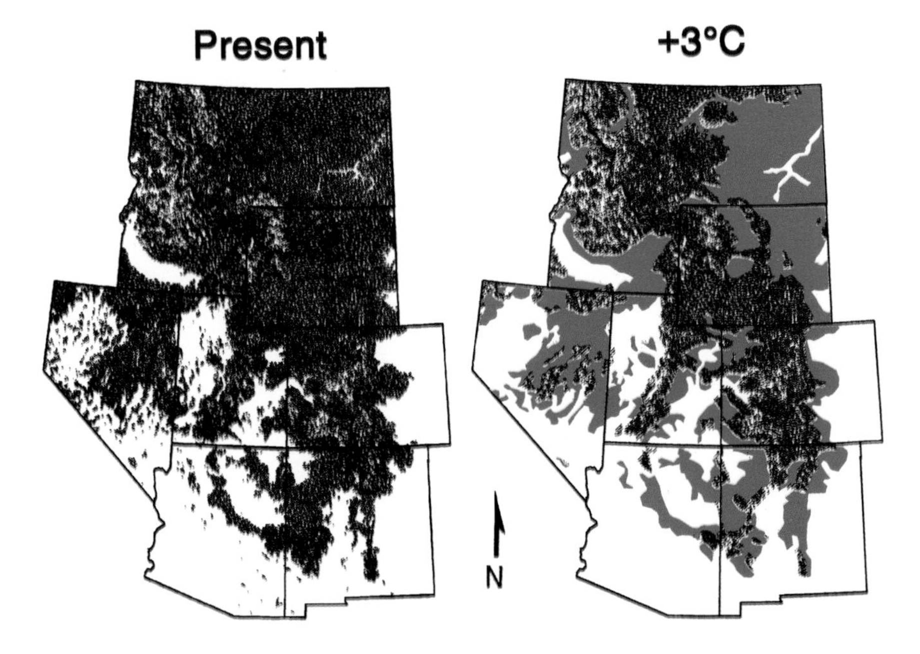

Figure 3.5 Present day potential distribution of trout in the Rocky Mountain region as limited by the 22°C isopleth of mean July air temperature (left panel; shaded area). Also shown is the future potential distribution of trout (right panel; dark shaded area) and habitat loss (red area) after a 3°C increase in mean July air temperature. There would be a predicted 50% loss of geographic range after such a temperature increase (Keleher and Rahel 1996).

Hydrological factors that influence stream fish assemblages, such as the timing of spring runoff and the stability of base flows, can be predicted from landscape characteristics by using a GIS. Wiley et al. (1997) developed a groundwater index that approximates groundwater flow into a stream and allows fisheries biologists to identify reaches likely to support trout populations. The index is raster based and involves multiplying values of topographic slope (derived from a DEM) by hydraulic conductivity (derived from a map of surficial geology; Figure 3.6). High Groundwater Index values indicate stream reaches having high rates of groundwater inputs that stabilize streamflow and keep water temperatures cool—conditions important for maintaining

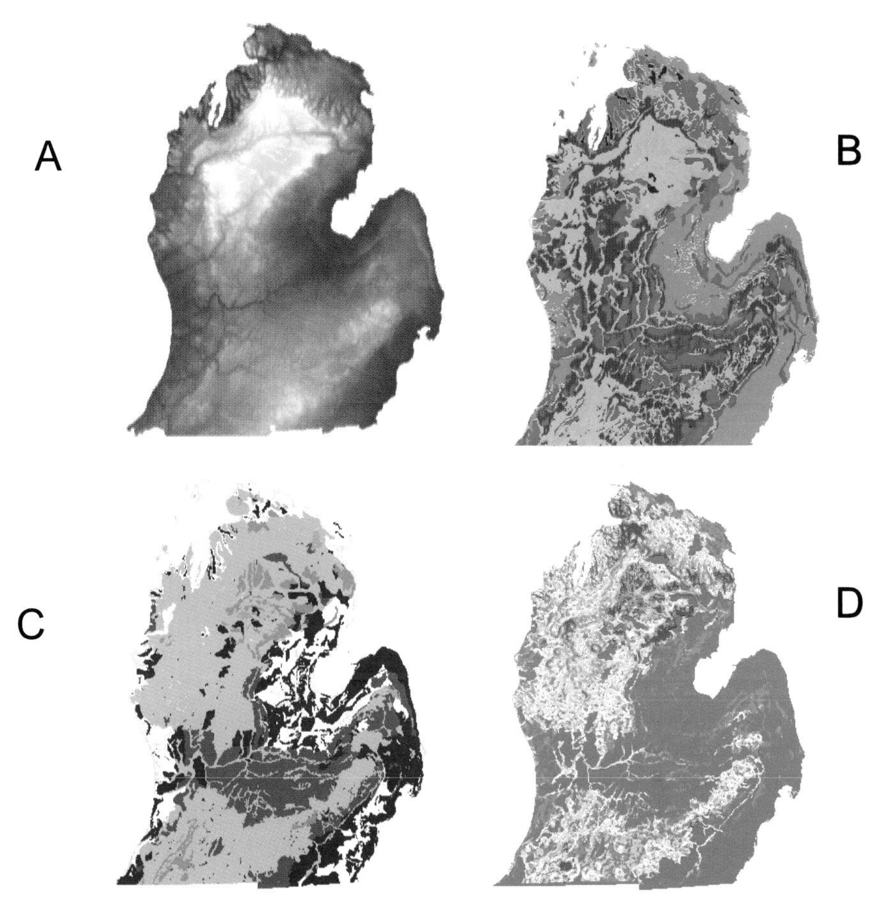

Figure 3.6 An example of using a raster-based GIS approach to model potential groundwater loading to stream channels in Michigan's lower peninsula (Wiley et al. 1997). The Groundwater Potential Map (D) is computed by multiplying values of topographic slope derived from a digital elevation map (A; U.S. Geological Survey, 1 arc-sec) by hydraulic conductivity (C) that is derived from a map of surficial geology (B). Areas with high groundwater potential (shown in red) have large inputs of groundwater into streams that stabilize flows and maintain cool water temperatures favorable for trout populations.

trout populations. Gresswell et al. (1997) found that two-thirds of the variation in the timing of cutthroat trout *Oncorhynchus clarki* spawning migrations among tributaries to Yellowstone Lake could be explained by regression equations involving basin area and aspect. These explanatory variables were calculated in a GIS and appear to influence spawning migrations through their effect on the timing of spring runoff, which cues cutthroat spawning migrations. Spring runoff and cutthroat spawning occurred earliest in tributaries with a small drainage basin and a southeast orientation.

How geologic and geomorphic features are related to the occurrence of fish species can be examined readily in a GIS framework. Nelson et al. (1992) examined relations between trout distribution and watershed geology derived from GIS coverages. They found that trout were essentially restricted to the sedimentary geologic district and were rare or absent in volcanic and detrital geologic districts. Streams in the sedimentary geologic district were higher in elevation, had more gravel substrate, and had less eroded banks, and this resulted in better trout habitat compared with streams in the other two geologic districts.

Estimating fish abundance over extensive geographic regions is another application of GIS in fisheries. Toepfer et al. (2000) developed a multistage approach for estimating the abundance of stream fishes by using a GIS. They mapped channel units (e.g., riffles, runs, and pools) in a river and used information on habitat preference of the leopard darter to assign each channel unit to a habitat suitability class. Then they determined the average abundance of leopard darters in a sample of each suitability class. By using a GIS, they estimated the total abundance of leopard darters throughout the entire river by multiplying the average abundance per suitability class by the area of river in each suitability class and summing across classes. Fisher et al. (2001) used a similar method to estimate the abundance of smallmouth bass *Micropterus dolomieu* in an Oklahoma stream (Figure 3.7). This approach is useful where fish-habitat relations are well known and habitat maps can be generated relatively easily (e.g., from aerial photos or by floating on a river), but fish populations are difficult to sample because of logistical, financial, or access constraints.

One of the most common uses of GIS is to find intersections among various data layers. This approach was used by Dauble and Johnson (1999) to direct search efforts for chinook salmon *Oncorhynchus tshawytscha* redds in the lower Snake River. Previous workers had identified the ranges of depth, velocity, and substrate used by spawning chinook salmon. Dauble and Johnson used a GIS to overlay maps of bathymetry, near-bed water velocity, and dominant substrate in the lower Snake River to find areas containing the appropriate combination of these features that could provide spawning habitat for chinook salmon. Their GIS-derived map narrowed the area of the river that had to be searched and increased the efficiency of locating redds with an underwater videocamera system. The authors reported the GIS-based filtering technique was helpful because the survey areas were large, the time interval for detecting redds was short, and

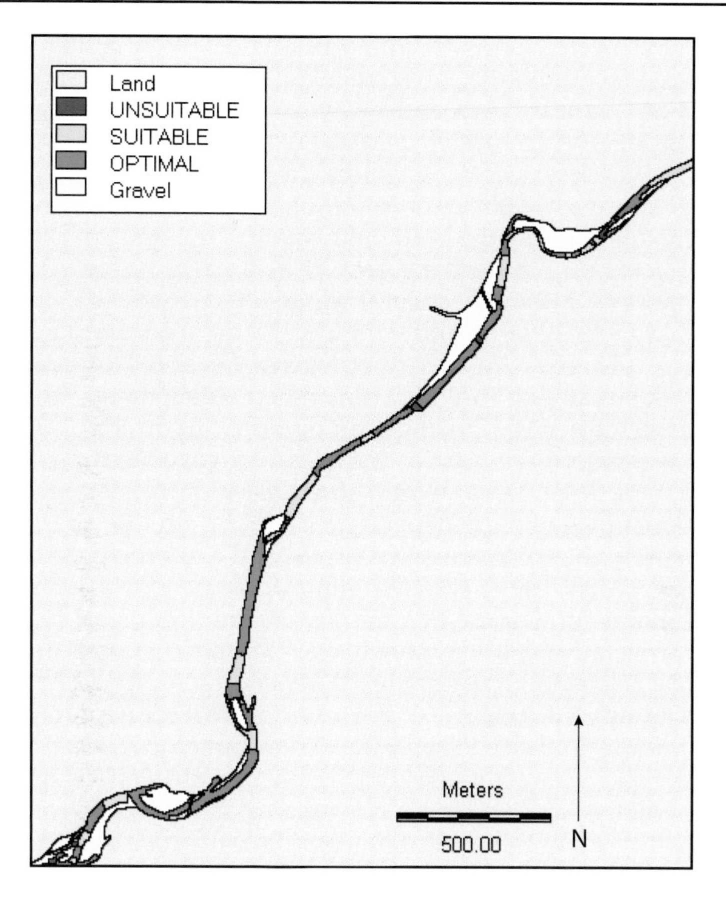

Figure 3.7 Smallmouth bass habitat suitability in the Baron Fork of the Illinois River, Oklahoma (modified from Fisher et al. 2001).

the environmental conditions frequently were challenging. Webb and Bacon (1999) described a similar application of GIS for prioritizing survey efforts, estimating natural spawning distribution, and identifying areas suitable for stocking of salmon in Scotland.

Modeling with GIS is used to extrapolate field measurements (e.g., stream depth or velocity) and integrate complex ecological information (watershed land use, stream habitat, fish abundance) over space and time. Cartographic modeling in GIS uses Boolean operators with two or more data layers to identify areas with certain properties. One common use of cartographic modeling is to identify suitable habitats from environmental variables that are related to the survival of an organism. For example, cartographic modeling has been used to model suitable habitats of stream fish such as Formosan landlocked salmon *Oncorhynchus masou formosanus* (Tsao et al. 1996; also known as cherry salmon), smallmouth bass (Figure 3.7; Fisher et al. 2001), and leopard darters (Toepfer et al. 2000). Rule-based models, or expert systems, linked to GIS produce spatially explicit information based on rules programmed into the computer. These systems have been used with GIS to evaluate habitat and environmental risk and recommend management solutions to environmental problems. For example, Lam and Swayne (1991) developed an expert system (RAISON) to model regional fish species richness, and White and Kapuscinski (1995) developed a prototype decision support system (UrbanFish) for the conservation of stream fish communities in urban planning. Statistical and mathematical modeling typically involves the export of GIS-derived data to external programs that generate the models, although several models (e.g., AGNPS, Engle et al. 1993; SWAT, Srinivasan and Arnold 1994) are integrated with GIS software.

There are many examples (see Section 3.3.2) of GIS-based statistical modeling in which watershed characteristics derived from GIS coverages (e.g., geology, topography, hydrology, land use, riparian vegetation) are used as independent variables to predict stream fish characteristics that would otherwise require extensive field sampling. Examples of such dependent variables are the presence–absence, or abundance of particular species, and assemblage characteristics such as species diversity or standing stocks of sport fish. Statistical models may be univariate or multivariate (multiple and logistic regression, principal components analysis, discriminant function analysis, canonical correspondence analysis) depending on whether the data are continuous or categorical. For example, Maret et al. (1997) used multivariate analyses to relate fish community metrics to GIS-derived watershed characteristics (e.g., land use, stream order, watershed size, and drainage density) and instream habitat measurements. Mathematical models that simulate population processes (growth, survival, mortality) can be coupled with GIS to predict outcomes of these processes in a spatial or an aspatial manner. Jessup (1998) developed a mathematical model that simulated population dynamics of brown trout and used it to relate GIS-derived watershed characteristics to trout habitat quality.

3.3.2 Effects of Watershed Land Use on Fish Populations and Assemblages

A major challenge facing fisheries managers is to understand how land use influences fish populations and assemblages. Understanding these relationships is necessary before we can propose changes that will mitigate harmful effects of current land-use practices. Geographic information systems are a useful tool for calculating the proportion of the watershed in different land-use categories and then relating land use to habitat conditions and fish abundance. Studies in the midwestern United States have shown that habitat quality and the health of stream fish assemblages are negatively related to the amount of agricultural land upstream of a sampling site (Allan et al. 1997; Wang et al. 1997). By clipping coverages for various subsets of the watershed, one can easily examine issues of scale in fish-habitat relations. For example, Allan et al. (1997) calculated the extent of agricultural land use upstream of sampling sites at four spatial scales, ranging from local (a 30-m buffer zone extending 150 m upstream of the sample site) to regional (area of agricultural land in the entire basin upstream of the sample site). The health of fish assemblages (measured by the Index of Biotic Integrity) was most strongly correlated with the amount of agricultural land at the largest spatial scale, suggesting that regional rather than local land use was the primary factor influencing fish assemblages. Other studies that used GIS to relate the well being of fish assemblages to anthropogenic disturbance and land use in the drainage basin include Maret et al. (1997) and Waite and Carpenter (2000).

Geographic information systems also can be used to develop cell-based models that quantify how different land-use scenarios affect sediment and nutrient delivery to streams. Allan et al. (1997) described a model that uses information on land use, soils, hydrography, and topography to simulate runoff, sediment, and nutrient transport from the catchment to a downstream point in response to a storm of a specified magnitude. The model was used to predict the change in runoff and water quality in the River Raisin basin of Michigan for various changes in urban, agricultural, or forested land cover within the watershed. Discharge volume, sediment yield, and total nitrogen and phosphorus yield all decreased as more land was converted to forest but increased as land was converted to urban or agricultural uses.

A cell-based approach also was used by Kelly (2001) to estimate sediment delivery into streams as a result of forest clearing in an Oregon watershed. Patches of cleared forest were identified from Landsat Thematic Mapper sensor data and mapped onto the watershed. A DEM consisting of cells with 30-m resolution was clipped to conform to the boundaries of each patch so that erosion could be considered only across a cleared patch. The GIS surface functions were used to calculate the direction of flow for each cell based on the relative elevation in eight cells surrounding the target cell. The accumulated flow to each cell was calculated by summing the number of cells that flow into each down-slope cell. This flow eventually drained into adjacent waterways, and it was assumed that the accumulated flow from a group of cells draining to a common pour point would be a good proxy for the sediment delivery coming from those cells. The geographic extent and underlying topography of a logged patch determined the amount of the patch that would drain into particular streams. When the analysis was applied to the entire watershed, the author was able to estimate which patches of cleared forest would be most problematic in delivering sediment to streams and also which streams would be receiving the most sediment from cleared patches. This approach could be used to model where forest clear-cuts should be placed on the landscape in order to minimize erosion problems and to identify which streams should be monitored because they would be most likely to experience increased sediment loads.

In an example of using GIS to model fish response to watershed modification, Jessup (1998) simulated brown trout population dynamics and related it to watershed land-use disturbance and imperviousness, which were obtained from a GIS database. Model results indicated that brown trout populations would take 12–15 years to recover after major highway construction in watersheds of moderate imperviousness (<15% of watershed covered with roofs or pavement) but may not recover in more densely developed watersheds.

3.3.3 Analysis of Spatial and Temporal Changes in Fish Distributions and Aquatic Habitats

Geographic information systems can be a powerful tool for analyzing spatial and temporal changes in fish distributions and aquatic habitats. For example, Keleher and Rahel (1996) used GIS to predict the loss of salmonid habitat across the western United States under several global warming scenarios. Salmonids were found to be limited to regions where mean July air temperatures were less than or equal to 22°C. A regression model based on site elevation and latitude was used to plot the distribution of mean July air temperatures across the western United States. By using a GIS, the geographic area having mean July air temperatures less than or equal to 22°C was calculated for current conditions, as well as for various warming scenarios (Figure 3.5). Results indicated that increases of 1°C, 3°C, or 5°C in mean July air temperature could decrease the geographic area suitable for salmonids by 17%, 50%, or 72%, respectively. In another study of potential climate change impacts on fish, Rahel et al. (1996) predicted that warming would not only reduce the total stream habitat suitable for salmonids but would fragment surviving populations into increasingly small and isolated enclaves. In this case, a GIS was used to calculate the total stream distance remaining in each enclave after a given amount of stream warming (Figure 3.8).

Oberfoell (1992) examined seasonal largemouth bass *Micropterus salmoides* movement and habitat use in the upper Mississippi River by using GIS. Radio tracking locations for largemouth bass were overlain on a habitat map so that seasonal movements and habitat use could be quantified. Fish were found to move up to 14 km to overwintering areas and to show a marked increase in use of vegetated areas during the spawning period.

Hilderbrand et al. (1999) found a GIS valuable in portraying seasonal changes in the location and area of habitat types in a Virginia stream. As discharge decreased from spring to summer, total stream surface area decreased and the pool-riffle sequencing changed. Interestingly, the total area of pool habitat actually increased as deep, fastwater run habitat shrank into a series of small, interconnected pools. Hilderbrand et al. (1999) cautioned that streamwide habitat assessments of fish habitat quality will vary depending on the flow stage.

Tiffan et al. (2002) further explored this issue by using a GIS to quantify rearing habitat for juvenile chinook salmon at different flow rates downstream of a hydroelectric dam. A 33-km reach of the Columbia River in Washington, was divided into 25-m^2 cells. For each

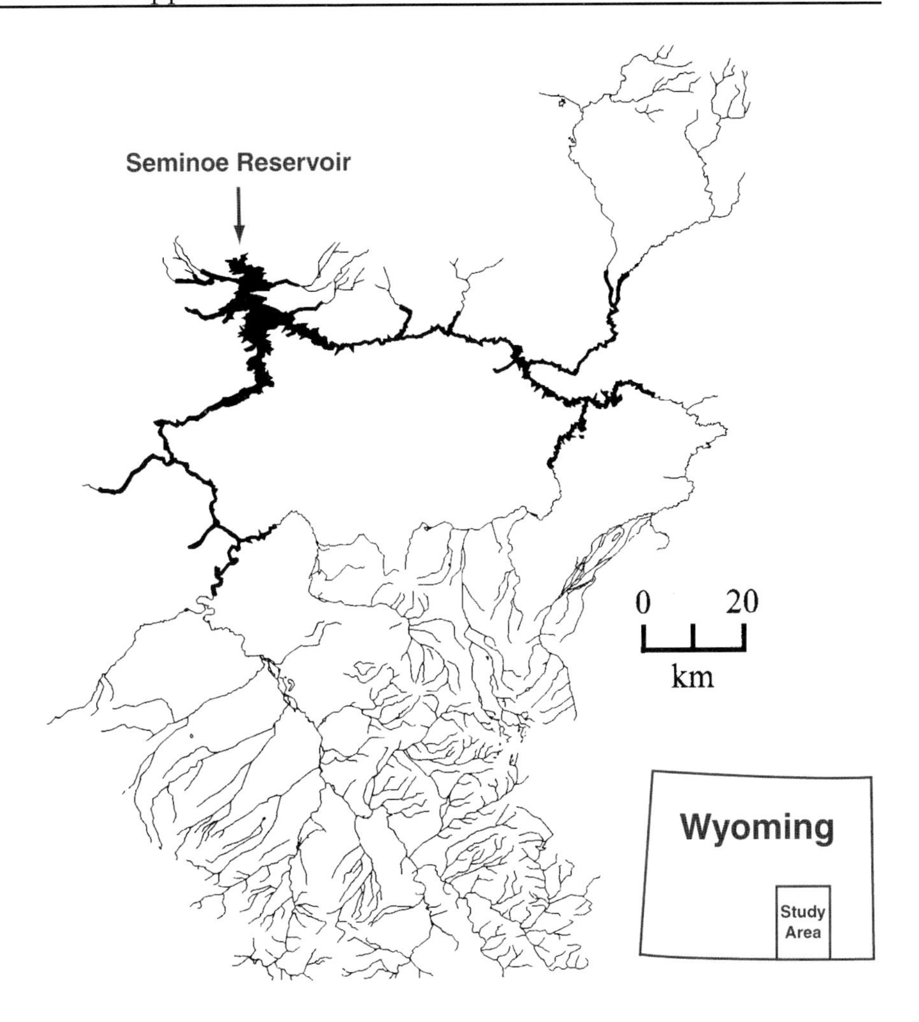

Figure 3.8 Projected fragmentation of stream habitat for coldwater fish in the North Platte River in Wyoming, upstream of Seminoe Reservoir due to climate warming. Stream reaches that would become too warm for trout with a 3°C increase in maximum July water temperature are indicated by the dark black shading. The length of stream with water temperatures suitable for trout would decline from 4,128 to 3,581 km and remaining habitat would become fragmented into small, isolated enclaves (Rahel et al. 1996).

cell, water velocity and depth were derived from a hydrodynamic model for a series of flow rates ranging from 1,416–11,328 m^3/s. The water depth for each cell was used to calculate lateral slope, which is the cell's water depth divided by the distance to shore and multiplied by 100. A logistic regression model indicated that the probability of juvenile salmon occurrence in a cell increased as the lateral slope or the water velocity decreased. Based on information about lateral slope and water velocity stored in the GIS, the authors used the logistic model to calculate the number of cells in the study reach predicted to have juvenile salmon (based on a probability of occurrence ≥ 0.5). This information was then translated into the total area of river predicted to be suitable habitat, and the process was repeated for all flow rates. The results indicated that the amount of suitable rearing habitat would decline as flow rate increased, with the sharpest declines occurring over the range of 1,416–4,814 m^3/s. In addition, the GIS was used to calculate the area of remnant pools in the study reach that resulted from daily flow fluctuations associated with peak power generation. Salmon that become stranded in these pools often die due to heat stress or predation. The authors demonstrated that minimizing the daily fluctuations in flow and thus minimizing stranding of fish was an important consideration for protecting chinook salmon in this portion of the Columbia River.

Changes in stream geomorphic channel units, channel morphology, watershed land use, or fish populations over time can be quantified with GIS. Detecting change with GIS is the process of comparing spatially explicit databases from two different time periods to determine the location and spatial extent of the changes (Johnston 1998). Terrestrial ecologists have used satellite imagery and aerial photography in combination with GIS to evaluate historical changes in land use, vegetation, wetlands, and urbanization patterns in North America (Sisk 1998). Geomorphologists have long been interested in changes in stream channels through time (Knighton 1998). They have documented these changes with remotely sensed images since the 1940s and are currently pioneering the use of multispectral, fine-scale digital imagery for mapping and classifying geomorphic channel units (Wright et al. 2000). Fisheries biologists have traditionally used time- and labor-intensive field surveys to inventory stream habitat types and sample fish populations (Dolloff et al. 1993). Remotely sensed surveys of stream-channel units not only reduce the time required to conduct inventories but also provide a spatially explicit, digital database that can be compared between time periods to evaluate habitat changes (Torgersen et al. 2001; O'Connor and Kennedy 2002; Whited et al. 2002).

3.3.4 Fish Conservation and Management

Many natural resource management agencies are putting information on fish distribution and aquatic habitat features into spatially referenced databases. Combining this information with other spatially referenced databases provides fisheries biologists with a powerful tool for making management and conservation decisions. For example, managers can readily produce maps of species distributions and can update these maps frequently as new information becomes available. Managers can query GIS databases to find combinations of features important for fish habitat such as where suitable current speeds, water depths, and substrates overlap to produce ideal spawning conditions for salmon (Dauble and Johnson 1999). Or, managers can ask which species are likely to be found at a particular location on a river and thus direct development activities away from sites harboring threatened or endangered species.

A GIS can play an important role in identifying areas worthy of conservation attention. Overlays of distribution maps for threatened or endangered species can be used to identify areas containing a high number of such species. Such areas could then be considered as high priority sites for establishing aquatic preserves. Angermeier and Bailey (1992) used this approach to identify watersheds in Virginia that contained high numbers of threatened or endangered species of fish and mussels. In Missouri, individual species models were constructed for 217 fish species and used in map overlays to determine the number of rare, threatened, and endangered fish species in streams across the state (Figure 3.9). A similar analysis was done for mussels and crayfish, and the resultant maps provided guidance as to where initial conservation efforts should be concentrated. When maps showing areas with a high concentration of threatened and endangered species are combined with maps of land ownership, one can determine which conservation hotspots are already protected (e.g., located within national forests) and which may need enhanced protection (e.g., private lands subject to impending development). This procedure has been referred to as "gap analysis" because it identifies "gaps" in the conservation network (Scott et al. 1993). As part of the Gap Analysis Program (GAP), many states in the United States are in the process of undertaking analyses in freshwater habitats, and this is an area where GIS will play a major role in the conservation of aquatic species, in-

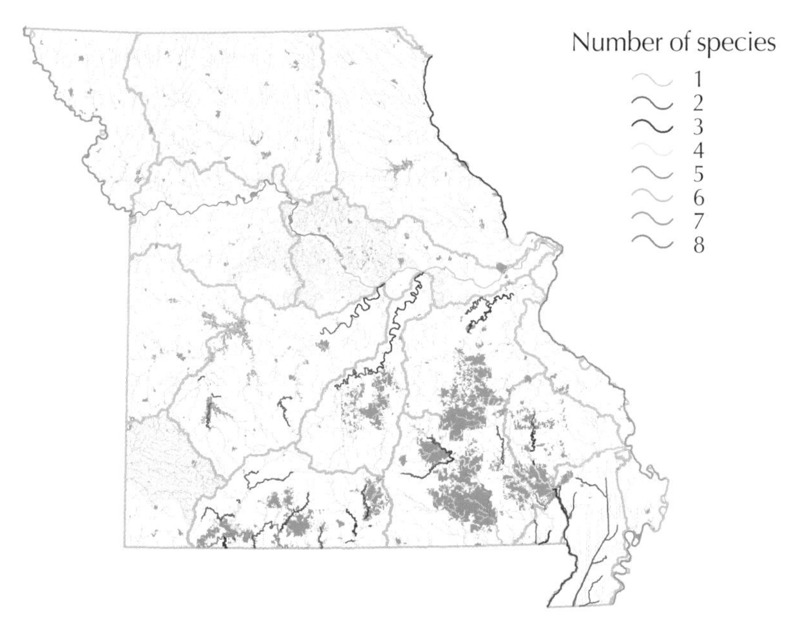

Figure 3.9 The number of globally rare, threatened, and endangered fish species in streams and rivers of Missouri. Status categories are based on assessments by NatureServe (2003). Data are from the Missouri Aquatic GAP Project (MoRAP 2003). The light gray lines represent the major streams, for context, the green areas are public lands, and the white areas (with dark gray outlines) are ecological drainage units.

cluding fish, mussels, and crayfish (USGS 2003). An example of a gap analysis for fishes in Missouri is presented in Figure 3.10.

In some cases, it may be desirable to identify areas worthy of protection even though data on species distributions are not available

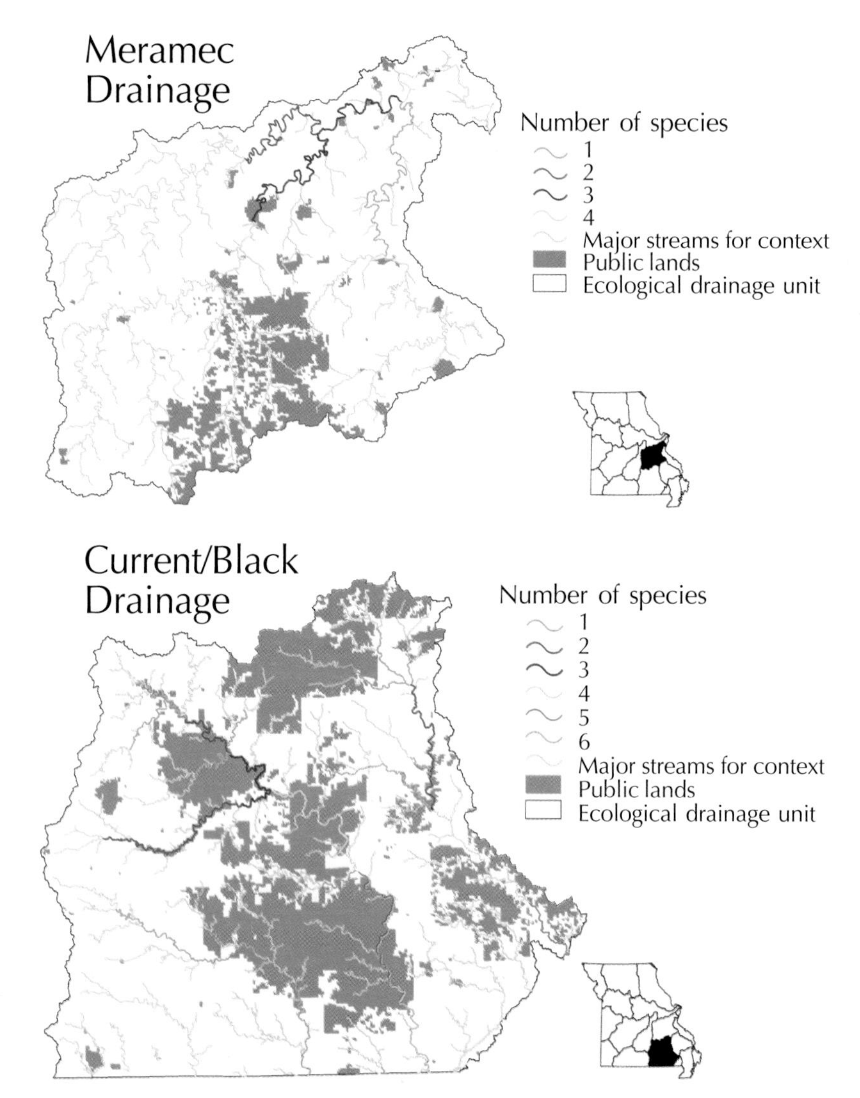

Figure 3.10 The number of globally rare, threatened, and endangered fish species in streams and rivers of Missouri in relation to the location of public lands. Species status categories are based on assessments by NatureServe (2003). In the Meramec Drainage, riverine sites with the largest number of rare, threatened, and endangered species are not located in areas protected by public ownership of the adjacent lands. In the Current/Black Drainage, at least some reaches with large numbers of rare, threatened, and endangered species flow through public lands where conservation efforts are easier to implement than on private lands. Data are from the Missouri Aquatic GAP Project (MoRAP 2003).

or are incomplete. In this situation, GIS can be used to model potential species distribution patterns based on an understanding of species habitat requirements. Lunetta et al. (1997) used this approach to identify watersheds in western Washington having the highest potential as salmonid habitat. They used a GIS to calculate the slope of stream reaches throughout each watershed and to quantify the type of land cover adjacent to the streams. They then queried the GIS database to find low-gradient reaches (<4% slope) in areas with old-growth forests. These areas were considered to provide the best habitat for salmonids because low-gradients coupled with large woody debris inputs from the riparian zone created abundant pool habitats favored by salmonids. Only about 2% of the 164,083 km of stream analyzed contained this combination of habitat features. An obvious management recommendation is to preserve these areas as important salmonid habitat. Another management recommendation is to seek protection for low-gradient stream reaches that currently have riparian areas with mid-successional stage vegetation. Such areas constituted almost 5% of the study reaches, and their value as salmonid habitat might increase as forest succession proceeds to the climax stage in the coming decades.

O'Brien-White and Thomason (1997) also used a GIS to assess the status of fish habitat among streams in a South Carolina watershed. Each stream reach was given a point score for seven factors related to the health of aquatic communities. These scores were calculated from GIS coverages of protected areas, fish assemblage status, water quality, riparian condition, land use, impoundments, and ditches. For each stream reach, points for all factors were summed to give a composite score that was placed into one of three classes representing high-, moderate-, and low-quality fish habitat. Maps showing the distribution of fish habitat quality across the drainage allowed managers to prioritize sampling efforts and locate mitigation and restoration sites.

In a similar vein, Tsao et al. (1996) developed a stream classification system to identify potential reintroduction sites that could support a self-sustaining population of Formosan landlocked salmon. Field data describing habitat features for stream reaches were incorporated into a GIS, and the total stream area in various habitat types such as pools or riffles was calculated. Knowledge of habitat requirements for each life history stage was used along with GIS to determine the amount of suitable habitat in each stream for fry, juvenile, and adult salmon. The results allowed managers to identify streams with suitable habitat for all life-history stages and, thus, the greatest potential for sustaining reintroduced populations of Formosan landlocked salmon.

To identify areas important for the conservation of native fish species in the Guadiana River in Portugal, Filipe et al. (2002) used multivariate logistic regression in a GIS framework. At 149 sites, fish assemblages were sampled, and large-scale habitat variables such as geomorphology, riparian vegetation, and reach location in the drainage basin were measured. Tracking this information in a GIS allowed the authors to develop logistic models that predicted the probability that a given species would be present at each of the 149 sampling sites. When the sites and their associated probability of having a given fish species were projected onto a map, it was easy to identify portions of the drainage that would be most likely to contain the species. These would be areas where habitat preservation efforts could be directed. The authors noted that perhaps the greatest value of the modeling effort was in identifying sites where a fish species was expected to be found but was absent. If reasons for the absence of the species could be identified, that is, whether it was due to pollution or obstruction of seasonal movement patterns, then managers could formulate rehabilitation measures to allow the species to reoccupy those areas.

Moyle and Randall (1998) found GIS helpful in calculating a series of broad-scale habitat variables used to characterize Sierra Nevada watersheds in California. They used square landscape units (pixels) of 1 ha and calculated measures of human impact such as the percentage of hectares in each watershed containing a water diversion or the percentage of hectares containing a road. They also measured the biological integrity of each watershed based on factors such as the number of native species of frogs and fish species, and the presence of anadromous fishes. Watersheds with low biotic integrity had been highly altered by human activity as indicated by a high percentage of pixels containing dams, water diversions, or roads.

3.4 CONCLUSIONS

Geographic information systems technology is the state of the art for describing large-scale habitat features in watersheds. Prior to the advent of GIS, locating, describing, and delineating watershed features such as boundaries, elevations, stream lengths, geologic formations, soil types, and land-cover features was done manually on paper maps. With GIS, these same processes can be done automatically and in considerably less time (Band 1986; Lanka et al. 1987; Jensen and Dominique 1988; Nelson et al. 1992). The challenge for future GIS applications in watershed science is the availability and acquisition of spatial data at different spatial scales. Creation of the National Spatial Data Infrastructure (NSDI) in the United States in 1994, whose development was coordinated by the Federal Geographic Data Committee, has greatly facilitated the distribution of geospatial data. The NSDI defines the technologies, policies, and people necessary to promote sharing of geospatial data throughout all levels of government, the private and nonprofit sectors, and the academic community (FGDC 2003). Geospatial data are distributed through the National Geospatial Data Clearinghouse via the Internet. Our ability to ask and answer questions about fisheries-watershed relationships by using GIS will improve as more geospatial data becomes available.

Geographic information systems technology has great potential for modeling fish distributions in relation to large-scale habitat features and human alterations of watersheds. Historic and current alterations, such as land-use conversions (Sisk 1998) and streamflow regime modification (Poff et al. 1997), as well as potential future alterations, such as global warming, may greatly influence the distribution of stream fishes (Matthews and Zimmerman 1990; Keleher and Rahel 1996; Rahel et al. 1996). Geographic information systems integrated with modeling (see Section 3.3.1) will continue to be the tool of choice for landscape ecologists and fisheries biologists as emphases shift from describing land and water alterations to restoring ecosystem structure and function.

One of the emerging applications in remote sensing is the use of airborne scanners and sensors to map water depths, geomorphic channel units, and large woody debris in streams (Gilvear et al. 1995; Lyon and Hutchinson 1995; Wright et al. 2000; Torgersen et al. 2001; Marcus 2002; Whited et al. 2002). Although refinement of the imagery classification techniques is needed, remote sensing technology and multispectral imagery have great potential for mapping habitats in streams and rivers.

The importance of the Internet as a tool for searching and acquiring spatial data for streams and rivers cannot be understated. Internet2 is a consortium of over 180 universities working in partnership with industry and government to develop and deploy advanced network applications and technologies such as digital libraries, virtual laboratories, and distance-independent learning that are not possible with the present-day Internet. With the advent of Internet2, access to and processing of geospatial data should increase enormously.

3.5 REFERENCES

Aldrich, R. C. 1979. Remote sensing of wildland resources: a state-of-the-art review. U.S. Forest Service General Technical Report RM-71.

Allan, J. D., D. L. Erickson, and J. Fay. 1997. The influence of catchment land use on stream integrity across multiple spatial scales. Freshwater Biology 37:149–161.

Angermeier, P. L., and A. Bailey. 1992. Use of a geographic information system in the conservation of rivers in Virginia, USA. Pages 151–160 *in* P. J. Boon, P. Calow, and G. E. Petts, editors. River conservation management. Wiley, New York.

Band, L. E. 1986. Topographic partitioning of watersheds with digital elevation models. Water Resources Research 22:15–24.

Baxter, C. V., C. A. Frissell, and F. R. Hauer. 1999. Geomorphology, logging roads, and the distribution of bull trout spawning in a forested river basin: implications for management and conservation. Transactions of the American Fisheries Society 128:854–867.

Bozek, M. A., and F. J. Rahel. 1991. Assessing habitat requirements of young Colorado River cutthroat trout by use of macrohabitat and microhabitat analyses. Transactions of the American Fisheries Society 120:571–581.

Chrisman, N. 1997. Exploring geographic information systems. Wiley, New York.

Clingenpeel, J. A., and B. G. Cochran. 1992. Using physical, chemical and biological indicators to assess water quality on the Ouachita National Forest utilizing basin area stream survey methods. Arkansas Academy of Science Proceedings 46:33–35.

Dauble, D. D., and R. L. Johnson. 1999. Fall chinook salmon spawning in the tailraces of Lower Snake River hydroelectric projects. Transactions of the American Fisheries Society 128:672–679.

DeMers, M. N. 1997. Fundamentals of geographic information systems. Wiley, New York.

Dolloff, C. A., D. G. Hankin, and G. H. Reeves. 1993. Basinwide estimation of habitat and fish populations in streams. U. S. Forest Service General Technical Report SE-83.

Dunham, J. B., and B. E. Rieman. 1999. Metapopulation structure of bull trout: influences of physical, biotic and geometrical landscape characteristics. Ecological Applications 9:642–655.

Engle, B. A., R. Srinivasan, and C. C. Rewerts. 1993. A spatial decision support system for modeling and managing agricultural non-point source pollution. Pages 231–237 *in* M. F. Goodchild, B. O. Parks, and L. T. Steyaert, editors. Environmental modeling with GIS. Oxford University Press, New York.

ESRI (Environmental Systems Research Institute, Inc.). 1992. Cell-based modeling with GRID 6.1. Hydrological and distance modeling tools. ESRI, Supplement, Redlands, California.

ESRI (Environmental Systems Research Institute, Inc.). 2001a. ArcGIS spatial analyst: advance GIS spatial analysis using raster and vector data. ESRI, White paper, Redlands, California.

ESRI (Environmental Systems Research Institute, Inc.). 2001b. Linear referencing and dynamic segmentation in ArcGIS 8.1. ESRI, Technical paper, Redlands, California.

Fausch, K. D., C. E. Torgersen, C. V. Baxter, and H. W. Li. 2002. Landscapes to riverscapes: bridging the gap between research and conservation of stream fishes. BioScience 52:483–498.

FGDC (Federal Geographic Data Committee). 2003. National spatial data infrastructure. U.S. Geological Survey, FGDC. Available: *www.fgdc.gov/nsdi/nsdi.html* (September 2003).

Filipe, A. F., I. G. Cowx, and M. J. Collares-Pereira. 2002. Spatial modeling of freshwater fish in semi-arid river systems: a tool for conservation. River Research and Applications 18:123–136.

Fisher, W. L., P. E. Balkenbush, and C. S. Toepfer. 2001. Development of stream fish population sampling surveys and abundance estimates using GIS. Pages 253–265 *in* T. Nishda, P. J. Kailola, and C. E. Hollingworth, editors. Proceedings of the first international symposium on GIS in fishery science. Fishery Research Group, Saitama, Japan.

Fisher, W. L., and C. S. Toepfer. 1998. Recent trends in geographic information systems education and fisheries research applications at U. S. universities. Fisheries 23(5):10–13.

Frissell, C. A., W. L. Liss, C. E. Warren, and M. D. Hurley. 1986. A hierarchical framework for stream habitat classification: viewing streams in a watershed context. Environmental Management 10:199–214.

Gardner, B., P. J. Sullivan, and A. J. Lembo, Jr. 2003. Predicting stream temperatures: geostatistical model comparison using alternative distance metrics. Canadian Journal of Fisheries and Aquatic Sciences 60:344–351.

Giles, R. H., and L. A. Nielsen. 1992. The uses of geographic information systems in fisheries. Pages 81–94 *in* R. H. Stroud, editor. Fisheries management and watershed development. American Fisheries Society, Symposium 13, Bethesda, Maryland.

Gilvear, D. J., T. M. Waters, and A. L. Milner. 1995. Image analysis of aerial photography to quantify changes in channel morphology and instream habitat following placer mining in interior Alaska. Freshwater Biology 34:389–398.

Goodchild, M. F. 2000. GIS and transportation: status and challenges. Geoinfomatica 4:127–139.

Gresswell, R. E., W. J. Liss, G. L. Larson, and P. J. Bartlein. 1997. Influence of basin-scale physical variables on life history characteristics of cutthroat trout in Yellowstone Lake. North American Journal of Fisheries Management 17:1046–1064.

Heuvelink, G. B. M. 1998. Error propagation in environmental modelling. Taylor and Francis, Bristol, Pennsylvania.

Hilderbrand, R. H., A. D. Lemly, and C. A. Dolloff. 1999. Habitat sequencing and the importance of discharge in inferences. North American Journal of Fisheries Management 19:198–202.

Hunsaker, C. T., and D. A. Levine. 1995. Hierarchical approaches to the study of water quality in rivers. BioScience 45:193–203.

Isaak, D. J., and W. A. Hubert. 1997. Integrating new technologies into fisheries science: the application of geographic information systems. Fisheries 22(1):6–10.

Isaak, D. J., and W. A. Hubert. 2001. A hypothesis about factors that affect maximum summer stream temperatures across montane landscapes. Journal of the American Water Resources Association 37:351–366.

Jensen, S. K., and J. O. Dominque. 1988. Extracting topographic structure from digital elevation data for geographic information system analysis. Photogrammetric Engineering and Remote Sensing 54:1593–1600.

Jessup, B. K. 1998. A strategy for simulating brown trout population dynamics and habitat quality in an urbanizing watershed. Ecological Modelling 112:151–167.

Johnson, L. B., and S. H. Gage. 1997. Landscape approaches to the analysis of aquatic ecosystems. Freshwater Biology 37:113–132.

Johnston, C. A. 1998. Geographic information systems in ecology. Blackwell Scientific Publications, Oxford, UK.

Johnston, C. A., and R. J. Naiman. 1990. The use of a geographic information system to analyze long-term landscape alteration by beaver. Landscape Ecology 4:5–19.

Keleher, C. J., and F. J. Rahel. 1996. Thermal limits to salmonid distributions in the Rocky Mountain region and potential habitat loss due to global warming: a geographic information system (GIS) approach. Transactions of the American Fisheries Society 125:1–13.

Kelly, N. M. 2001. Spatial pattern of forest clearing and potential sediment delivery in a north-western Oregon watershed. Pages 281–294 *in* T. Nishda, P. J. Kailola, and C. E. Hollingworth, editors. Proceedings of the first international symposium on GIS in fishery science. Fishery Research Group, Saitama, Japan.

Knighton, D. 1998. Fluvial forms and processes: a new perspective. Oxford University Press, New York.

Lam, D. C. L., and D. A. Swayne. 1991. Integrating database, spreadsheet, graphics, GIS, statistics, simulation models, and expert systems: experiences with the RAISON system on microcomputers. Pages 429–459 *in* D. P. Loucks and J. R. da Costa, editors. Decision support systems: water resources planning. Springer-Verlag, NATO, ASI Series G26, Heidelberg, Germany.

Lanka, R. P., W. A. Hubert, and T. A. Wesche. 1987. Relations of geomorphology to instream habitat and trout standing stocks in small Wyoming streams. Transactions of the American Fisheries Society 116:21–28.

Lunetta, R. S., B. L. Cosentino, D. R. Montgomery, E. M. Beamer, and T. J. Beechie. 1997. GIS-based evaluation of salmon habitat in the Pacific Northwest. Photogrammetric Engineering and Remote Sensing 63:1219–1229.

Lyon, J. G., and W. S. Hutchinson. 1995. Application of a radiometric model for evaluation of water depths and verification of results with airborne scanner data. Photogrammetric Engineering and Remote Sensing 61:161–166.

Maidment, D. R. 2002. Arc Hydro: GIS for water resources. ESRI Press, Redlands, California.

Marcus, W. A. 2002. Mapping of stream microhabitats with high spatial resolution hyperspectral imagery. Journal of Geographical Systems 4:113–126.

Marcus, W. A., R. A. Marston, C. R. Colvard, Jr., and R. D. Gray. 2002. Mapping the spatial and temporal distributions of large woody debris in rivers of the Greater Yellowstone Ecosystem, USA. Geomorphology 44:323–335.

Maret, T. R., C. T. Robinson, and G. W. Minshall. 1997. Fish assemblages and environmental correlates in least-disturbed streams of the Upper Snake River basin. Transactions of the American Fisheries Society 126:200–216.

Matthews, W. J., and E. G. Zimmerman. 1990. Potential effects of global warming on native fishes of the southern Great Plains and the Southwest. Fisheries 15(6):26–32.

MoRAP (Missouri Resource Assessment Partnership). 2003. Missouri resource assessment partnership. MoRAP. Available: *www.cerc.usgs.gov/morap/* (September 2003).

Moyle, P. B., and P. J. Randall. 1998. Evaluating the biotic integrity of watersheds in the Sierra Nevada, California. Conservation Biology 12:1318–1326.

Narumalani, S., Z. Yingchun, and J. R. Jensen. 1997. Application of remote sensing and geographic information systems to the delineation and analysis of riparian buffer zones. Aquatic Botany 58:393–409.

NatureServe. 2003. NatureServe: a network connecting science with conservation. NatureServe. Available: *http://natureserve.org/* (September 2003).

Nelson, R. L., W. S. Platts, D. P. Larsen, and S. E. Jensen. 1992. Trout distribution and habitat in relation to geology and geomorphology in the North Fork Humbolt River drainage, northeastern Nevada. Transactions of the American Fisheries Society 121:405–426.

Nibbelink, N. P. 2002. Spatial patterns in fish distribution and density: the influence of watershed-based habitat gradients. Doctoral dissertation. University of Wyoming, Laramie.

Oberfoell, G. A. 1992. The use of geographic information systems to develop applications for the analysis of fisheries telemetry data. U.S. Fish and Wildlife Service, Environmental Management Technical Center, Special Report 92-2037, Onalaska, Wisconsin.

O'Brien-White, S., and C. S. Thomason. 1997. Evaluating fish habitat in a South Carolina watershed using GIS. Proceedings of the Annual Conference Southeastern Association of Fish and Wildlife Agencies 49(1995):153–166.

O'Connor, W. C. K., and R. J. Kennedy. 2002. A comparison of catchment based salmonid habitat survey techniques in three river systems in Northern Ireland. Fisheries Management and Ecology 9:149–161.

Parsons, M. G., J. R. Maxwell, and D. Heller. 1982. A predictive fish habitat index using geomorphic parameters. Pages 85–91 *in* N. B. Armantrout, editor. Acquisition and utilization of aquatic habitat inventory information. American Fisheries Society, Western Division, Bethesda, Maryland.

Poff, N. L., J. D. Allan, M. B. Bain, J. R. Karr, K. L. Prestegaard, B. D. Richter, R. E. Sparks, and J. C. Stromberg. 1997. The natural flow regime: a paradigm for river conservation and restoration. BioScience 47:769–784.

Porter, M. S., J. Rosenfeld, and E. A. Parkinson. 2000. Predictive models of fish species distribution in the Blackwater Drainage, British Columbia. North American Journal of Fisheries Management 20:349–359.

Radko, M. A. 1997. Spatially linking basinwide stream inventories to arcs representing streams in a geographic information system. U.S. Forest Service General Technical Report INT-GTR-345.

Rahel, F. J., C. J. Keleher, and J. L. Anderson. 1996. Potential habitat loss and population fragmentation for cold water fish in the North Platte River drainage of the Rocky Mountains: response to climate warming. Limnology and Oceanography 41:1116–1123.

Rahel, F. J., and N. P. Nibbelink. 1999. Spatial patterns in relations among brown trout (*Salmo trutta*) distribution, summer air temperature, and stream size in Rocky Mountain streams. Canadian Journal of Fisheries and Aquatic Sciences 56(Supplement 1):43–51.

Rieman, B. E., and J. D. McIntyre. 1995. Occurrence of bull trout in naturally fragmented habitat patches of varied size. Transactions of the American Fisheries Society 124:285–296.

Roper, B. B., and D. L. Scarnecchia. 1995. Observer variability in classifying habitat types in stream surveys. North American Journal of Fisheries Management 15:49–53.

Rossi, R. E., D. J. Mulla, A. G. Journel, and E. H. Franz. 1992. Geostatistical tools for modeling and interpreting ecological spatial dependence. Ecological Monographs 62:277–314.

Roth, N. R., J. D. David, and D. L. Erickson. 1996. Landscape influences on stream biotic integrity assessed at multiple spatial scales. Landscape Ecology 11:141–156.

Scott, J. M., F. Davis, B. Csuti, R. Noss, B. Butterfield, C. Groves, H. Anderson, S. Caicco, F. D'erchia, T. C. Edwards, Jr., J. Ulliman, and R. G. Wright. 1993. Gap analysis: a geographic approach to protection of biological diversity. Wildlife Monographs 123:1–41.

Sisk, T. D., editor. 1998. Perspectives on the land use history of North America: a context for understanding our changing environment. U.S. Geological Survey, Biological Resources Division, Biological Science Report USGS/BRD/BSR-1998-0003.

Srinivasan, R., and J. G. Arnold. 1994. Integration of a basin-scale water quality model with GIS. Water Resources Bulletin 30:453–462.

Tiffan, K. F., R. D. Garland, and D. W. Rondorf. 2002. Quantifying flow-dependent changes in subyearling fall chinook salmon rearing habitat using two-dimensional spatially explicit modeling. North American Journal of Fisheries Management 22:713–726.

Toepfer, C. S. 1997. Population and conservation biology of the threatened leopard darter. Doctoral dissertation. Oklahoma State University, Stillwater.

Toepfer, C. S., W. L. Fisher, and W. D. Warde. 2000. A multi-stage approach to estimate stream fish abundance using geographic information systems. North American Journal of Fisheries Management 20:634–645.

Torgersen, C. E., R. N. Faux, B. A. McIntosh, N. J. Poage, and D. J. Norton. 2001. Airborne thermal remote sensing for water temperature assessment in rivers and streams. Remote Sensing of Environment 76:386–398.

Torgersen, C. E., D. M. Price, H. W. Li, and B. A. McIntosh. 1999. Multiscale thermal refugia and stream habitat associations of chinook salmon in northeastern Oregon. Ecological Applications 9:301–319.

Tsao, E. H., Y. S. Lin, E. P. Bergersen, R. B. Behnke, and C. R. Chiou. 1996. A stream classification system for identifying reintroduction sites of Formosan landlocked salmon. Acta Zoologica Taiwanica 7:39–59.

USFS (U.S. Forest Service). 1994. A cumulative effects analysis of silvicultural best management practices using basin area stream survey methods. U.S. Forest Service, Ouachita National Forest, Report, Hot Springs, Arkansas.

USGS (U.S. Geological Survey). 2003. National Gap Analysis Program. USGS. Available: *www.gap.uidaho.edu* (September 2003).

Vannote, R. L., G. W. Minshall, K. W. Cummins, J. R. Sedell, and C. E. Cushing. 1980. The river continuum concept. Canadian Journal of Fisheries and Aquatic Sciences 37:130–137.

Waite, I. R., and K. D. Carpenter. 2000. Associations among fish assemblage structure and environmental variables in Willamette Basin streams, Oregon. Transactions of the American Fisheries Society 129:754–770.

Wang, L., J. Lyons, P. Kanehl, and R. Gatti. 1997. Influences of watershed land use on habitat quality and biotic integrity in Wisconsin streams. Fisheries 22(6):6–12.

Webb, A. D., and P. J. Bacon. 1999. Using GIS for catchment management and freshwater salmon fisheries in Scotland: the DeeCAMP project. Journal of Environmental Management 55:127–143.

Weibel, R., and B. P. Buttenfield. 1992. Improvement of GIS graphics for analysis and decision-making. International Journal of Geographical Information Systems 6:223–245.

White, G. M., and A. R. Kapuscinski. 1995. Urban planning for the conservation of stream ecosystems: critique of a prototype decision support tool. Pages 928–938 *in* J. M. Power, M. Strome, and T. C. Daniels, editors. Proceedings of decision support 2001. American Society of Photogrammetry and Remote Sensing, Bethesda, Maryland.

Whited, D., J. A. Stanford, and J. S. Kimball. 2002. Application of airborne multispectral digital imagery to quantify riverine habitats at different base flows. River Research and Applications 18:583–594.

Wiley, M. J., S. L. Kohler, and P. W. Seelbach. 1997. Reconciling landscape and local views of aquatic communities: lessons from Michigan trout streams. Freshwater Biology 37:133–148.

Wright, A., W. A. Marcus, and R. Aspinall. 2000. Evaluation of multispectral, fine scale digital imagery as a tool for mapping stream morphology. Geomorphology 33:107–120.

Chapter 4

Geographic Information Systems Applications in Reservoir Fisheries

CRAIG P. PAUKERT AND JAMES M. LONG

4.1 INTRODUCTION

Geographic information systems (GIS) are a rapidly emerging technology in freshwater fisheries science (Isaak and Hubert 1997). Although GIS technology has been used widely in terrestrial applications and increasingly has been used in marine environments, reservoir applications have lagged behind other freshwater applications. Four general questions can be answered with GIS (ESRI 1990) that apply to reservoir systems. (1) What characteristics (i.e., attributes) are associated with a particular location (e.g., What reservoir bottom topography is associated with fish spawning areas?)? (2) Which locations meet specific criteria (e.g., Which reservoir areas have a slope <5% and a gravel substrate?)? (3) Are there any spatial patterns related to that location (e.g., Do fish use these areas more during periods of higher inflows or water levels?)? (4) What happens if a particular event occurs (e.g., How much spawning habitat is lost if a reservoir is drawn down 3 m?)?

Although there has been extensive research on fish-habitat associations in reservoirs, there has been little attention given to the use of GIS in reservoir fisheries assessment. The value of GIS is the ability to analyze complex spatial patterns simultaneously. For example, coverages of reservoir attributes (depth, slope, vegetation coverage) can be overlaid with telemetry locations to determine habitat preferences of a fish. From this, biologists can query certain attributes and ask specific questions such as: Do fish use vegetated areas with slopes less than 5% during the spring? Examples of GIS applications in reservoir fisheries involve examination of habitat (Cross 1991; Parsely and Beckman 1994; Irwin and Noble 1996; Long 2000), vegetation (Remillard and Welch 1993; Cunningham 2000), fish movement (Rogers and Bergersen 1996; Zigler et al. 1999), and fish yield (DeSilva et al. 2001).

Reservoirs are hybrids between rivers and lakes, incorporating characteristics of each and, thus exhibit high spatial heterogeneity (Noble et al. 1994). Reservoirs often vary longitudinally in regard to primary productivity, physical habitat, and fish distribution (Buynak et al. 1989;

Thornton et al. 1990; Noble et al. 1994; Long 2000). Reservoirs also exhibit horizontal and vertical zonation (Thornton et al. 1990; Noble et al. 1994), which may affect fish distribution and movement. Fish distributions in reservoirs may be positively correlated with productivity, prey abundance, or other measures of suitable habitat (e.g., temperature) that vary spatially (e.g., Siler et al. 1986; Irwin et al. 1997; Phillips et al. 1997; Wilkerson and Fisher 1997; Paukert and Fisher 2000). Population dynamics of fish (i.e., recruitment and growth) are influenced by the physical and biological features of reservoirs (Gasaway 1970; Kimmel and Groeger 1986; Guy and Willis 1995). A GIS is a unique tool for analyzing these complex spatial patterns of high habitat heterogeneity and their influence on reservoir fisheries.

Water-level fluctuations strongly impact reservoir fisheries. These fluctuations change the physical habitat (littoral zone and spawning) used by fish (Ploskey 1986). Fluctuating water levels can severely affect fish behavior, distribution, recruitment, and growth (Ploskey 1986; Rogers and Bergersen 1996; Sammons and Bettoli 2000). Reservoir drawdowns expose littoral zones that, when inundated again, provide increased productivity and food for fishes (Ploskey 1986), assuming the drawdown is not prolonged enough to reduce macroinvertebrate abundance (Johnson and Andrews 1974). The timing of water-level fluctuations also is an important factor influencing reservoir fishes (Willis 1986), especially in regard to spawning and recruitment (Ploskey 1986; Gipson and Hubert 1993; Kohler et al. 1993; Sammons and Bettoli 2000).

The objectives of this chapter are to examine the techniques and challenges of using GIS in reservoirs and to provide examples of the use of GIS in reservoir fisheries management and research. We conclude with our view of the future of GIS in reservoir fisheries applications.

4.2 DATA COLLECTION TECHNIQUES IN RESERVOIRS

4.2.1 Physicochemical Features

Many of the spatial data collection techniques used in reservoirs are similar to those used in other water bodies. Aerial photography or satellite imagery can be used to delineate the reservoir boundary. Alternatively, the shoreline of a small reservoir can be demarcated by foot or by boat with a global positioning systems (GPS) receiver (Wilson 2002). Water depth and substrate classification may be determined with a sonar unit positioned directly under a GPS receiver (Rukavina 1998). For large reservoirs, LIDAR sensors (similar to radar but uses laser light instead of radio waves) mounted on an aircraft may be used to determine water depth over a large scale (Muirhead and Cracknell 1986). Side-scanning sonar can be used to classify shoreline features and sediment size (Pratson and Edwards 1996; Jansson and Erlingsson 2000), and water velocities in reservoirs can be determined with acoustic Doppler technology (Lemmin and Rolland 1997) and georeferenced in a GIS.

Topographic maps created before reservoir impoundment can provide insight into bottom features and depths. Although bottom contours change because of sedimentation and the modified flow regimes, these maps can provide a useful starting point to determine general reservoir characteristics, such as water depth and shoreline slope. Accurate depth estimates should be determined by direct measurement because water elevation (height above mean sea level, msl) is dependent on location in the reservoir. The elevation of the water in the upper end typically is higher than in the lower end. Reservoir water elevation usually is measured at the dam and is not transferable to the entire reservoir. However, accurate measures of depth from a topographic map may be possible by using methods that adjust for the longitudinal differences in water surface elevation, such as backwater curves (Chow et al. 1988).

Sources of data for the physical features of reservoirs include federal (e.g., U.S. Geological Survey, USGS), state (e.g., natural resource departments), local (e.g., city planning departments) agencies, nonprofit groups (e.g., local lake planning commission), and schools (e.g., university library). It may be advantageous to obtain similar data from different sources. For example, a local organization and a state agency may both have aerial photos but taken at different times and water levels, making it possible to examine how water level affects the physical environment of the reservoir. However, because most of the time and effort involved in a GIS application may be data acquisition (Isaak and Hubert 1997), careful planning with clear objectives is needed.

Limnological data typically needs to be acquired from either onsite or remote field measurements. Remote sensing has been used to determine reservoir turbidity (Nellis et al. 1998), chlorophyll *a* (Ritchie et al. 1994), emergent vegetation (Narumalani et al. 1997), and submergent vegetation (Malthus and George 1997). However, reduced visibility through the water column and difficulty in precision of locations due to lack of physical landmarks (Lehmann and Lachavanne 1997) make remote sensing and GIS applications for these variables difficult in aquatic systems. Currently, on-site measurements, coupled with GPS readings, are probably the most effective methods for determining limnological features of reservoirs (Gubala et al. 1994).

One of the greatest challenges in mapping the reservoir limnological environment is accurately characterizing its three-dimensional (3-D) aspects. Variations in the limnology of a reservoir can occur longitudinally (offshore–nearshore; upper reservoir–lower reservoir) and vertically (e.g., surface to bottom). Although researchers have been collecting 3-D data for some time, few have fully incorporated the 3-D aspect into their analyses (but see Chen et al. 1998).

4.2.2 Fish Locations and Distributions

The most common use of GIS in reservoir fishery applications is onsite measurement of fish locations. Fisher and Toepfer (1998) found, in a survey of universities when GIS is used in fisheries research, the most widely used application was mapping fish distribution data, along with aquatic habitat. Telemetry studies now use GIS to plot and analyze individual fish location data in reservoirs (e.g., Rogers and Begersen 1996; Wilkerson and Fisher 1997). Although research on the spatial aspect of catch rate data exists in marine environments (e.g., Fox and Starr 1996; Zheng et al. 2001; Ruckert et al. 2002), reservoir applications are in their infancy.

Analysis of historical data with GIS can provide insights into changes in distribution patterns. Currently, GIS use in reservoirs in this area is limited, but Conover and Grady (2000) used GIS to describe historic distribution of paddlefish *Polyodon spathula* and their movement in rivers and reservoirs of the Mississippi River basin. Mietz (2003) used GIS to determine if historical electrofishing sampling was conducted in proportion to the available shoreline habitat in the Glen Canyon Dam tailwaters and the Grand Canyon (Colorado River, Arizona). The most likely cause for this limited use is that the technology has been implemented only recently and historic data are time consuming to enter into GIS databases. Nonetheless, the utility of GIS will presumably become more apparent as current reservoir fish distribution data are entered for use in future analyses.

4.3 DATA ANALYSIS TECHNIQUES IN RESERVOIRS

Potential GIS analyses in reservoirs range from the micro-level to macro-level scale and from individuals to populations and communities. The most basic GIS analyses include data overlays, buffering, and point interpolation, although incorporation of geostatistics is becoming more prominent in fisheries.

4.3.1 Overlays

Overlay operations have been the mainstay of GIS for reservoir fisheries analysis. Data layers, or coverages in a GIS, contain different information and, by overlaying them, spatial relationships among variables are revealed. For example, Rogers and Bergersen (1996) used an overlay analysis in GIS to examine the spatial coincidence of largemouth bass *Micropterus salmoides* and northern pike *Esox lucius* in relation to vegetation type in a lake. They mapped the vegetation types by using GPS and then transferred the data as a coverage into a GIS. Individual largemouth bass and northern pike were implanted with ultrasonic transmitters, fish locations were plotted on a map, and the information was transferred as a coverage into a GIS. By overlaying the fish distributions coverage onto the vegetation type coverage, they were able to determine locations of certain vegetation types used by each fish species.

While an overlay of fish data with habitat data may be commonplace, habitats in reservoirs may change due to water-level fluctuations. For example, a researcher conducting a telemetry project might use GPS to determine a fish's position in the reservoir and then compute the distance from shore for each individual fish by using GIS with a map depicting the shoreline at the full conservation pool elevation. The resulting computations will be accurate only if the locations were recorded when the reservoir was at full conservation pool elevation. At all other elevations, the distances computed will be inaccurate. The degree of inaccuracy will depend on the local shoreline slope (low slopes will have more error than high slopes). In this case, researchers should relate their fish data in the proper context to their habitat data and not rely on any one habitat map.

4.3.2 Buffers

Buffers can determine proximity of an object to specified features. Rogers and Bergersen (1996) used buffers to determine near-shore areas of a lake and then queried the GIS to determine what percentage of fish locations, based on telemetry, fell within the near-shore boundary. Because longitudinal variation of fishery parameters is a common phenomenon in reservoirs (Siler et al. 1986; Buynak et al. 1989), buffers can be used to relate these phenomena to their relative position in the reservoir. For example, Long (2000) modeled relative abundance (i.e., catch per unit effort, CPUE) of black bass *Micropterus* spp. at shoreline sampling sites throughout a reservoir as a function of distance to the dam by using buffers.

4.3.3 Interpolation

Sampling surveys often measure a variable of interest at points throughout an area and then use this information to infer conditions at nonsampled locations. Interpolation is a means to estimate and visualize the entire surface area of interest. The two most common approaches to interpolation are inverse distance weighting (IDW) and spline (Lam 1983). Interpolation can be performed in vector GIS with points, arcs, or lines, or in raster GIS with individual cells. For example, the Oklahoma Department of Wildlife Conservation (ODWC) monitors abundance of shad *Dorosoma* spp. with hydroacoustics at sample points in reservoirs (G. Summers, ODWC, personal communication). The point data are entered into a GIS, and interpolation is performed to estimate abundance throughout the reservoir. Spatial trends in shad abundance are discerned from the interpolated map. Similar research also is being conducted in South Carolina on Lake Jocassee (D. Coughlan, Duke Power, personal communication).

Kriging is a specific method of interpolation that is little used in reservoir fisheries work, likely because this method is more complex computationally than other methods (Lam 1983; Guan et al. 1999). Kriging is a statistical approach to IDW interpolation (instead of a mathematical approach), consisting of a suite of generalized least-squares algorithms based on linear regression (Goovaerts 1997; Clark and Harper 2000). Although all interpolation methods produce estimates for nonsampled areas, kriging provides variances of those estimates in relation to spatial structure (Oliver and Webster 1990; Atkinson

1996). Therefore, kriging allows for estimation of map precision. Kriging has been shown to be one of the most precise interpolation methods (Guan et al. 1999). Regardless of the interpolation method used, precision and accuracy can be severely affected by the original data and complexity of the surface being interpolated (Lam 1983; Wilson 2002).

4.4 CHALLENGES OF USING GIS IN RESERVOIRS

Reservoirs provide uncommon challenges for GIS use and analysis. In this section, we discuss spatial, temporal, and habitat change challenges when using GIS for reservoir fisheries research and management.

4.4.1 Spatial Challenges

Perhaps the single most important spatial challenge in reservoirs is the effect of water-level fluctuations on fish habitat. For example, annual water-level fluctuations in Bull Shoals Reservoir, Arkansas–Missouri, averaged 6.9 m from 1968 to 1973 (Aggus and Elliot 1975). When water levels fluctuate, environmental conditions at sampling sites change. Irwin and Noble (1996) noted that the amount of gravel substratum declined as water levels decreased in embayments of B. E. Jordan Reservoir, North Carolina, whereas Parsely and Beckman (1994) suggested that lowering dam discharges reduces available spawning habitat for white sturgeon *Acipenser transmontanus* in the Columbia River. For studies that do not concurrently measure environmental and fisheries characteristics along the shoreline, steps must be taken to ensure that fisheries data correspond to environmental data at the same water level. For example, substrate types classified on one day might not be available at a later date when the site is sampled for fish if there is a change in water level. Subsequent analyses between substrate type and catch rates, therefore, would be in error. Fortunately, daily water-level data usually are easily obtainable from reservoir operators (e.g., U.S. Army Corps of Engineers,USACOE), often over the Internet and they can be used to correctly match environmental data with fisheries data.

4.4.2 Temporal Challenges

Geographic information systems are well suited for modeling spatial data but are not well suited to modeling data that are time dependent. A researcher conducting an electrofishing survey might spend twice as much time in one cove versus another of the same size because of differences in habitat complexity (e.g., maneuvering around standing timber versus maneuvering through submersed vegetation). Although these samples are equal in size, temporally, they are different. Therefore, results from statistical analyses will depend on whether the researcher used time versus space to depict catch rates (i.e., number per electrofishing hour versus number per electrofishing meter). Although the choice of how to analyze these data is more of a sampling and analysis issue, how to properly portray them in a spatial setting is a GIS issue.

4.4.3 Habitat Changes

While many aspects of reservoir habitat are relatively static (e.g., substrate type, shoreline slope), many others are dynamic (e.g., water temperature, turbidity). Even variables that seem static may change. In reservoirs, this is often the case for fish cover. Reservoir managers often add trees or other forms of artificial cover into reservoirs (Summerfelt 1993). These forms of cover should be recorded when mapping reservoirs and the GIS maps updated when new additions are made. Keeping maps updated is even more difficult when groups other than the management authority (e.g., nonprofit groups or angling groups) or individuals add artificial cover or increase shoreline development. Building relationships with these groups and individuals is a good way to obtain information to keep maps updated.

A GIS has the potential to be a valuable tool for determining sites for cover additions. Cross (1991) described how GIS could be used to select sites in the drawdown zone of Trinity Reservoir, California, for habitat enhancement for largemouth bass and smallmouth bass *Micropterus dolomieu*. Coupled with a reservoir elevation model, the amount of time that an individual habitat enhancement would be available (i.e., flooded) to fish could be calculated. In this example, GIS made it possible for placing cover in areas that provided maximum benefit to the targeted fish populations.

4.5 EXAMPLES OF GIS APPLICATIONS IN RESERVOIRS

The objective of this section is to provide specific examples of how GIS has been used in reservoirs. These examples provide an array of different applications. Our hope is that these examples will spur interest and expansion of GIS applications in reservoir environments in the future.

4.5.1 Use of GIS to Determine the Impact of Hydrologic Operations on Fall Chinook Salmon Spawning and Rearing Habitat in John Day Reservoir

The reduction of salmonid habitat in the northwestern United States has been attributed to habitat loss due to dam construction and reservoir operations. The USGS evaluated the effects of these dam operations on fall chinook salmon *Oncorhynchus tshawytscha*, a species listed as endangered under the Endangered Species Act (Dauble 2000). The objective of this study was, in part, to evaluate the restoration and recovery of fall chinook salmon in John Day Reservoir (Washington–

Oregon). The amount of spawning and rearing habitat in John Day Reservoir at three different water-level regimes, normal pool (80.5 m above msl), spillway crest (66.7–70.4 m above msl), and natural river (48.5–48.8 m above msl), was determined by GIS. A GIS also was used to explore alternatives to fall chinook salmon restoration efforts: dam removal and reservoir drawdown. In addition, the model incorporated three simulated discharges (2,832, 4,418, and 8,496 m^3/s [100,000, 156,000, and 300,000 ft^3/s]) for each of the three water levels. For space considerations, we will discuss only the 4,418 m^3/s discharge with a reservoir drawdown option.

Spatial data were obtained from several different sources and incorporated into the fall chinook spawning and rearing habitat models. Channel features (e.g., shoreline, bars, islands, etc.) were collected from quadrangle sheets, land survey maps, 1931 War Department hydrographic maps, digital line graph maps, digital elevation models (DEM), and aerial photos. All data were georeferenced to the same projection and coordinate system.

Water flow data and classification of riffle, pools, and runs were estimated by using mean water column velocities derived from a two-dimensional hydrologic model. The Froude number was used to delineate if an area was a pool, a riffle, or a run. Riverbed elevation was based on a 1994 survey by the USACOE and a 30-m resolution DEM. Suitable fall chinook spawning and rearing habitat was based on existing fisheries surveys and the literature. Suitable spawning habitat was defined as reservoir areas with mean water column velocities of 0.4–2.0 m/s and water depths of 0.3–9.0 m. Riffle habitat also was considered suitable for spawning. Rearing habitat was estimated by catch data from electrofishing and concurrent habitat surveys. By using multivariate analyses (e.g., discriminant function and principal components analyses), the reservoir was classified into 10-m × 10-m cells. A probability function was used to estimate the probability of at least 10 juvenile fall chinook salmon being collected for each cell. If the probability was 0.7 or greater, then that cell was considered a high-probability rearing habitat.

At 4,418 m^3/s, the river was classified as 99.9% pool habitat under normal pool levels, 93% under spillway crest, and 76% under natural river flow conditions. The area of potential spawning habitat was lowest under normal pool conditions and highest under natural flow conditions. Based on the GIS models using water depth, velocity, and mesohabitat (i.e., pool-riffle-run), the total area meeting the criteria for fall chinook spawning in John Day Reservoir was 607 ha (3% of the total wetted area) under normal pool conditions, 2,650 ha (23% of the total wetted area) under spillway crest, and 4,600 ha (51% of the total wetted area) under natural flow conditions (Figure 4.1).

Compared with a normal pool scenario, a reservoir drawdown simulating natural flow conditions would increase the amount of potential fall chinook spawning habitat by nearly 4,000 ha in John Day Reservoir.

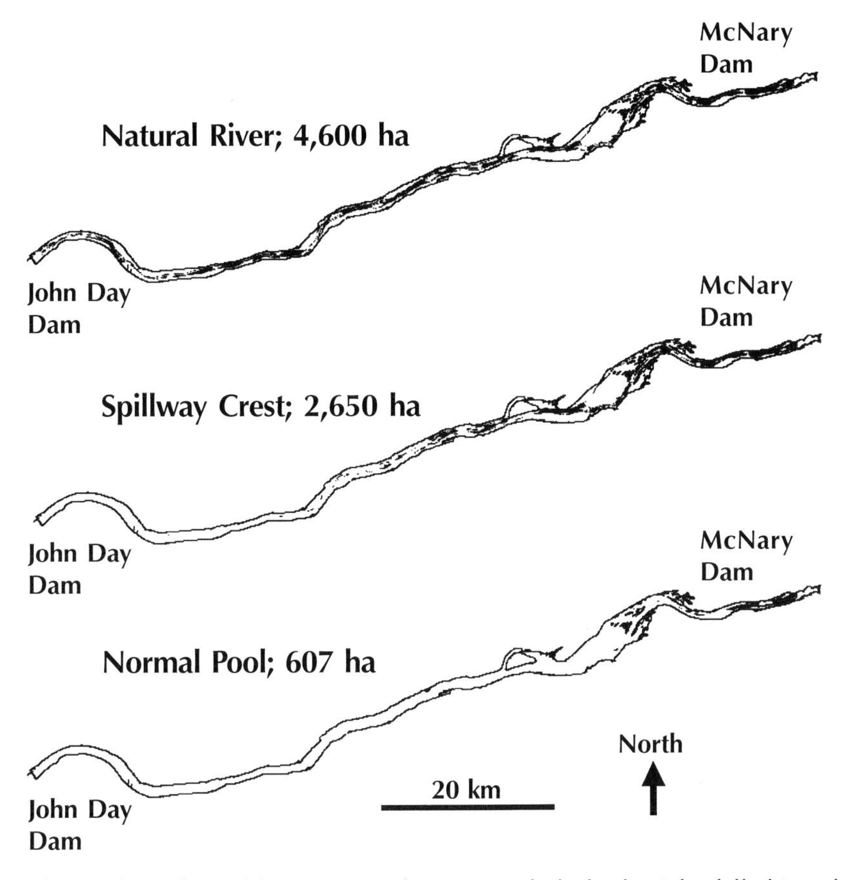

Figure 4.1 Total area (ha) of suitable spawning locations (dark shading) for fall chinook salmon in John Day Reservoir, Washington–Oregon, at 4,418 m^3/s at three water-level scenarios: natural river, spillway crest, and normal pool. Modified from Dauble (2000).

Fall chinook rearing habitat was highest under the natural river flow conditions. Under a flow regime of 4,418 m^3/s, there were 255 ha of high probability (≥0.7) rearing habitat at normal pool water levels. Under the spillway crest scenario, 289 ha had high probabilities of fall chinook rearing, whereas the natural river water levels had 356 ha classified as high probability of rearing habitat (Table 4.1).

By using GIS technology and spatial layers depicting hydrology, geomorphology, and fish catch rate, the authors determined that the natural flow regime scenario at 4,418 m^3/s increased potential fall chinook spawning and rearing habitat. Partial (i.e., spillway crest) or complete (natural flow) drawdown of John Day Reservoir provided increased spawning and rearing habitat for fall chinook salmon. The natural flow regime yielded the largest area of spawning habitat and had slightly higher rearing habitat when compared with the spillway crest scenario. Under normal pool, fall chinook spawning and rearing habitat were at their lowest levels across all three scenarios. However, the authors cautioned that these scenarios predicted where

Table 4.1 Area (ha) and percent of high-probability and low-probability fall chinook salmon rearing habitat in John Day Reservoir at 4,418 m^3/s at three scenarios: normal pool, spillway crest, and natural river. Total usable habitat is the high-probability and low-probability habitat combined. Table modified from Dauble (2000).

Variable	Normal pool	Spillway crest	Natural river
High probability (ha)	255	289	356
Low probability (ha)	279	239	300
Total reservoir area (ha)	19,779	11,748	9,055
Total usable area (ha)	534	528	565
Percent high probability	1.3	2.5	3.9
Percent total usable area	2.7	3.7	5.6

spawning could occur. Where spawning actually occurred would be influenced by additional factors such as substrate, spawning bed slope, and spawning bed scour (Dauble 2000).

The use of GIS in this analysis provided spatial data (e.g., mesohabitat, water velocity, etc.) that were incorporated into a predictive model. The authors used GIS to help visually display their data, as well as analyze spatial data with relative ease. Refining the model by incorporating additional variables would be relatively simple because existing data already were built into the database. Without GIS, the analysis of these data would be much more time intensive. An additional advantage to using GIS is the visual display of relatively complex relationships. Based on a series of layers and criteria, the authors created relatively simple maps, tables, and figures that the public and administrators could understand without knowing the intricacies of a GIS.

4.5.2 Use of GIS to Determine Habitat Selection of Northern Pike and Largemouth Bass in a Small Colorado Reservoir

Rogers and Bergersen (1996) used GIS to evaluate habitat use in Lake Ladora, a 25-ha reservoir in Colorado. Aquatic macrophytes dominate Lake Ladora, and the authors hypothesized that both largemouth bass and northern pike would select for these macrophytes.

A vegetation map of the reservoir was created by measuring vegetation along 24 transects. The data acquisition system consisted of a boat-mounted computer integrated with a compass and speedometer. When the transects were run, the dominant vegetation was identified continuously along the transects. After the transects were run, the reservoir was partitioned into 10-m × 10-m cells and the remainder of the lake that was not mapped was interpolated to these 10-m × 10-m cells by using the closest four points (i.e., nearest neighbor) to classify vegetation.

Six largemouth bass and four northern pike were implanted with ultrasonic transmitters and located weekly by canoe. When a fish was located, its coordinates were mapped by using aerial photos of the reservoir with a meter grid overlay. In the laboratory, these coordinates were converted to Universal Transverse Mercator coordinates to facilitate transfer to a GIS.

The authors also evaluated the use of nearshore habitats by using GIS. The reservoir perimeter was digitized into a GIS, and a 20-m buffer around the perimeter was generated. Fish locations were separated by season: spring (March–May), summer (June–August), fall (September–November), and winter (December–February). Fish locations during these seasons were overlaid onto the buffer, and GIS was used to query the proportion of locations within the 20-m buffer.

Statistical analyses consisted of logistic regression for the habitat selection data. In this analysis, largemouth bass and northern pike binary (i.e., presence or absence) location maps were integrated into the aquatic macrophyte coverage maps. In addition, Ivlev's electivity index also was used to evaluate habitat selection. A chi-square analysis was used to determine if the proportion of fish locations within the 20-m buffer differed from random habitat use (e.g., was the proportion of fish locations higher or lower than the proportion of the lake within the 20-m buffer).

Lake Ladora consisted primarily of vegetated areas. The GIS-derived vegetation map indicated large areas of common water milfoil *Myriophyllum sibiricum*, American pondweed *Potamogeton nodosus*, and muskgrass *Chara* spp. with some deepwater and silt areas (Figure 4.2).

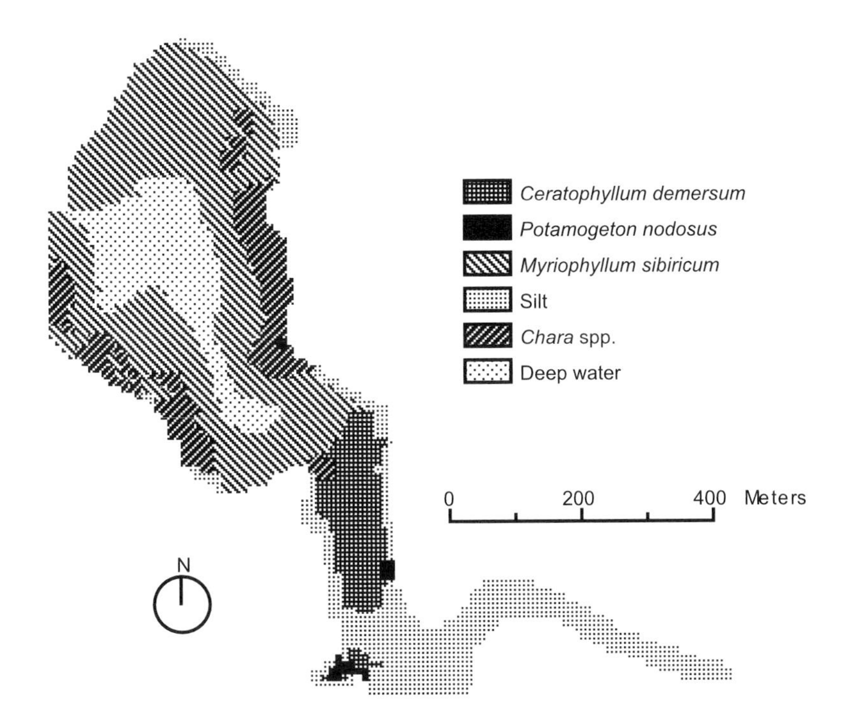

Figure 4.2 Dominant vegetation cover in Lake Ladora, Colorado, as measured by 24 transects and subsequent interpolation into 10-m × 10-m cells. Reprinted from Rogers and Bergersen (1996).

Based on Ivlev's electivity index, largemouth bass selected *Potamogeton nodosus* and northern pike selected the silted areas of the lake. Both northern pike and largemouth bass avoided *Myriophyllum sibiricum*, *Chara* spp., and deep water (Figure 4.3).

Lake Ladora largemouth bass typically used nearshore habitats in higher proportions than random use in spring and summer, and in lower proportions than random use in winter. Northern pike used nearshore habitats more than expected in summer and less than expected in winter (Figure 4.4). In general, northern pike were nearshore in summer and offshore during winter.

The authors surmised that the Lake Ladora largemouth bass results were not surprising because *Potomogeton nodosus* may provide a cover that these fish use to ambush prey. The avoidance of other plant species by largemouth bass may be attributed to the fact that these species grew in dense mats that may not provide adequate room for navigation. This hypothesis was further supported because largemouth bass were found near shore where habitat was more complex. The winter offshore migration of largemouth bass was probably to locate deep, warmer water.

The northern pike results were slightly uncharacteristic for this species because they selected for open-water silted areas instead of

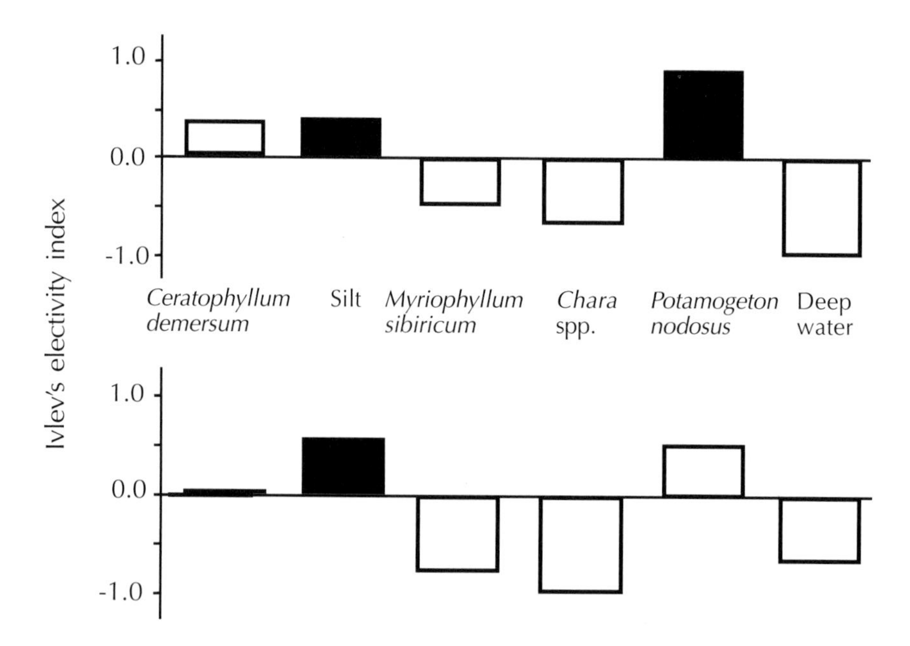

Figure 4.3 Ivlev's electivity index of largemouth bass (top) and northern pike (bottom) located during summer 1992 in Lake Ladora, Colorado. Available proportion of vegetation was assessed by using a vegetation map interpolated by using GIS. Positive values indicate selection for that cover type and negative vales represent avoidance for that cover type. Solid bars represent significant ($P < 0.001$) selection using logistic regression. Modified from Rogers and Bergersen (1996).

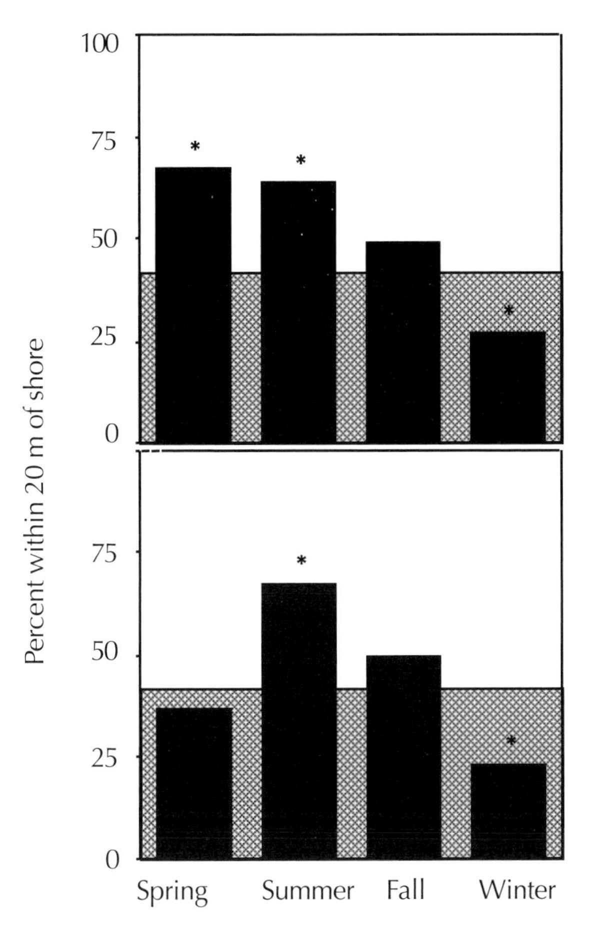

Figure 4.4 Seasonal selection for nearshore (i.e., within 20 m from shore) habitats by largemouth bass (top) and northern pike (bottom) in Lake Ladora, Colorado. The gray area represents the proportion of available habitat within 20 m from shore, whereas the solid bars represent the proportion of fish locations within 20 m from shore. Bars with asterisks indicate significant ($P < 0.05$) differences between the proportion of observed and available habitat use. Modified from Rogers and Bergersen (1996).

vegetation. The authors attributed this to the silted area being the coolest portion of the lake in summer. The selection of nearshore areas in summer by northern pike was attributed to these fish being located in the narrow arm on the southeast section of the lake (Figure 4.2). These fish might have been selecting for the coolwater areas with silt substrates, which happened to be near shore. This GIS analysis was relatively simple, and additional layers of water depth, water temperature, vegetation density, forage density, and so on, would provide a more complex model and may increase the predictive power of the analysis. Nonetheless, this example shows the utility of GIS in the analysis of telemetry data in reservoirs.

4.5.3 Use of GIS to Randomly Select Electrofishing Sampling Sites in Skiatook Lake, Oklahoma

Random sampling generally is regarded as necessary to obtain data for rigorous statistical analysis. The use of GIS can facilitate this process in reservoirs with prior collection of data on resource availability (Wilde and Fisher 1996). Long (2000) used GIS to summarize habitat availability data in order to randomly select potential electrofishing sampling sites along the shoreline of Skiatook Lake, Oklahoma, in his study of black bass resource use. The overall objective was to assess differential resource use by largemouth bass, spotted bass *Micropterus punctulatus,* and smallmouth bass.

Skiatook Lake is a 4,266-ha flood-control impoundment of Hominy Creek in north-central Oklahoma, which was formed in 1984 by the USACOE. The lake perimeter consists largely of steep, bedrock substrata with little aquatic vegetation. The upper end of the reservoir is more turbid and has higher primary productivity (Fisher et al. 2000).

To produce a GIS map for selecting sampling sites, all of the 7.5-min topographic maps (mapped in 1966; photo-revised in 1983) encompassing the boundaries of Skiatook Lake were scanned, edited, and exported to a GIS (ArcView, Environmental Systems Research Institute, Redlands, California) (Figure 4.5A). The entire shoreline of the lake was surveyed, and areas were classified according to substrate (i.e., sand, gravel, cobbler, boulder) and cover types (i.e., emergent vegetation, flooded standing timber, fallen timber). Substrate and cover types were added to the GIS map as line features based on visual landmarks and were considered potential sampling sites (Figure 4.5B).

Because Skiatook Lake had longitudinal variation in water quality, it was stratified into four areas: Hominy Creek, Bull Creek, mid-lake, and lower lake (Figure 4.5C). Hominy and Bull creeks were located in the upper reservoir and represented the two major tributaries into Skiatook Lake. The mid-lake region was a transitional zone where the water received from the upper reservoir area was mixed. The lower-lake region was the area where the reservoir widened and most of the suspended material settled, resulting in higher water clarity. Sampling sites (uniquely identifiable combinations of cover and substrate types) were randomly selected from all potential sites identified within each of the four strata (Figure 4.5D). Numbers drawn from a random number table, linked to the GIS database tables, indicated the unique sites to be sampled. This process was repeated in spring and fall from 1997 through 1999 so that 10% to 20% of the entire shoreline was sampled in each time period. By using GIS, a map of the selected sampling sites was produced each year and season to facilitate navigation.

The scheme used in this study may not be appropriate for every reservoir sampling situation. The objective of this study was to determine relative abundance of black bass while controlling for the effects of substrate and cover type. Texas Parks and Wildlife Department (TPWD) currently uses GIS routinely in their reservoir sampling surveys (F. Janssen, TPWD, personal communication). There, reservoirs

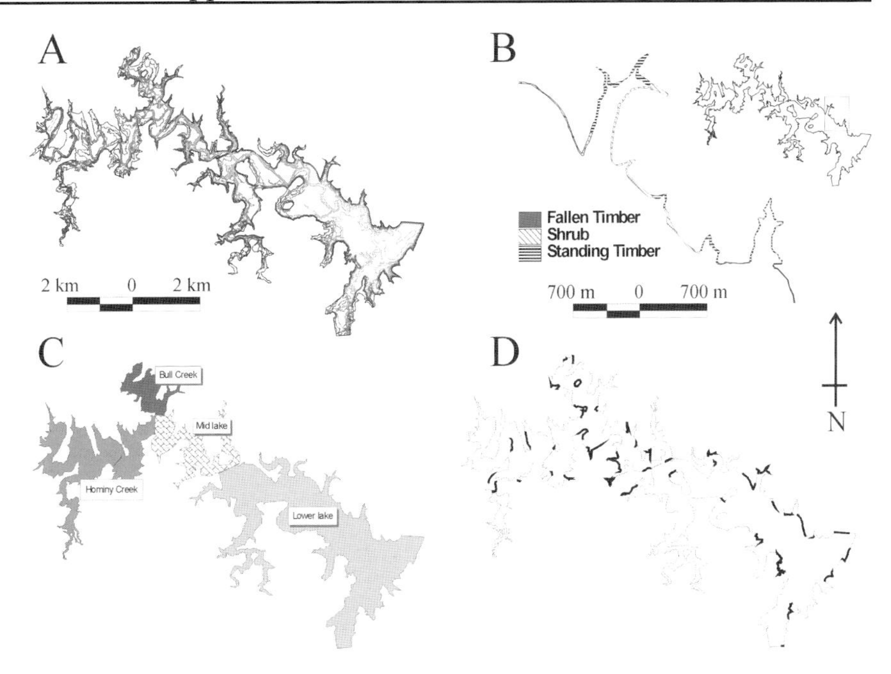

Figure 4.5 Diagram illustrating the three major overlays used to construct a map of potential electrofishing sampling sites in Skiatook Lake, Oklahoma. First, a U.S. Geological Survey topographical map was converted into digital format (A), then shoreline sampling sites (combinations of substrate and cover type) were added to the map from on-site reconnaissance (B, example of the cover type theme, polygons are shown for clarity). Skiatook Lake was then divided into four strata (C) from which sampling sites were randomly selected (D, spring 1997 electrofishing sampling sites). Modified from Long (2000).

are divided into 10-s grids and a predetermined number of stations (grids) are randomly selected for each gear type during a sampling year. All grids within a reservoir are available for gill netting stations, while electrofishing and trap netting stations are selected from grids that are adjacent to the shoreline. For pelagic species, shoreline habitat would not be useful in determining a random sampling scheme but water temperature or dissolved oxygen may be, and these can be modeled and used to develop a sampling design in GIS. Another alternative is to overlay a grid encompassing the entire water body and randomly select individual cells for sampling (e.g., Wilkerson and Fisher 1997; Paukert and Willis 2002). As long as the variable of interest (i.e., sample sites) has a spatial component, GIS can be a useful tool to facilitate sample site selection.

4.5.4 Evaluating the Effectiveness of Revegetation Efforts in Lake Chambers, Oklahoma, by Using GIS

The amount and type of aquatic vegetation present in a reservoir is one of the most important factors affecting fish populations. Aquatic vegetation concentrates fish (Crowder and Cooper 1979; Killgore et al. 1989; Chick and McIvor 1994), influences species composition (Xie et al. 2000), alters predator efficiency (Savino and Stein 1982, 1989), and influences recruitment (Durocher et al. 1984). Because reservoirs are artificial systems and often have large water-level fluctuations, vegetation usually is sparse (Summerfelt 1993). However, reservoir fisheries often can be improved through efforts to reestablish aquatic vegetation.

Before 1991, electrofishing surveys performed by the ODWC indicated increasing abundance of largemouth bass in Lake Chambers, Oklahoma (Stahl 1995). In 1991–1992, the water level in Lake Chambers was lowered for spillway repairs. This resulted in the extirpation of aquatic vegetation, and subsequent electrofishing surveys showed depressed numbers of largemouth bass through 1995 (Stahl 1995). The reduction in largemouth bass abundance was attributed to the loss of aquatic vegetation, and two native species of vegetation, coontail *Ceratophyllum demersum* and northern water milfoil *Myriophyllum spicatum*, were reintroduced into the reservoir in 1997.

Cunningham (2000) evaluated the effectiveness of this revegetation effort by using GIS. The overall study objective was to assess if GIS could be used as an efficient tool to determine the effectiveness of reservoir revegetation.

Lake Chambers is a 36.6-ha recreational impoundment in northwest Oklahoma. To evaluate the effectiveness of the revegetation effort, Cunningham (2000) first constructed a map of the lake and then determined which areas of the lake were suitable for vegetation growth. Variables collected to use in determining suitability were water depth, clarity, shoreline slope, and substrate (Figure 4.6).

To obtain an accurate map of the shoreline, Cunningham (2000) collected GPS data by following the shoreline at normal pool elevation (708.2 msl). These data were differentially corrected, resulting in locational data accuracy of 3 m or less. A bottom topography map of the lake was obtained by attaching a depth sounder to the GPS unit and by obtaining depth information at known locations ($N = 385$). These point data were interpolated to produce a smooth surface of the lake bottom (DEM). From the DEM, a moving window analysis (Burroughs 1987) was used to calculate shoreline slope. The depth data were used to infer areas of light penetration (water clarity) because Secchi depth data indicated that light would likely penetrate to the bottom year round in areas 51 cm deep or less, variably in areas 52–182 cm, and never in areas deeper than 183 cm. To obtain a map of substrate types, soils were collected from 35 sites on Lake Chambers, analyzed in a laboratory, and interpolated among substrate sampling points to estimate substrate types for the entire lake. Because soil productivity was negatively related to percent sand, and percent sand was easier to measure, percent sand was used as the interpolation variable.

To obtain a map of areas suitable for growth of aquatic macrophytes, depth, slope, water clarity, and substrate-type data were

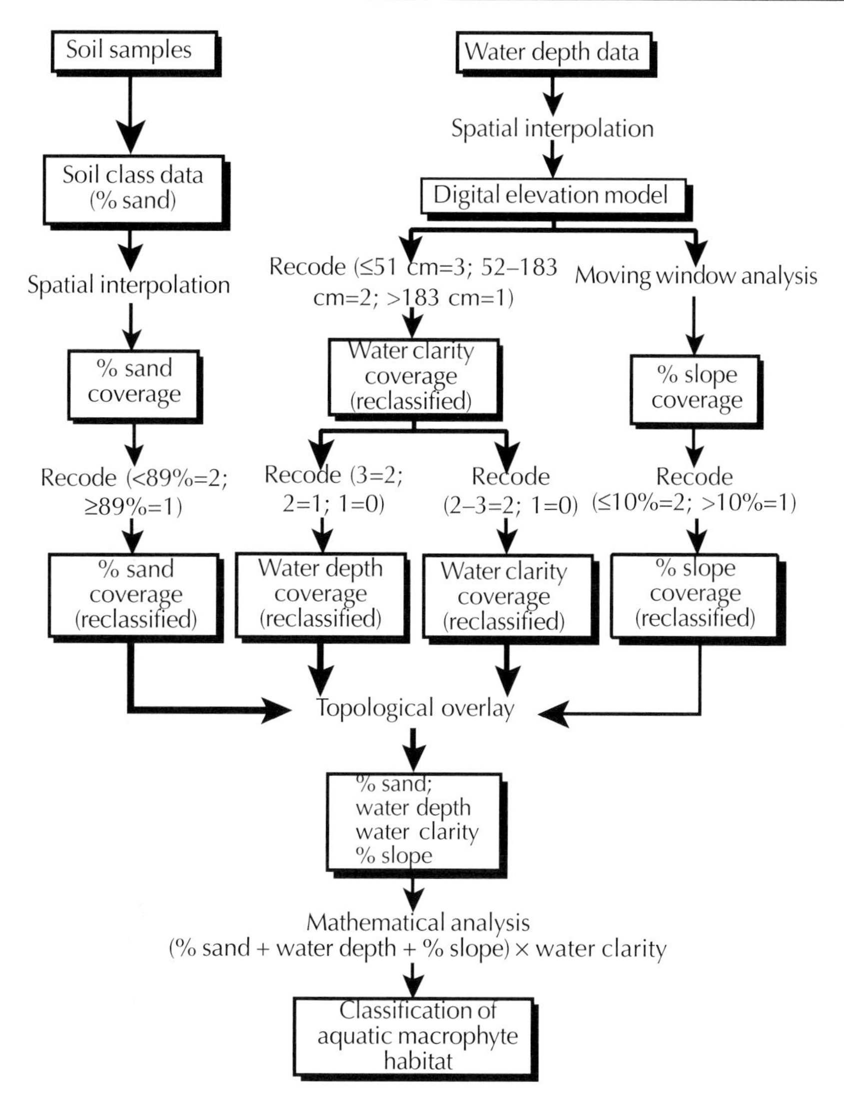

Figure 4.6 Flow chart diagram showing how GIS and spatial data were used to evaluate the effectiveness of revegetation efforts in Lake Chambers, Oklahoma (Cunningham 2000).

classified into various categories representing "good" and "poor" (Figures 4.6 and 4.7). Depths 51 cm or less were classified as good, depths from 52 to 183 cm were classified as poor, and depths more than 183 cm were classified as unsuitable. These values were then re-coded to produce good, poor, and unsuitable habitats for the water-clarity coverage and the water-depth coverage. Percent slope values 10% or less were considered good and values higher than 10% were poor. Percent sand values less than 89% were good and 89% or higher were poor. Combining all the coverages with an overlay analysis allowed a categorization of the entire lake into areas of excellent, good, moderate, and poor habitat for aquatic macrophyte

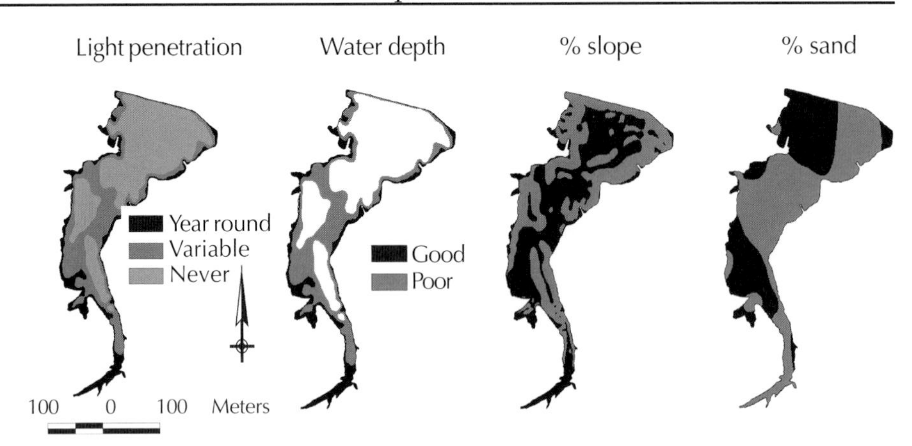

Figure 4.7 Map showing areas of suitability for aquatic macrophyte establishment in Lake Chambers, Oklahoma, based independently on water clarity, water depth, percent slope, and percent sand. All maps are shown at the same scale (Cunningham 2000). The white areas indicate unsuitable land.

growth (Figure 4.8). This overlay procedure was performed with the mathematical equation:

$$(\% \text{ sand} + \text{water depth} + \% \text{ slope}) \times \text{water clarity.}$$

Cunningham (2000) tested the accuracy of this model by randomly selecting 66 sites to survey for presence or absence of the two aquatic macrophyte species reintroduced into the reservoir (Figure 4.8). Only 6 of the 66 sites surveyed contained aquatic macrophytes, and the only species found was northern water milfoil. Five of the positive sites were contained in areas classified as excellent, and the final positive site was classified as good (Figure 4.8).

The use of GIS in this application was considered a valuable tool in evaluating the revegetation effort in Lake Chambers (Cunningham 2000). The general lack of macrophytes occurring in the reservoir was attributed to low water clarity. Lack of knowledge about the ecology of specific plant species was considered to be a problem in accurately assessing revegetation success. In addition, it was not known if one of the variables measured was more important than the others. In this example, all variables affecting macrophyte establishment were weighted equally. One of author's recommendations was that future research focus on gaining a better understanding of the role of various habitat variables on individual plant species establishment.

4.6 CONCLUSION AND FUTURE OF GIS APPLICATIONS IN RESERVOIRS

Our conclusions are separated into three sections: current knowledge and status of GIS applications in reservoirs, improvements and needs in reservoir GIS applications, and what we believe will be the future

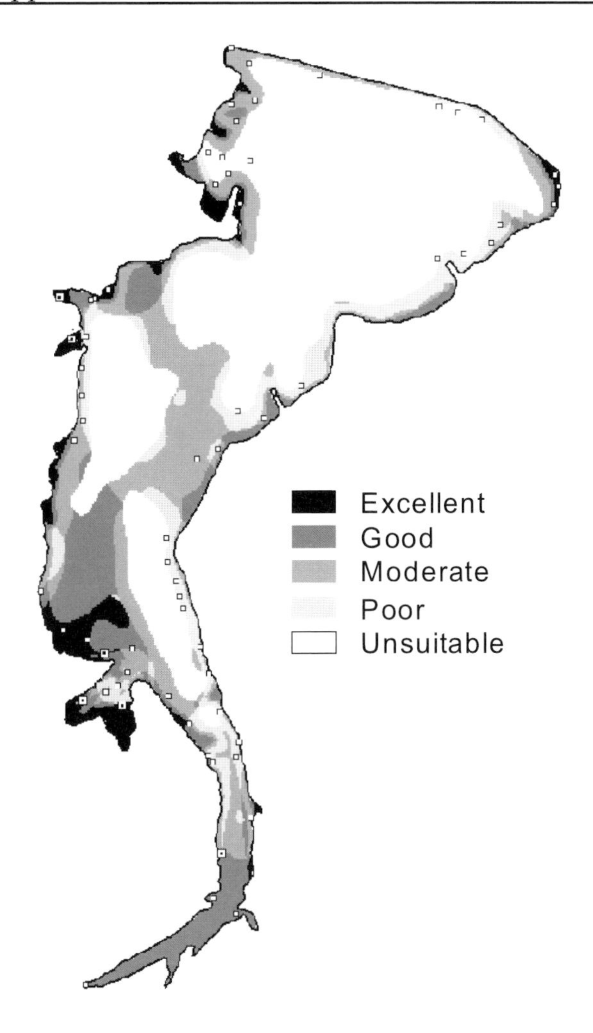

Figure 4.8 Map showing areas of suitability for aquatic macrophyte establishment in Lake Chambers, Oklahoma, based on mathematical overlay of water clarity, water depth, slope, and percent sand coverages. Squares indicate areas that were searched for aquatic macrophyte establishment and squares with dots indicate areas where aquatic macrophytes were found. Modified from Cunningham (2000).

role of GIS in reservoir fisheries. Although these sections are separated, they are not necessarily mutually exclusive. For example, future directions and improvement of GIS applications in reservoirs are based on the current state of these applications.

4.6.1 Current Knowledge and Status of Reservoir GIS Applications

Applications of GIS in reservoir fisheries management and research are increasing. The first fisheries GIS publication occurred in 1984, and the number of publications has increased substantially since the mid-1990s (Meaden 2001). Much of the early fisheries GIS literature focused on lentic systems and marine environments. Even today reservoir applications are limited. However, as more fisheries professionals realize the efficiency of GIS for analyzing habitat and managing complex reservoir environments (Irwin and Noble 1996), the role of GIS will certainly increase.

The technology for GIS applications in reservoir fisheries is readily available, and fisheries students and professionals are seeking training to use it. Education and knowledge of GIS by fisheries professionals is increasing. In a recent survey of 24 universities, 96% indicated their fisheries students were taking at least one introductory course in GIS, with about half of those students taking courses directly related to GIS and natural resources (Fisher and Toepfer 1998). Based on their survey results, the authors surmised that GIS is becoming an integral part of fisheries research and management. As new biologists with a GIS background enter the fisheries profession, GIS applications will undoubtedly increase.

4.6.2 Improvement Needed in Reservoir GIS Applications

Although our understanding of GIS technology has improved over the last two decades, many improvements are still needed, particularly in applications in aquatic and reservoir environments. Some of the primary improvements needed for reservoir applications are: (1) better collection and visualization of 3-D data, (2) better agency and government cooperation, (3) awareness of landscape-level influences on reservoir characteristics, (4) increased awareness and implementation of error classification, (5) increased knowledge of spatial analysis, and (6) better quantification of water-quality data.

Collection of 3-D data is essential in reservoir fisheries applications. The depth component is integral in determining fish distribution and ecology. Three-dimensional GIS modeling has been difficult in larger water bodies such as reservoirs (Mason et al. 1994). This is primarily because the interactions between water flow and other parameters are complex (e.g., temperature, water quality parameters, fish distributions), the computing power and algorithms needed to analyze 3-D data are complex and computer-intensive, and there are few ways to incorporate time-series analysis with a 3-D GIS (Chen et al. 1998). Both computing power and algorithms have become more efficient, and analysis of spatial data continually are being refined. However, classification of these 3-D habitats is needed to improve reservoir GIS models.

It is imperative that governmental agencies and countries cooperate and share spatial information. Availability of spatial data is more of a problem in countries outside the United States, where the governments may have more control over the data. For example, information on the land use around some Sri Lankan reservoirs was classified by the government and could not be used by outside individuals and agencies (DeSilva et al. 2001). In addition, cooperation by agencies who use GIS is needed so they do not "reinvent the wheel." The Federal Geographic Data Committee (FGDC) currently provides a source of spatial data and metadata that has been produced in coordination with policies and standards provided by the U.S. Government. The goal of FGDC is to provide geographic information to the public and minimize the duplication of data collection by various agencies. Because of the larger scale of reservoirs and their multiple uses, many agencies may have GIS databases that are of use to fisheries managers and researchers. Agencies need to cooperate with organizations such as the FGDC and determine what information they have and share it with others.

With the increasing interest in ecosystem-based management, landscape-level influences on reservoirs need to be addressed further. The analysis of reservoir fisheries should not stop at reservoir boundaries. Reservoirs themselves may be considered landscapes and may be managed at multiple scales (Noble et al. 1994; Irwin et al. 2002). Watershed and landscape-level attributes are necessary to better understand reservoir fisheries ecology.

The classification of errors is a much needed and often neglected attribute of GIS applications. Johnson (1990) suggested that there are three types of errors: (1) overt errors, (2) manuscript errors, and (3) processing errors. The use of outdated data sources or improper resolution or scale causes overt errors, whereas manuscript errors are caused by poor georeferencing or inappropriate interpolation techniques. Processing errors are caused by inexact algorithms and operator error. All these errors are compounded as you increase the complexity (e.g., more overlays) of the GIS database. Although researchers often recognize these errors, they rarely are calculated or reported (but see Paukert and Willis 2002; Wilson 2002). It is common to report estimates of variance or error with traditional statistical tests, yet this is still not the case with GIS applications. The user of the spatial information needs to be informed about the certainty of the data presented.

There needs to be improvement in the use of spatial analysis techniques for reservoir GIS applications. Aquatic ecologists typically have lagged behind terrestrial ecologists in the use of spatial analysis procedures (Johnson 1990). Indeed, reservoir applications have lagged behind other aquatic applications such as marine applications. As GIS education and software increase, spatial analysis techniques in reservoirs must follow.

Finally, there needs to be better quantification of water quality and water chemistry parameters in reservoirs. This information typically has been very difficult to quantify because of its dynamic nature. However, remote sensing (e.g., satellite imagery) technology can now collect "instantaneous" information. The next step is to use that information to estimate primary productivity (e.g., chlorophyl concentration), nutrients (e.g., phosphorous), or other physical characteristics (e.g., depth and turbidity). With this technology, fisheries professionals can further their understanding of reservoir limnology and the associated ecology of the fishes.

4.6.3 Future Directions of Reservoir GIS Applications

Geographic information systems applications in reservoirs are in their infancy. As alluded to earlier in this chapter, 3-D maps of reservoirs will become increasingly important as technology and information becomes available and cost decreases. The human dimensions aspect of reservoir fisheries management is becoming increasingly popular, and GIS can be an integral tool in analysis of this information. A simple example would be by using zip codes in angler survey data and incorporating this information into a GIS to identify the geographic locale of the users. In addition, with multiple-use management of reservoirs the norm, spatial information relating location of various users (e.g., anglers, swimmers, water skiers) within a reservoir may be expedited by using GIS.

The analysis of fish population dynamics by using GIS in reservoirs has received limited attention. Other than with telemetry studies, the use of GIS to understand spatial relationships of fish has been limited. In particular, species interactions (e.g., competition and predation) are often inferred in fisheries, but GIS technology is rarely used to analyze them. However, determining sampling locations, differences in relative abundance estimates, and estimating population size are all relatively common uses of GIS in reservoirs. Site selection by using GIS has not been widely used for reservoir sampling design. Wilkerson and Fisher (1997) and Long (2000) both used GIS technology to determine sample sites for telemetered fish and electrofishing sampling, respectively, in Oklahoma reservoirs. In addition, Mietz (2003) used GIS to analyze historical electrofishing sampling in the Glen Canyon Tailwaters (i.e., Glen Canyon, Colorado River, Arizona) to determine if sampling was in proportion to available habitat. Use of GIS technology to map relative abundance (i.e., CPUE) of fish in reservoirs is nearly nonexistent. These CPUE data can be overlaid with other attributes (e.g., water quality, depth, user-group distribution) to better understand the interaction of fish with their environment and users.

An emerging and promising application is the use of GIS to calculate population estimates. Different habitats can be incorporated into a GIS, and population estimates can be calculated within these habitats. This information then could be used to extrapolate the estimate across the total area of all the habitats in the reservoir (or embayment) to create a habitat-based population estimate. This technique has been used in streams (Toepfer et al. 2000) and could be applied to reservoirs. In reservoirs, the use of sonar and GPS technology, coupled with GIS, has been used to assess reservoir habitat (Pratson and Edwards 1996; Jansson and Erlingsson 2000), and potentially could be used as a tool to estimate relative and absolute abundance of fishes in these systems.

In conclusion, there is a great potential for using GIS to address fisheries management issues in reservoirs. Developing this potential will require increased education and experience among fisheries biologists and a willingness to cooperate with other resource professionals in developing appropriate databases.

4.7 REFERENCES

Aggus, L. R., and G. V. Elliot. 1975. Effects of food and cover on year-class strength of largemouth bass. Pages 317–322 *in* R. H. Stroud and H. Clepper, editors. Black bass biology and management. Sport Fishing Institute, Washington, D.C.

Atkinson, P. M. 1996. Optimal sampling strategies for raster-based geographical information systems. Global Ecology and Biogeography Letters 5:271–280.

Burroughs, P. A. 1987. Principles of geographic information systems for land resources assessment. Clarendon Press, Oxford, UK.

Buynak, G. L., L. E. Kornman, A. Surmont, and B. Mitchell. 1989. Longitudinal differences in electrofishing catch rates and angler catches of black bass in Cave Run Lake, Kentucky. North American Journal of Fisheries Management 9:226–230.

Chen, X., A. J. Brinicombe, B. M. Whiting, and C. Wheeler. 1998. Enhancement of three-dimensional reservoir quality modeling by geographic information systems. Geographical and Environmental Modeling 2:125–139.

Chick, J. H., and C. C. McIvor. 1994. Patterns in the abundance and composition of fishes among beds of different macrophytes: viewing a littoral zone as a landscape. Canadian Journal of Fisheries and Aquatic Sciences 51:2873–2882.

Chow, V. T., D. R. Maidment, and L. W. Mays. 1988. Applied hydrology. McGraw-Hill, New York.

Clark, I., and W. V. Harper. 2000. Practical geostatistics 2000. Eccosse North America LLC, Columbus, Ohio.

Conover, G. A., and J. M. Grady. 2000. Mississippi River basin paddlefish research coded-wire tagging project, 1998 annual report. Mississippi Interstate Cooperative Resource Association, Bettendorf, Iowa.

Cross, D. 1991. Reservoir, GIS, and bass habitat. U.S. Forest Service General Technical Report RM 207:121–125.

Crowder, L. B., and W. E. Cooper. 1979. Structural complexity and fish–prey interactions in ponds: a point of view. Pages 2–10 *in* D. L. Johnson and R. A. Stein, editors. Response of fish to habitat structure in standing water. American Fisheries Society, North Central Division, Special Publication 6, Bethesda, Maryland.

Cunningham, K. 2000. Use of geographical information systems to predict the success of revegetation efforts on Oklahoma lakes. Oklahoma Department of Wildlife Conservation, Federal Aid in Sport Fish Restoration, Project F-50-R, Job 12, Oklahoma City.

Dauble, D. 2000. Assessment of the impacts of development and operation of the Columbia River hydroelectric system on main-stem riverine processes and salmon habitat. Report to Bonneville Power Administration, Contract 1998AC08104, Project 199800402, Portland, Oregon.

DeSilva, S. S., U. S. Amarasinghe, C. Nissanka, W. D. D. Wijesooriya, and M. J. J. Fernando. 2001. Use of geographical information systems as a tool for predicting fish yield in tropical reservoirs: case study on Sri Lankan reservoirs. Fisheries Management and Ecology 8:47–60.

Durocher, P. P., W. C. Proine, and J. E. Kraai. 1984. Relationship between abundance of largemouth bass and submerged vegetation in Texas reservoirs. North American Journal of Fisheries Management 4:84–88.

ESRI (Environmental Systems Research Institute, Inc.). 1990. Understanding GIS: the ArcInfo method. ESRI, New York.

Fisher, W. L., J. M. Long, and R. G. Hyler. 2000. Evaluation of a differential harvest regulation on black bass populations in Skiatook Lake, Oklahoma. Oklahoma Department of Wildlife Conservation, Federal Aid in Sport Fish Restoration, Grant F-41-R, Final Report, Oklahoma City.

Fisher, W. L., and C. S. Toepfer. 1998. Recent trends in geographic information systems education and fisheries research application at U.S. universities Fisheries 23(5):10–13.

Fox, D. S., and R. M. Starr. 1996. Comparison of commercial fishery and research catch data. Canadian Journal of Fisheries and Aquatic Sciences 53:2681–2694.

Gasaway, C. R. 1970. Changes in the fish population in Lake Francis Case in South Dakota in the first 16 years of impoundment. U.S. Bureau of Sport Fisheries and Wildlife, Technical Paper 56, Washington, D.C.

Gipson, R. D., and W. A. Hubert. 1993. Spawning-site selection by Kokanee along the shoreline of Flaming Gorge Reservoir, Wyoming–Utah. North American Journal of Fisheries Management 13:475–482.

Goovaerts, P. 1997. Geostatistics for natural resources evaluation. Oxford University Press, New York.

Guan, W., R. H. Chamberlain, B. M. Sabol, and P. H. Doering. 1999. Mapping submerged aquatic vegetation with GIS in the Caloosahatchee estuary: evaluation of different interpolation methods. Marine Geodesy 22:69–91.

Gubala, C. P., C. Branch, N. Roundy, and D. Landers. 1994. Automated global positioning system charting environmental attributes: a limnologic case study. The Science of the Total Environment 148:83–92.

Guy, C. S., and D. W. Willis. 1995. Population characteristics of black crappies: a case for ecosystem-specific management. North American Journal of Fisheries Management 15:754–765.

Irwin, E. R., J. R. Jackson, and R. L. Noble. 2002. A reservoir landscape for age-0 largemouth bass. Pages 61–72 *in* D. P. Philipp and M. S. Ridgway, editors. Black bass: ecology, conservation, and management. American Fisheries Society, Symposium 31, Bethesda, Maryland.

Irwin, E. R., and R. L. Noble. 1996. Effects of reservoir drawdown on littoral habitat: assessment with on-site measures and geographic information systems. Pages 324–331 *in* L. E. Miranda and D. R. DeVries, editors. Multidimensional approaches to reservoir fisheries management. American Fisheries Society, Symposium 16, Bethesda, Maryland.

Irwin, E. R., R. L. Noble, and J. R. Jackson. 1997. Distribution of age-0 largemouth bass in relation to shoreline landscape features. North American Journal of Fisheries Management 17:882–893.

Isaak, D. J., and W. A. Hubert. 1997. Integrating new technologies into fisheries science: the application of geographic information systems. Fisheries 22(1):6–10.

Jansson, M. B., and U. Erlingsson. 2000. Measurement and quantification of a sedimentation budget for a reservoir with regular flushing. Regulated Rivers: Research and Management 16:279–306.

Johnson, J. N., and A. K. Andrews. 1974. Growth of white crappie and channel catfish in relation to variations in means annual water level of Lake Carl Blackwell, Oklahoma. Proceedings of the Annual Conference Southeastern Association of Game and Fish Commissioners 27(1973):767–776.

Johnson, L. B. 1990. Analyzing spatial and temporal phenomena using geographic information systems: a review of ecological applications. Landscape Ecology 4:31–43.

Killgore, K. J., R. P. Morgan, and N. B. Rybicki. 1989. Distribution and abundance of fishes associated with submersed aquatic plants in the Potomac River. North American Journal of Fisheries Management 9:101–111.

Kimmel, B. L., and A. W. Groeger. 1986. Limnological and ecological changes associated with reservoir aging. Pages 103 to 109 *in* G. E. Hall and M. J. Van Den Avyle, editors. Reservoir fisheries management: strategies for the 80's. American Fisheries Society, Southern Division, Reservoir Committee, Bethesda, Maryland.

Kohler, C. C., R. J. Sheehan, and J. J. Sweatman. 1993. Largemouth bass hatching success and first-winter survival in two Illinois reservoirs. North American Journal of Fisheries Management 13:125–133.

Lam, N. S. 1983. Spatial interpolation methods: a review. American Cartographer 10:129–149.

Lehmann, A., and J. B. Lachavanne. 1997. Geographic information systems and remote sensing in aquatic botany. Aquatic Botany 58:195–207.

Lemmin, U., and T. Rolland. 1997. Acoustic velocity profiler for laboratory and field studies. Journal of Hydraulic Engineering 123:1089–1098.

Long, J. M. 2000. Population dynamics and interactions of three black bass species in an Oklahoma reservoir as influenced by environmental variability and a differential harvest regulation. Doctoral dissertation. Oklahoma State University, Stillwater.

Malthus, T. J., and D. G. George. 1997. Airborne remote sensing of macrophytes in Cefni Reservoir, Anglesey, UK. Aquatic Botany 58:317–332.

Mason, D. C., M. A. O'Conaill, and S. B. M. Bell. 1994. Handling four-dimensional geo-referenced data in environmental GIS. International Journal of Geographic Information Systems 2:191–215.

Meaden, G. J. 2001. GIS in fisheries science: foundations for the new millennium. Pages 3 to 29 *in* T. Nishida, P. J. Kailola, and C. E. Hollingworth, editors. Proceedings of the first international symposium on GIS in fishery science. Fishery GIS Research Group, Saitama, Japan.

Mietz, S. W. 2003. Evaluating historical electrofishing distribution in the Colorado River, Arizona, based on shoreline substrate. Master's thesis. Northern Arizona University, Flagstaff.

Muirhead, K., and A. P. Cracknell. 1986. Airborne LIDAR bathymetry. International Journal of Remote Sensing 7:597–614.

Narumalani, S., J. R. Jensen, J. D. Althausen, S. G. Burkhalter, and H. E. Mackey, Jr. 1997. Aquatic macrophyte modeling using GIS and logistic multiple regression. Photogrammetric Engineering and Remote Sensing 63:41–49.

Nellis, M. D., J. A. Harrington, and J. Wu. 1998. Remote sensing of temporal and spatial variations in pool size, suspended sediment, turbidity, and Secchi depth in Tuttle Creek Reservoir, Kansas: 1993. Geomorphology 21:281–293.

Noble, R. L., J. R. Jackson. E. R. Irwin, J. M. Phillips, and T. N. Churchill. 1994. Reservoirs as landscapes: implications for fish stocking programs. Transactions of the North American Wildlife and Natural Resources Conference 59:281–288.

Oliver, M. A., and R. Webster. 1990. Kriging: a method of interpolation for geographical information systems. International Journal of Geographic Information Systems 4:313–332.

Parsely, M. J., and L. G. Beckman. 1994. White sturgeon spawning and rearing habitat in the Lower Columbia River. North American Journal of Fisheries Management 14:812–827.

Paukert, C. P., and W. L. Fisher. 2000. Abiotic factors affecting summer distribution and movement of male paddlefish *Polyodon spathula*, in a prairie reservoir. Southwestern Naturalist 45:133–140.

Paukert, C. P., and D. W. Willis. 2002. Seasonal and diel habitat selection by bluegills in a shallow natural lake. Transactions of the American Fisheries Society 131:1131–1139.

Phillips, J. M., J. R. Jackson, and R. L. Noble. 1997. Spatial heterogeneity on abundance of age-0 largemouth bass among reservoir embayments. North American Journal of Fisheries Management 17:894–901.

Ploskey, G. R. 1986. Effects of water level changes in reservoir ecosystems, with implications for fisheries management. Pages 86–97 *in* G. E. Hall and M. J. Van Den Avyle, editors. Reservoir fisheries management: strategies for the 80's. American Fisheries Society, Southern Division, Reservoir Committee, Bethesda, Maryland.

Pratson, L. F., and M. H. Edwards. 1996. Introduction to advances in seafloor mapping using sidescan sonar and multibeam bathymetry data. Marine Geophysical Research 18:601–605.

Remillard, M. M., and R. A. Welch. 1993. GIS technologies for aquatic macrophyte studies: modeling applications. Landscape Ecology 8:163–175.

Ritchie, J. C., F. R. Schiebe, C. M. Cooper, and J. A. Harrington, Jr. 1994. Chlorophyll measurements in the presence of suspended sediment using broad band spectral sensors aboard satellites. Journal of Freshwater Ecology 9:197–206.

Rogers, K. B., and E. P. Bergersen. 1996. Application of geographic information systems in fisheries: habitat use by northern pike and largemouth bass. Pages 315–323 *in* L. E. Miranda and D. R. DeVries, editors. Multidimensional approaches to reservoir fisheries management. American Fisheries Society, Symposium 16, Bethesda, Maryland.

Ruckert, C., J. Floeter, and A. Temming. 2002. An estimate of horse mackerel biomass in the North Sea, 1991–1997. ICES Journal of Marine Science 59:120–130.

Rukavina, N. A. 1998. RoxAnn survey of Lake Ontario nearshore sediments at Metro Toronto Region Conservation Area's (MTRCA) East Point development site, Scarborough, Ontario. Environment Canada, National Water Research Institute, Burlington, Saskatchewan.

Sammons, S. M., and P. W. Bettoli. 2000. Population dynamics of a reservoir sport fish community in response to hydrology. North American Journal of Fisheries Management 20:791–800.

Savino, J. F., and R. A. Stein. 1982. Predator–prey interaction between largemouth bass and bluegills as influenced by simulated, submersed vegetation. Transactions of the American Fisheries Society 111:255–266.

Savino, J. F., and R. A. Stein. 1989. Behavior of fish predators and their prey: habitat choice between open water and dense vegetation. Environmental Biology of Fishes 24:287–293.

Siler, J. R., W. J. Foris, and M. C. McInerny. 1986. Spatial heterogeneity in fish parameters within a reservoir. Pages 122–136 *in* G. E. Hall and M. J. Van Den Avyle, editors. Reservoir fisheries management: strategies for the 80's. American Fisheries Society, Southern Division, Reservoir Committee, Bethesda, Maryland.

Stahl, J. W. 1995. Fish management surveys and recommendations for Lake Chambers. Oklahoma Department of Wildlife Conservation, Federal Aid in Sport Fish Restoration. Grant F-44-D, Oklahoma City.

Summerfelt, R. C. 1993. Lake and reservoir habitat management. Pages 231–261 *in* C. C. Kohler and W. A. Hubert, editors. Inland fisheries management in North America. American Fisheries Society, Bethesda, Maryland.

Thornton, K. W., B. L. Kimmel, and F. E. Payne. 1990. Reservoir limnology: ecological perspectives. Wiley, New York.

Toepfer, C. S., W. L. Fisher, and W. D. Warde. 2000. A multistage approach to estimate fish abundance in streams using geographic information systems. North American Journal of Fisheries Management 20:634–645.

Wilde, G. R., and W. L. Fisher. 1996. Reservoir fisheries sampling and experimental design. Pages 397–409 *in* L. E. Miranda and D. R. DeVries, editors. Multidimensional approaches to reservoir fisheries management. American Fisheries Society, Symposium 16, Bethesda, Maryland.

Wilkerson, M. L., and W. L. Fisher. 1997. Striped bass distribution, movements, and site fidelity in Robert S. Kerr Reservoir, Oklahoma. North American Journal of Fisheries Management 17:677–686.

Willis, D. W. 1986. A review of water level management in Kansas reservoirs. Pages 110–114 *in* G. E. Hall and M. J. Van Den Avyle, editors. Reservoir fisheries management: strategies for the 80's. American Fisheries Society, Southern Division, Reservoir Committee, Bethesda, Maryland.

Wilson, S. K. 2002. Relation of habitat to fish community characteristics in small South Dakota impoundments. Master's thesis. South Dakota State University, Brookings.

Xie, S. G., Y. B. Cui, T. L. Zhang, R. L. Fang, and Z. J. Li. 2000. The spatial pattern of the small fish community in the Biandantang Lake—a small shallow lake along the middle reach of the Yangtze River, China. Environmental Biology of Fishes 57:179–190.

Zheng, X., G. J. Pierce, and D. G. Reid. 2001. Spatial patterns of whiting abundance in Scottish waters and relationships with environmental variable. Fisheries Research 50:259–270.

Zigler, S. J., M. R. Dewey, and B. C. Knights. 1999. Diel movement and habitat use by paddlefish in Navigation Pool 8 of the Upper Mississippi River. North American Journal of Fisheries Management 19:180–187.

Chapter 5

Geographic Information Systems Applications in Lake Fisheries

Carolyn N. Bakelaar, Peter Brunette, Paul M. Cooley, Susan E. Doka, E. Scott Millard, Charles K. Minns, and Heather A. Morrison

5.1 INTRODUCTION

Geographic information systems (GIS) have emerged as a pivotal technology in the scientific study and management of renewable natural resources. Early evolution and applications of the technology centered on terrestrial resources (forestry, land-use planning, and wildlife). The next phase added rivers and streams, largely in the context of watershed analysis, modeling, and planning (see Chapter 3). The current phase of growth, finally, is encompassing lakes and oceans. This sequence of events has been shaped to some extent by the "visibility" of the ecosystems being mapped. In contrast to the terrestrial situation, our inability to "see" (directly observe or view) all parts of aquatic ecosystems, especially the deeper portions, has posed many challenges. Advances in remote-sensing technology (light, infrared, laser, and radar) are increasing our ability to penetrate the murky depths. The sophistication of acoustic technology, coupled with global positioning systems (GPS) for transect and swath mapping of biota and habitat features, and with radiotelemetry for tracking the movements of biota, can be combined with GIS to enhance our understanding and management of lake fish resources.

Scientific knowledge and understanding of renewable natural resource systems expanded in the 20th century with the realization that the spatial properties of ecosystems and their biota were important. The rapid growth of computer hardware and software in the last 25 years has spawned, in turn, the development of highly capable and user-friendly GIS. This new technology tended, at first, to be used to perform tasks in the "old" ways, for example, marking sampling locations and transects on maps. Now, as familiarity with GIS is increasing, new ways of storing, analyzing, and visualizing data are emerging that take fuller advantage of the technology. Geographic information systems now allow spatial aspects to be explicitly integrated with other natural resource concerns (e.g., the behavior and physiology of organisms, modeling ecosystem processes). These technical developments are rapidly enabling information needs to be met for aquatic ecosystem science and management.

In this chapter, we cover the following subjects:

• the main themes of lake resource management with special reference to the central importance of fish and fisheries;

• the scientific and technical issues encountered when applying GIS to lakes and associated wetlands;

• a survey of current literature emphasizing the various capabilities of GIS and the variety of applications of GIS to lakes and their biota;

• a series of case studies illustrating the potential range of GIS applications in lakes; and

• an assessment of emerging trends and potentials for enhancing scientific understanding and the management of fish and fisheries in lakes.

5.2 MANAGING LAKE RESOURCES

The issues and challenges in managing lake resources encompass four interconnected topics:

(1) *Determining factors.* What factors determine the composition and productivity potential of fish resources in lakes?

(2) *Process ecology.* How do habitat features shape the spatial and temporal distributions and dynamics of lake fish stocks?

(3) *Fisheries management.* How should fish exploitation be managed to ensure the sustainability and socioeconomic benefits of harvests in lakes?

(4) *Habitat management.* What conservation and protection measures are required in lakes to ensure the continued diversity and productivity of fish resources?

Each of these questions has many facets and GIS is playing an increasing role in providing the answers.

5.2.1 Determining Factors

The biological productivity of lakes is determined by the inputs of nutrients and solar energy, by the array of species present, and by morphometric and other habitat characteristics (Steedman and Regier

1987). The allocations of biomass and productivity among phytoplankton, macrophyte communities, zooplankton, benthos, fish, and other biota are determined by allometric consumption, growth, reproductive, and mortality relationships (Pauly 1980), and by biotic interactions that are predominantly trophic within constraints set by environmental tolerances of biota.

The horizontal and vertical distribution of biomass and productivity of any life stage, species, or assemblage rarely is uniform or random. Geographic information systems can be useful in organizing multispatial data sets to enable modeling and visualization of spatial patterns; GIS also can aid in extrapolation from point and line transect surveys to whole lake estimates of total biomass or production. Spatial relationships can be compared through multilayer overlays of lake characteristics such as bathymetry, fetch, or slope. Brosse et al. (1999a, 1999b) provide an example of using GIS to identify determining factors for fish in Lake Pareloup, France. The abundance of roach *Rutilus rutilus* was related to geospatial features such as depth, slope, and distance from shore, as well as limnological parameters such as temperature and oxygen.

5.2.2 Process Ecology

The activities (e.g., feeding, reproducing, and moving) of individual fish are strongly influenced by characteristics and configurations of habitat features. The aggregate outcomes of individual activities are expressed spatially and temporally in population process rates, such as birth, death, immigration and emigration (Pulliam and Danielson 1991), or growth. Habitat features are multidimensional and may include aspects of depth, substrate, cover (both vegetative and other), exposure, light, temperature, circulation, food availability, and predator abundance. Some features, such as temperature and vegetation, vary through time, and the spatiotemporal dynamics can be represented in GIS. The impact of the habitat features on population processes can depend on the size and shape of habitat patches, on the edges or ecotones connecting patches, on the spatiotemporal proximity of different patches, and on the hierarchy of patches at different scales (Horne and Schneider 1994; Minns et al. 1996a). The ability to formulate, analyze, and interpret these process–habitat linkages is greatly enhanced by the use of GIS to produce maps of features and to quantify new metrics. Different species and life stages may be responding to different combinations and scales of habitat features, and GIS capabilities make it possible to accommodate a wide variety of representations once the basic layers have been mapped at adequate resolutions. Geographic information systems allow us to move away from static, paper maps of habitat and productive capacity based on fixed categorizations toward a more spatiotemporally dynamic view of the interactions of habitat features and biotic processes. For example, Mason et al. (1995) showed how spatial data on habitat conditions and prey abundance could be used in conjunction with standard bioenergetic models to predict the temporal dynamics of growth potential for salmonine fishes in Lakes Michigan and Ontario.

5.2.3 Fisheries Management

Many commercial fisheries are being exploited at or beyond sustainable levels (FAO 1999). Geographic information systems can aid in detecting shifts in spatial patterns of effort and yield that can occur as stocks expand and contract (MacCall 1990). Reviewing past and emerging approaches to fisheries management, Caddy (1999) showed how spatial information plays a major role in the management measures needed to sustain or recover fish stocks. One approach actively being investigated throughout the world is the use of protected areas or reserves as a fishery management tool (Guénette et al. 1998). A GIS should be an essential element in the design of reserves and the evaluation of their effectiveness.

Most of the literature addressing fisheries management concerns marine applications, although the issues are much the same in lakes. As in oceans, overexploitation may easily occur in lakes as a result of commercial, recreational, or subsistence fishing. Instances where GIS can be applied include identifying conservation or restoration reserves; spatial management of competing uses of waterways (e.g., fisheries versus transportation); and mapping distribution–stock size relationships. Often different fisheries, for example, sport, commercial, and subsistence fisheries, or gill netters and trawlers, target the same or different species occurring in the same regions of a lake. The interactions among these fisheries have spatial aspects that can be examined by using GIS to help resolve conflicts, to assess bycatch, and to study the distributions of biomass, effort, and harvest. Selgeby (1982) documented the decline of lake herring *Coregonus artedi* in portions of Lake Superior due to overexploitation. As the stock declined, the fishery receded into six small areas. With the benefit of hindsight, it is now clear that a GIS-based analysis of the spatial distributions of fishing effort, catch per unit effort, and total catch might have allowed a conservation management strategy to emerge.

5.2.4 Habitat Management

Freshwater fish are declining globally, in part because their "invisibility" underwater results in an attitude of "out of sight, out of mind" (Tudge 1990). These declines are a result of alteration and destruction of habitat, exotic species introductions, entrainment and impingement mortalities, nutrient and organic enrichment, chemical contaminants, and overexploitation. Consequently, there is increasing interest in conservation of freshwater fish populations.

Two types of conservation activities may occur in lakes, one reactive and one proactive. Geographic information systems analyses and modeling can be used for both types. Reactive conservation often is applied belatedly, after a species has become endangered or threatened. Many fish species occurring in lakes are considered to be at risk in North America (Williams and Miller 1990) and elsewhere (Bruton 1995). In these reactive cases, GIS could be used to map the remaining distribution and to identify important habitats that must be protected or restored if a species is to recover. Sanctuaries may be designated to minimize the pressures on remaining stocks.

A proactive form of conservation is emerging whereby conservation plans attempt to divert development pressure away from areas and habitat features known to be critical for the main fish resources in the lake (Morrison et al. 2001). Proactive conservation is anticipatory, precautionary, and preventative. Geographic information systems map classification, analysis, and modeling can play a central role in the integration and assessment of the many issues that must be considered in conservation planning. The application of a gap analysis program (GAP) in terrestrial ecosystems (Scott et al. 1993; Jennings 2000) provides an indication of the potential for using GIS in proactive conservation efforts. A GAP analysis relies on overlays to combine maps of land cover, predicted species' distributions, and land management status to highlight biodiversity hot spots and priority areas for conservation and restoration efforts. The goal of proactive conservation is to avoid the need for reactive actions. Baban (1999) illustrated the use of GIS with remote sensing to establish a quantitative framework for lake and watershed management in the Norfolk Broads region of the United Kingdom. Maps of surface water quality derived from remote sensing were combined with maps of shorelines, land use, and bathymetry as a basis for delineating management units.

5.3 TECHNIQUES AND CHALLENGES OF APPLYING GIS IN LAKE STUDIES

The main challenges in the use of GIS for management of lakes and their fisheries involve the collection, analysis, and representation of spatial data. These challenges are organized under six topics: data acquisition, representation, analysis and modeling, statistical issues, spatial modeling, and visualization.

5.3.1 Data Acquisition

In North America, the infrastructure for the assembly, organization, and distribution of geospatial information is still emerging, and it is a massive undertaking. The infrastructure for accessing and distributing free data (especially those acquired with public funding) is more advanced in the United States than in Canada. Lake data are limited due in part to the "invisibility" or submerged nature of many lake features. Federal, state, and provincial agencies are developing geospatial information systems and including physical, chemical, and biological data for lakes where it is available. To assemble the geospatial data for a particular study may require data from several organizations: landscape elements from one, climate from another, and bathymetry from a third.

5.3.2 Representation

5.3.2.1 ***Projections and Coordinates.*** Selection of coordinates and a projection for lake applications of GIS usually are straightforward because an equal area format is preferred to avoid visual distortions. Large lakes (e.g., the Great Lakes) may cross reference zones (e.g., Lake Ontario spans Universal Transverse Mercator zones 17 and 18). Therefore, representation of the area may be distorted as proximity to the contact boundary decreases. It is preferable to use a projection that suits the specific requirements of the study (e.g., area or distance analysis). However, the final output may be displayed with a more visually pleasing or familiar projection. The horizontal datum and coordinates used in analysis may be different than those used to collect the data. This discrepancy is not a difficult issue to overcome because most GIS software includes transformation models that convert coordinate system and datum ellipsoids. It should be noted that geodetic transformations may differ between Canada and the United States.

5.3.2.2 ***Temporal Variability.*** Temporal variability is a recurring problem in the development of GIS databases for lakes. Water levels vary seasonally and annually. Consequently, the shoreline and bathymetry of the lake change. Ideally, GIS would provide a seamless digital terrain model of both the lake bottom and the surrounding landscape, with the means to choose any lake water level and obtain updated depth contour maps. The extremes of annual and seasonal water-level changes can be modeled three dimensionally to represent seasonally exposed and inundated lake bottom. With vegetation, there is further variability related to (a) the seasonal succession and production dynamics of density, biomass, and composition; and (b) the longer-term patch dynamics of communities where a particular area may undergo a cycle of absence, invasion, abundance, and senescence (Scheffer et al. 1992).

Presentation of three-dimensional (3-D; x, y, z) and four-dimensional (x, y, z, time) information, such as thermal structure in a lake, is a challenge because two-dimensional (2-D) representation is the norm. The visualization aspects of this are discussed below. However, it will often be necessary to find ways to collapse higher dimensional data into two dimensions to allow them to be intersected with other two-dimensional features for modeling and analysis purposes. For example, three-dimensional bathymetry is routinely displayed as two-dimensional contour lines.

5.3.3 Analysis and Modeling

5.3.3.1 ***Data Analysis and Modeling.*** A variety of modeling tools can be used within a GIS framework. Geological sciences have created a set of spatial tools known as geostatistics, with kriging a mainstay for generating surface maps from point surveys in two and three dimensions. Some aquatic studies have made use of kriging in lakes. For example, Guillard et al. (1990) used kriging of acoustic fish count surveys in Lake Annec, France, to produce a fish density map for the whole lake. Essington and Kitchell (1999) examined the factors associated with movement of largemouth bass *Micropterus salmoides* in a Michigan lake by using random and null models with explicit consideration of spatial resolution. These papers reveal the potential for fuller analysis of spatial issues in ecological studies of fish. Studies that used spatial acoustic surveys of fish in lakes illustrate the growing use of artificial neural network models in aquatic ecological modeling (Brosse et al. 1999a, 1999b), particularly in France (Lek and Guégan 1999). Classification and regression trees (CART) overcome some of the limitations of traditional linear statistical modeling and allow spatial variables to be readily included in the analysis. Rejwan et al. (1999) provide an example of using CART to characterize the distribution of smallmouth bass *Micropterus dolomieu* nests in Lake Opeongo. The model combines temperature and spatial variables to predict nest density. Lamon and Stow (1999) used CART to examine how both fish-specific parameters and geographic location influenced organic contaminant levels in salmonids from various regions of Lake Michigan. There has been growing interest in the application of fuzzy set theory in multicriteria mapping, especially where the properties of patches may be characterized by distributions of values rather than a single or mean value. Minns and Moore (1995) used Bayesian methods to model the occurrence of fish species in watershed areas across Ontario as part of a study to predict regional changes in distribution and species richness given the potential impact of climate change on annual and seasonal temperatures. Hobbs (1997) provides an overview of Bayesian analysis and decision making for examining the impact of projected declines in water levels of coastal wetlands of the Great Lakes as a result of climate change.

5.3.3.2 ***Simulation Modeling.*** Given the long history of population modeling and simulation in fisheries management, the emergence of spatial models linking GIS representations of ecosystems to populations and communities was to be expected. While most of the effort has been directed to marine and stream fish stocks, some developments have taken place in lakes. Minns et al. (1996b) developed a model of the northern pike *Esox lucius* population in Hamilton Harbor on Lake Ontario as a basis for assessing the potential benefits of proposed habitat restoration measures. This work was an outgrowth of earlier studies mapping and measuring the supply of suitable habitat for pike (Minns et al. 1993a, 1993b). The model uses whole system measures of suitable habitat supply to affect density-dependent population processes such as growth and survival. A similar effort by Frisk et al. (1988) examined the potential impacts of lake level regulation on a whitefish stock in a Finnish lake.

Developments in the modeling of energy flows of individuals, populations, and ecosystems have added spatial considerations. Brandt et al. (1992) pioneered a spatial application of bioenergetic modeling in the Great Lakes and elsewhere by using spatial mapping of predator and prey abundance along with related environmental conditions to create growth potential maps, although not within a conventional GIS software environment. More recently, a spatial component, ECOSPACE, has emerged from the productivity-energy flow modeling approach of ECOPATH and ECOSIM (Walters et al. 1999). Spatially explicit individual-based models (IBM) (Tyler and Rose 1994) largely are superceding simpler approaches represented by spatial, habitat-based extensions of traditional stock–recruitment models (Beverton and Holt 1957; Ricker 1975).

Individual-based models were developed initially as a basis for examining the consequences of individual variability in process rates or performance (growth, foraging, mortality, reproduction; Tyler and Rose 1994). As the sophistication of IBM models and GIS has grown, spatially explicit IBMs have emerged as a major tool for the exploration of habitat–productivity links (Rose 2000). Spatially explicit IBMs bring together three major elements: individual variability, individual movement, and spatial heterogeneity. Such models can provide great insights into the spatial links between habitat and fish population dynamics, insights that are essential for the management of fisheries and the identification and conservation of critical and limiting habitats. Scheffer et al. (1995) have shown how IBMs might be used to represent statistical distributions of individual variability and movement where the simulation of all individuals in a stock may be too computationally complex. Further development of spatially explicit IBMs will require that GIS systems provide more dynamic modeling capability to represent individual movements and the aggregate accounting necessary for population simulation (Tischendorf 1997).

5.3.4 Statistical Issues

Concerns about the impact of autocorrelation in analysis and interpretation of spatial data and models have grown alongside longer-standing unresolved concerns about issues related to uncertainty in data and models. Legendre's (1993) overview provided some methods for coping with spatial autocorrelation in analysis and modeling, and a guide to the available software. In GIS applications, consideration of spatial autocorrelation still is uncommon, although Hinch et al. (1994) examined littoral fish–habitat links in small Ontario lakes. Because landscapes and ecosystems are not assembled randomly from arbitrary elements, spatial autocorrelation is inevitable and needs more attention when modeling with GIS. In general, explicit inclusion of spatial autocorrelation in statistical analyses designed to assess the strength of links between nonspatial variables results in a lower likelihood of significance at the usual benchmark probability levels. This poses particular difficulties for studies that depend on compilation of data from multiple sources for comparative analysis and modeling.

While there are many places where errors of precision and accuracy can arise in the assembly and manipulation of GIS databases, there are larger issues concerning errors that arise from the extrapolation and interpolation of limited data to generate maps. Those issues include how to express uncertainty in a map, how to carry uncertainty about input data into the generation of maps via models (propagation of error), and how to set intervals when classifying a map. In aquatic ecosystems, complete mapping of features is a rare luxury and often maps are generated from sparse networks of data points with ill-defined boundaries. Propagation of error can produce maps of dubious value, although straightforward methods have yet to emerge to deal with this problem (Haining and Arbia 1993). Thus, the maps obtained often are used in subsequent modeling, and the uncertainty of new maps is unmeasured and forgotten (Bailey 1988). The exception to this is kriging where standard deviation maps are routinely produced (Burrough 1987). When creating maps for display, GIS users often treat the selection of data classification breaks as trivial, although important statistical issues are involved in choosing intervals that are not misleading (Polfeldt 1999).

5.3.5 Spatial Modeling

Most GIS software provides a range of spatial modeling tools that enable ecological and ecosystem modeling and analysis. Use of these tools requires acceptance of certain assumptions and has implications for modeling and analysis. Burrough (1987) provided a formal GIS-oriented overview of spatial modeling and Berry (1987) described the basic "map algebra" capabilities of GIS. A simplified list of these tools includes (a) extrapolation and interpolation, (b) area overlays, (c) buffering, and (d) attributing points from maps. Each of these tools has potential application in lakes, though the limitations must be considered.

5.3.5.1 ***Extrapolation and Interpolation.*** In many lake GIS projects, maps are generated from incomplete point and line surveys. Bathymetry maps are derived from points collected along multiple line transects of echosounding traces. Contours are interpolated, within a shoreline boundary, at selected depth intervals. A substrate map will be derived from a set of point measures of composition. Maps can be derived by contouring, which requires that fixed contour intervals be selected, or by some form of tessellation where boundary areas are assigned around each observation (Bonham-Carter 1994). From a modeling viewpoint, tessellation often is preferred because the original data values continue to be used in map calculations, whereas modeling with contours requires that computations be done by using the contour interval range of values. Tessellation also enables some consideration of error propagation.

5.3.5.2 *Area Overlays.* These are the mainstays of GIS modeling, allowing information from several map layers to be brought together. For example, Minns et al. (1999a) overlaid maps of bathymetry, substrate, vegetation, and delimited wetlands as an important step to assigning habitat suitability classes in the littoral zone of Severn Sound, Canada. A downside of overlay is the excessive fragmentation that may occur, especially where numerous tiny polygon slivers are generated. In view of the inherent uncertainties of the polygon boundaries in the source maps, many of the slivers can contain misleading combinations of attributes or unusual index values. Sometimes overlays may produce an overly simplistic analysis and users must be prepared to explore more complex spatial or multivariate approaches.

5.3.5.3 *Buffering.* Ecotones and edges are important aspects of habitat use by fish in lakes (Willis and Magnuson 2000), and buffering tools allow proximity to macrophytes, stream mouths, and other discontinuities to be mapped. The challenge with buffering lies in having the field data to determine the size of the buffer radius or width.

5.3.5.4 *Attributing Points from Maps.* An intermediate step in many GIS-based modeling projects is assigning spatial attributes derived from base maps to point data sets. The point data sets may consist of limnological surveys of surface water quality, grid–point surveys of sediment characteristics, fish catch per effort from a multistation survey with trawls or gill nets, or a sequential set of positional observations derived from radio-tagged fish (e.g., Bégout-Anras et al. 1999). In each data set, the points are georeferenced by using GPS. When the points are incorporated into a GIS with existing maps showing the shoreline and bathymetry, attributes such as depth, slope, and fetch can be derived for each point.

5.3.6 Visualization

Visualization of lake data is different from the traditional mapping of terrestrial systems because organisms and physical habitat are dispersed in a three-dimensional space. The abundance of organisms and physical attributes can vary with multiple depths at each geographic location on a two-dimensional surface (Herbst and Bradley 1993; Mason et al. 1995). A detailed three-dimensional view may be unnecessary for shallow lakes and reservoirs, but in larger systems such as the North American Great Lakes, with maximum depths of 60 to 400 m, there are large volumes of aquatic habitat to evaluate. Only certain habitat characteristics vary in three-dimensional space, such as thermal structure, light intensity, oxygen content, nutrients, and the distribution of pelagic organisms (Schertzer and Sawchuk 1990; Kornijów and Moss 1998; Job and Bellwood 2000). Substrate type, benthic organisms, and aquatic vegetation densities are associated with the lake bottom and can be represented in two dimensions, similar to terrestrial mapping of soils and vegetation. Linking two- and three-dimensional characteristics is problematic. If three-dimensional features cannot be adequately represented in two-dimensional space by using aggregate statistics or metrics, alternate visualization methods need to be explored (Minns et al. 1999b).

The spatial resolution required for mapping the vertical component of a lake (water profile) usually is different from that of the horizontal component (surface area), depending on the habitat characteristics of interest. Temperature can vary rapidly over a few meters vertically at the thermocline. In contrast, a square kilometer grid may give adequate spatial resolution for representing the surface temperature of a lake except in nearshore areas and during thermal bar events when the horizontal temperature may change more rapidly over relatively small distances (Ullman et al. 1998).

Rarely is one species distributed ubiquitously or a habitat variable constant throughout the entire volume of a lake. Therefore, three-dimensional analysis and visualization in lakes can provide further insight into habitat associations and species distributions. Spatial patterns also vary temporally, adding a fourth dimension to the analysis (Stephenson 1990; Gaudreau and Boisclair 1998). Given the dynamic nature of species that inhabit lakes and the changes in physical habitat that occur over time and space, the analysis and visualization of fish habitat is truly a four-dimensional problem and alternate approaches are necessary to understand the underlying relationships that exist beyond the traditional two-dimensional area representation. The thermal habitat requirements of fish exemplify the four-dimensional problem as the second case study below indicates.

5.4 REVIEW OF APPLICATIONS OF GIS IN LAKES

Our view of GIS applications in lakes is presented in two parts. In this section, the literature is analyzed in relation to a simple classification scheme. In the next section, a series of case studies is presented to give some insight into the specific components of typical applications.

To obtain an overall picture of the scope and emphasis of GIS applications in lakes, we classified studies based on three aspects: system, target, and focus.

- The *system* refers to the lake context of the study and includes small lakes, large lakes (like the North American Great Lakes), regional populations of lakes, wetlands, and generic, where the study was not system specific.

- The primary *target* of the study is partitioned into fish, biota, habitat, macrophytes, or generic, where multiple taxa were considered.

- The *focus* is divided into two categories, mapping and modeling to indicate the main emphasis in the geospatial aspects of the study.

Those studies included in the survey are annotated in the reference list below. This simple classification was used to look for the main themes of GIS applications involving lakes.

A total of 60 studies were assessed (Table 5.1). Almost half the studies occurred in large lakes systems ($N = 29$), especially within the North American Great Lakes, followed by small lakes ($N = 14$) and lake regions ($N = 8$). In terms of target, habitat applications accounted for nearly half the sample, with fish and macrophytes being the next most prevalent. The focus was divided evenly between mapping ($N = 24$) and modeling applications ($N = 31$). Surprisingly, no studies were specifically directed to fisheries management in lakes.

Meaden and Kapetsky (1991) published an excellent compendium of information on the application of GIS and remote sensing in inland waters and aquaculture. Although the technical information about particular GIS and remote-sensing platforms has become dated, the report's broad overview still provides an excellent introduction to the concepts and potential applications. They include several case studies on aspects of lake ecosystems.

As the literature survey shows, habitat and macrophyte mapping and modeling account for a majority of the GIS applications in lakes. Once the horizontal and vertical physical, chemical, and biological characteristics of lakes gained attention, GIS applications inevitably followed (Wang and Xie 1994; Duel et al. 1995; MacLeod et al. 1995; de Vries 1996; Duel and Laane 1996; Minns and Bakelaar 1999; Minns et al. 1993a, 1993b, 1999b; Morrison et al. 2001; Dolan et al., abstract from the Proceedings of the 38th Conference of the International Association for Great Lakes Research, 1995; Koonce et al., unpublished paper from State of the Lakes Ecosystem Conference, 1998). Horizontal limnological and lower trophic level applications of GIS have emerged, especially in larger lakes where the greater distances result in pronounced environmental gradients (Johannsson 1995; Millard et al. 1999; Johannsson et al., abstract from the Proceedings of the 38th Conference of the International Association for Great Lakes Research, 1995). Vegetation draws much attention in lakes because it is seen to be important to fish and a major indicator of the health of the lake ecosystem. Thus, the spatial patterns and dynamics of macrophytes have generated many GIS studies (Jean and Bouchard 1991; Scheffer et al. 1992; Remillard and Welch 1993a, 1993b; Johnson 1994; Lehmann et al. 1994, 1997; Richardson and Hamouda 1995; Lehmann and Lachavanne 1997; Narumalani et al. 1997; Schmieder 1997; Williams and Lyons 1997; Gottgens et al. 1998; Burton and Prince, abstract from the Proceedings of the 38th Conference of the International Association for Great Lakes Research, 1995; Hanlon, abstract from Lake and Reservoir Management, 1995). Mapping and modeling studies of topography, bottom substrates, and chemical gradients have used GIS or geospatial approaches (Gubala et al. 1994; Brassard and Morris 1997; Cooley 1999; Haltuch and Berkman 1999; Schernewski et al. 2000; Gaugush, abstract from Lake and Reservoir Management, 1994; Hutchinson et al., abstract

Table 5.1 Frequency analysis of GIS-related studies involving fish and related aquatic topics in lakes. Studies were classified using a system–target–focus scheme as described in the text. In the references column, SL = small lake, LL = large lake, LR = lake region, WL = wetlands, and G = generic.

		System						
Target	Focus	Small lakes	Large lakes	Lake regions	Wet-lands	Generic	Total	References
Fish	Mapping	3	3	1			7	Guillard et al. 1990; Bégout-Anras et al. 1999; Brosse et al. 1999a (SL); Appenzeller 1995; Bernardo 1997; J. F. Koonce et al., Case Western Reserve University, paper from State of the Lakes Ecosystem Conference, 1998 (LL); B. Mower, Maine Department of Environmental Protection, abstract from Lake and Reservoir Management, 1995 (LR)
	Modeling	3	4	1		1	9	Brosse et al. 1999b; Essington and Kitchell 1999; Rejwan et al. 1999 (SL); Brandt et al. 1992; Minns et al. 1996b, 1999b; Gaff et al. 2000 (LL); Minns and Moore 1992 (LR); Maury and Gascuel 1999 (G)
Biota	Mapping		2				2	O. E. Johannsson et al., Canada Centre for Inland Waters, abstract from the Proceedings of the 38th Conference of the International Association for Great Lakes Research, 1995; L. Rudstam and C. Greene, abstract from the Proceedings of the 37th Conference of the International Association for Great Lakes Research, 1994 (LL)
	Modeling		3	2			5	Johannsson 1995; Minns et al. 1996c; R. Bailey et al., University of Western Ontario, abstract from the Proceedings of the 38th Conference of the International Association for Great Lakes

Table 5.1 Continued.

Target	Focus	System					Total	References
		Small lakes	Large lakes	Lake regions	Wet-lands	Generic		
Biota	Modeling							Research, 1995 (LL); Minns et al. 1990; Koutnik and Padilla 1994; Hrabik and Magnuson 1999 (LR)
Macro-phytes	Mapping	1	2	1	1	1	6	Remillard and Welch 1993b (SL); Jean and Bouchard 1991; Schmieder 1997 (LL); Johnson1994 (LR); C. Hanlon, South Florida Water Management District, abstract from Lake and Reservoir Management, 1995 (WL); Lehmann and Lachavanne 1997 (G)
	Modeling	3	2		1		6	Scheffer et al. 1992; Remillard and Welch 1993a; Narumalani et al. 1997 (SL); Lehmann et al. 1994, 1997 (LL); Gottgens et al. 1998 (WL)
Habitat	Mapping	2	8	2	1		13	Gubala et al. 1994, P. D. St. Onge and E. Bentzen, Trent University, abstract from Lake and Reservoir Management, 1995 (SL); Wang and Xie 1994; MacLeod et al. 1995; Haltuch and Berkman 1999; Minns and Bakelaar 1999; Minns et al. 1999a; K. Baron et al., National Oceanic and Atmospheric Administration, abstract from the Proceedings of the 37th Conference of the International Association for Great Lakes Research, 1994; D. M. Dolan et al., International Joint Commission, abstract from the Proceedings of the 38th Conference of the International Association for Great Lakes Research, 1995; J. Song and J. V. Depinto, State University of

Table 5.1 Continued.

Target	Focus	System: Small lakes	Large lakes	Lake regions	Wet-lands	Generic	Total	References
Habitat	Mapping							New York, Buffalo, abstract from the Proceedings of the 38th Conference of the International Association for Great Lakes Research, 1995 (LL); Jordan and Enlander 1990; Omernik et al. 1991; TNC 1994 (LR); T. M. Burton and H. H. Prince, Michigan State University, abstract from the Proceedings of the 38th Conference of the International Association for Great Lakes Research, 1995 (WL)
	Modeling	4	5	1	1		11	Cooley 1999; Schemewski et al. 2000; R. F. Gaugush, National Biological Survey, abstract from Lake and Reservoir Managment, 1994; N. J. Hutchinson et al., Ontario Ministry of the Environment, abstract from Lake and Reservoir Management, 1995 (SL); Minns et al. 1993a, 1993b; Duel et al. 1995; Duel and Laane 1996; Brassard and Morris 1997 (LL); Thierfelder 1998 (LR); Richardson and Hamouda 1995; Williams and Lyons 1997 (WL)
Generic	Mapping					1	1	Meaden and Kapetsky 1991(G)
	Modeling		1				1	de Vries 1996 (LL)
Total	Mapping	6	15	4	2	2	29	
	Modeling	10	15	4	2	1	32	

from Lake and Reservoir Management, 1995; Song and Depinto, abstract from the Proceedings of the 38th Conference of the International Association for Great Lakes Research, 1995).

Other applications have focused on fish resource management in a landscape context within regional populations of lakes. Within this context, relatively few offshore stations are sampled to represent the whole lake, while fish abundance and fishery effort and harvest indices are compiled on a whole lake basis. Various morphometry indices such as mean:maximum depth ratio and shoreline development are used to capture interlake differences. Examples include Jordan and Enlander (1990); Minns et al. (1990); Omernik et al. (1991); Minns and Moore (1992); Koutnik and Padilla (1994); Thierfelder (1998); Hrabik and Magnuson (1999); Mower (abstract from Lake and Reservoir Management, 1995); St. Onge and Bentzen (abstract from Lake and Reservoir Management, 1995).

Radiotelemetry tracking of fish and hydroacoustic enumeration of fish populations are inherently spatial. Hydroacoustic technology routinely is used to enumerate fish in lakes (Rudstam and Greene, abstract from the Proceedings of the 37th Conference of the International Association for Great Lakes Research, 1994). Typically, a network of transects is used to ensure wide coverage of the lake. Surprisingly, given the spatial nature of the data collected, GIS rarely has been used in the analysis of such data although the potential has been recognized (Rudstam and Magnuson 1985). For example, Brosse et al. (1999a, 1999b) analyzed a hydroacoustic survey of fish, mainly roach in Lake Pareloup, France, exploring links to several geospatial and limnological measures. Appenzeller (1995) used kriging to derive a fish biomass for Lake Constance from a hydroacoustic survey. Analysis of radiotelemetry studies of fish movement is greatly facilitated by the use of GIS (Bernardo 1997; Bégout-Anras et al. 1999; Essington and Kitchell 1999).

Although there are no studies directly involving fisheries management, there are several studies seeking to link habitat supply and dynamics to fish population dynamics and productivity (Brandt et al. 1992; Minns et al. 1996b, 1999b; Maury and Gascuel 1999; Walters et al. 1999; Gaff et al. 2000).

5.5 CASE STUDIES OF LAKE RESEARCH BY USING GIS

To illustrate the scope and potential for applying GIS to the assessment and management of lakes and their fisheries, we have assembled a series of case studies emphasizing different aspects that correspond with the four main areas outlined in Section 5.2.

5.5.1 Determining Factors 1: Gradient Patterns and Integral Productivity in the Great Lakes of North America

Fisheries management in lakes and elsewhere should be based on measures of ecosystem productivity as well as on records of effort and harvest. In the Great Lakes, this approach is complicated by the size of the lakes and the territorial boundaries and activities of nations, states, and provinces. Reliable data on productivity of entire fisheries and valued stocks are difficult and expensive to obtain. Consequently, predictors or surrogates of fish productivity offer promising alternatives in helping to set constraints on harvest and stocking practices.

In large, deep lakes, phytoplankton photosynthesis is the main source of primary production driving the food web. Empirical relationships between primary production and fish production in lakes have been documented previously (Downing et al. 1990). Other models, based on particle size spectral analysis, also provide methods for predicting fish production in lakes (Borgmann 1987; Leach et al. 1987). These alternative approaches depend on estimates of primary or secondary production, preferably on a whole-lake basis. Whole-lake assessment can provide data that establish empirical relationships between nutrients, primary production, zooplankton production, and fish production that can help fishery managers make sound, ecosystem-based decisions.

A GIS has been used in spatial phytoplankton production assessments on Lake Ontario and Lake Erie (Millard et al. 1996). Assessing phytoplankton productivity of whole lakes always has been a tradeoff in effort between spatial and seasonal sampling intensity. Millard et al. (1996) documented generally predictable patterns of seasonal productivity, and Millard et al. (1999) showed that spatial variability could be represented by a few key stations and temporal trends generally could be depicted with spring, summer, and fall lakewide cruises. Whole-lake carbon fixation was calculated for Lake Erie and Lake Ontario from spatially intensive surveys by extrapolating daily areal phytoplankton photosynthesis measured at each station to the area of a Theissen polygon unique to each station (Figure 5.1). Extrapolating depth profiles of primary production over lake areas without correction for morphometry can overestimate areal primary production by up to 20% in small lakes (Fee 1980). This correction is less important on a whole-lake basis in the Great Lakes but can be very important on a regional basis such as the shallow western basin of Lake Erie. Morphometric data for polygons were calculated by using a GIS to overlay the depth-contour and Theissen maps, and then cross-tabulate area by depth contour in each polygon. Total mass of carbon fixed in the lake for a day was obtained by summing the results for all polygons in the lake after correcting for depth (Figure 5.1).

By using this approach, Millard et al. (1999) estimated total whole-lake primary production for Lake Erie and Lake Ontario. Johannsson (1995), by using similar techniques, was able to produce a whole-lake estimate of *Mysis relicta*, a large zooplankter, in Lake Ontario. Both studies demonstrate how the application of basic GIS functions can take analysis beyond description to integration of information and prediction.

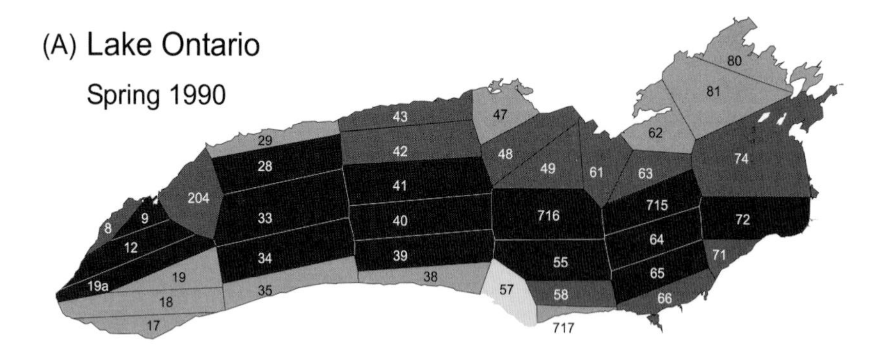

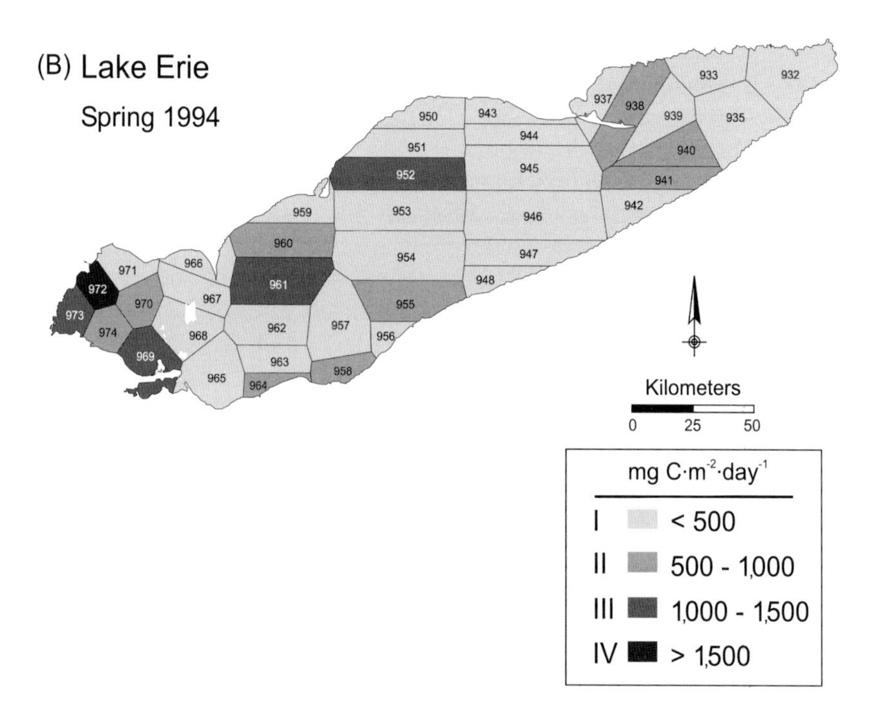

Figure 5.1 Maps of Lake Ontario (A) and Lake Erie (B) showing the stations (numbers) sampled on selected synoptic cruises in the springs of 1990 and 1994, respectively; the Thiessen polygons assigned to each station; and the polygons classified according to the level of daily areal phytoplankton photosynthesis computed with cloudless irradiance (mg $C \cdot m^{-2} d^{-1}$).

5.5.2 Determining Factors 2: Thermal Structure and Fish Habitat in Long Point Bay, Lake Erie

The purpose of this project was to investigate different methods of incorporating a time series of three-dimensional lake thermal structure into a GIS and to develop thermal habitat suitability indices (HSI) as a measure of thermal preference for fish (Minns et al. 1999b). This work is part of a larger project to build a habitat-based population model for yellow perch *Perca flavescens*. The flowchart for converting raw data to a three-dimensional thermal structure and then to habitat suitabilities shows the complexities of the process (Figure 5.2). Original data from different sources, such as satellite imagery, field vertical temperature profiles, and bathymetry were used to create three-dimensional temperature matrices (or cubes). These matrices were, in turn, converted into thermal habitat suitabilities, where an estimate of temperature suitability between 0 and 1 was evaluated for each cell within the matrix. The suitabilities were based on "optimal" temperature curves for different fish species at different life stages based on thermal preferences and growth rates (cf. Minns et al. 1999b for northern pike).

Thermal habitat suitability maps can be visualized in four ways: a two-dimensional image of an area from either a surface or lake bottom perspective; a layered two-dimensional approach showing

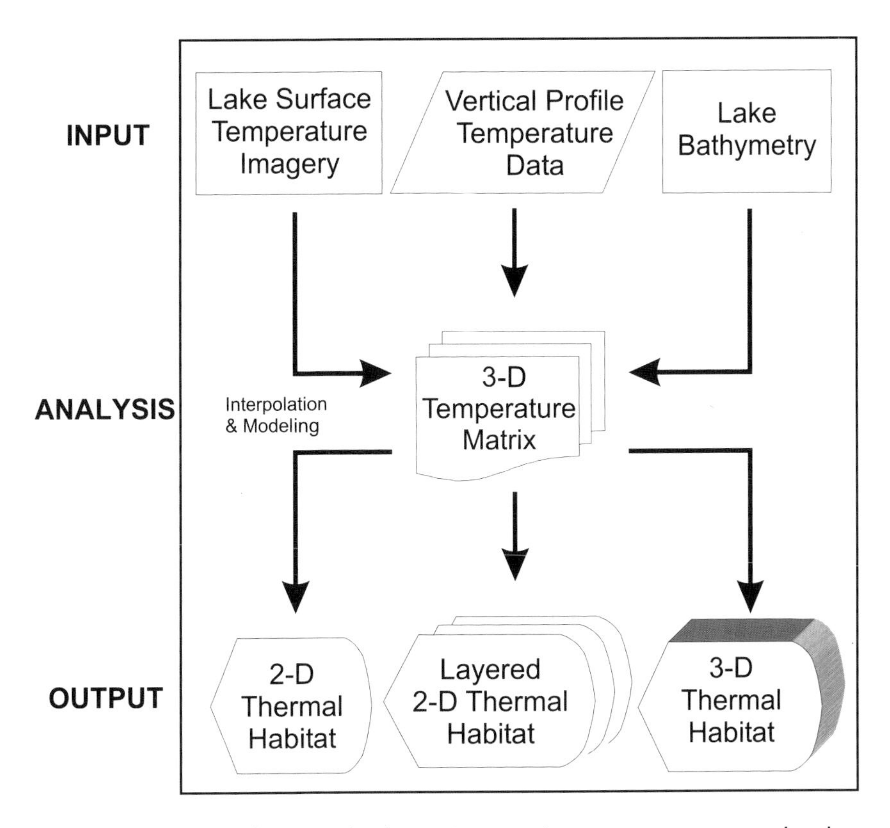

Figure 5.2 Flow chart illustrating the spatial information requirements necessary to develop a three-dimensional matrix of lake thermal structure, the subsequent conversion of temperatures to habitat suitabilities, and the possibilities for visualization of the finished thermal suitability maps.

thermal habitat at different depths or horizontal slices (Figure 5.3); a two-and-a-half-dimensional representation of the lake bottom topography (not shown); or a fully three-dimensional image or solid model that represents the entire lake volume (Figure 5.4). Each approach has pros and cons. The two-dimensional approach does not portray information about underlying depth distribution for each location, but it is easy to interpret each depth layer visually. A three-dimensional approach retains all the information, but it is difficult to manipulate and interpret the final image in one image. A GIS can be used at any point in the conceptual framework (Figure 5.2). To fully realize a time-series representation of the thermal regime in Long Point Bay, a movie may be created showing a series of one of the visualization types described above, tracking the seasonal progres-

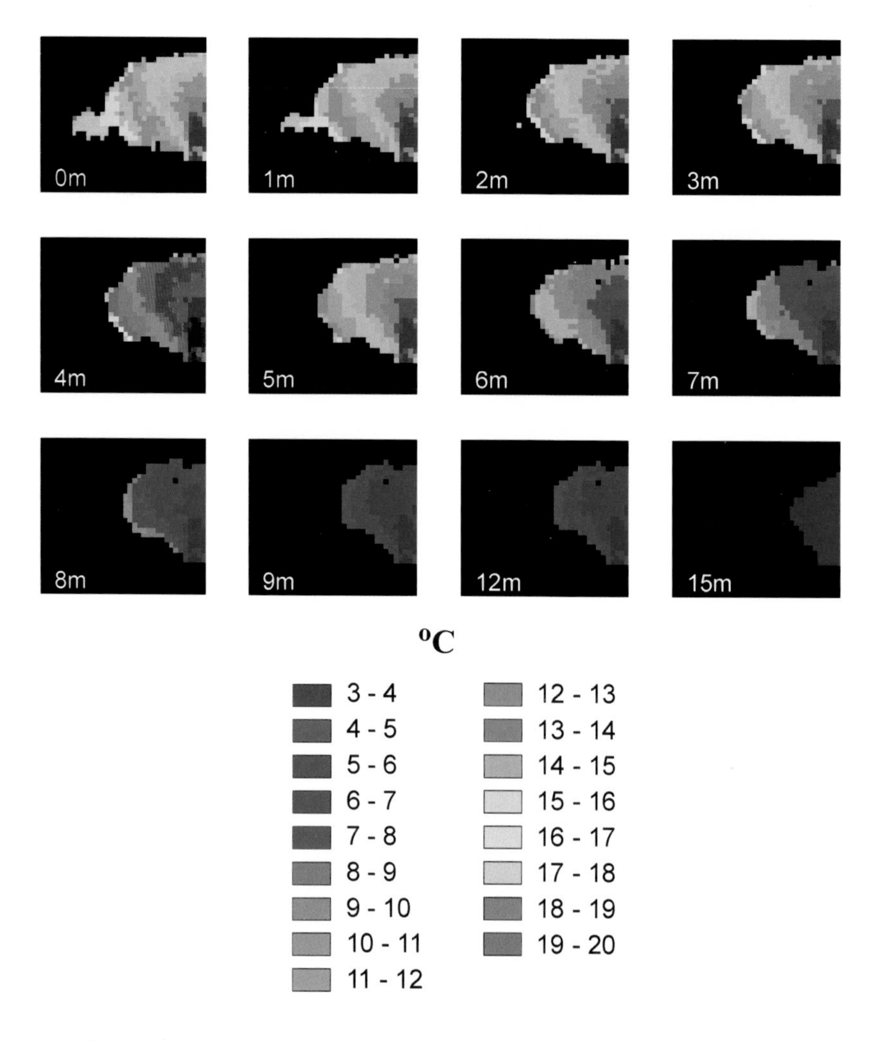

Figure 5.3 Layered two-dimensional approach to representing thermal habitat. Thermal structure of Long Point Bay, Lake Erie, on May 10, 1993, at different depth layers is represented.

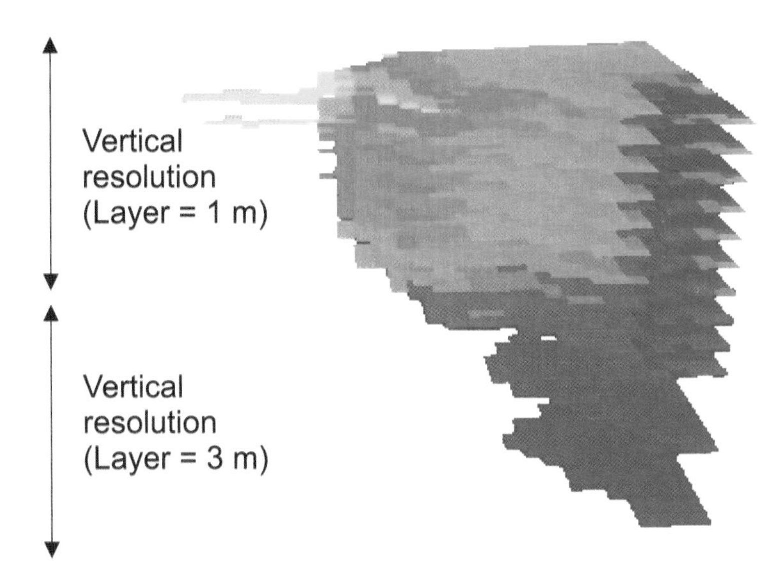

Figure 5.4 Stacked thermal layers illustrating the differences in spatial resolution used for horizontal and vertical representation. Three-dimensional thermal structure of Long Point Bay, Lake Erie, on May 10, 1993. (The figure was created by using three-dimensional Spatial Analyst extension of Environmental Systems Research Institute ArcView 3.01).

sion of warming and cooling, stratification and turnover. Fish must track suitable thermal habitat in conjunction with other habitat requirements, making it difficult to decide how four-dimensional (time plus three dimensions) thermal habitat suitability "maps" can be integrated with two-dimensional physical suitability maps.

5.5.3 Process Ecology 1: Habitat-Based Mapping and Modeling of Fish Population Dynamics and Productivity

"Habitat" may be defined as a place with specific physical, chemical, and biological conditions (sensu Odum 1971) occupied by an organism. Habitat-based modeling approaches usually show how the availability of physical habitat affects the population dynamics of a species. The key stages in habitat model development involve identification of habitat requirements at different life stages, quantification of the habitat, and formulation of links between habitat availability and the population's vital rates (i.e., growth, reproduction, and mortality; Figure 5.5). A GIS is used to quantify the area or volume of suitable habitat available within a defined region. By overlaying maps of relevant physical attributes, polygons that have particular attributes can be defined (Minns and Bakelaar 1999). The habitat attributes of each

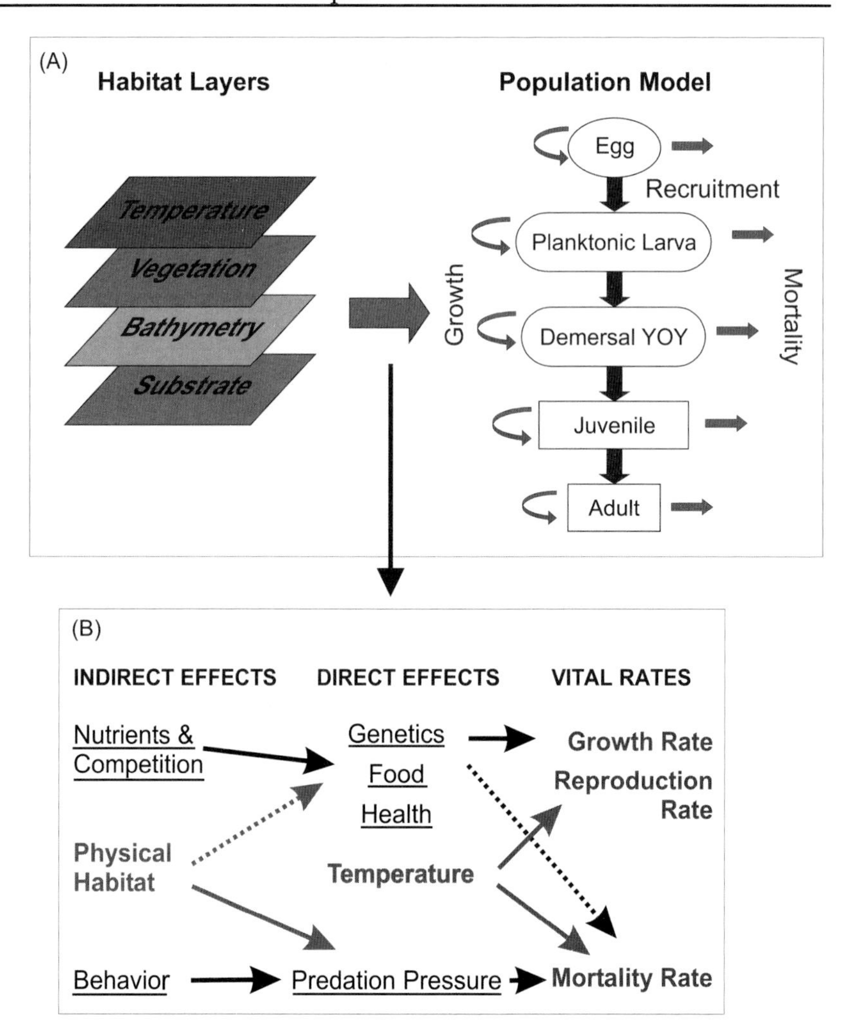

Figure 5.5 Diagram illustrating the connection between physical habitat and a population model. The effects of temperature, vegetation cover, bathymetry, and substrate (A) can vary with life stage. Temperature can affect growth, reproduction, and mortality rates directly, while the other physical habitat variables (vegetation, bathymetry, and substrate type) have indirect effects on these rates through their influence on fish health, food availability, and predation pressure (B). These effects can be confounded by other direct and indirect links to the vital rates of fish populations. Solid arrows indicate established links between variables; dashed arrows indicate more tenuous associations. Underlined variables are not considered explicitly in the current habitat models. YOY = young of the year.

polygon or summations across polygons can be linked to vital rates directly or indirectly through HSI models (Terrell 1984). Habitat suitability indexes range from 0 to 1 for each habitat variable, depending on the habitat's suitability for growth, reproduction, and survival.

As noted in Section 5.3, both individual- and population-based modeling approaches are used to examine the influence of habitat supply. The establishment of linkages between vital rates and habitat suitability is the crucial step in predicting changes in fish population dynamics and productivity when habitat modifications are made. Whether following a spatially explicit individual-based model or population-based approach, GIS maps and overlays of relevant habitat features are an essential input.

Of course, habitat characteristics vary over time as well as space and at different temporal scales. For example, bottom substrate composition may vary on an annual time scale, whereas surface water temperatures can vary on a diurnal time scale. It is, therefore, important to incorporate the biologically relevant time scales and subsequent changes in suitable habitat availability into habitat models (Eklov 1997). Currently, habitat-based population models have been developed for yellow perch in Long Point Bay, Lake Erie, and northern pike in Hamilton Harbor, Lake Ontario as regional test sites (Minns et al. 1996b, 1999b). In Long Point Bay, a remote-sensing inventory of nearshore physical habitat was integrated with bathymetric and offshore substrate data to provide a template for population modeling (Figure 5.6A; Minns et al. 1999b). Physical habitat suitability maps for northern pike (Figure 5.6B) were used to derive estimates of potential biomass and production from the model of Minns et al. (1996b). Ongoing work to map thermal habitat suitability for yellow perch (Figure 5.7) will allow consideration of physical habitat in a new population model as outlined in Figure 5.5. Such a model can be used to examine the influence of time-varying elements of habitat supply such as thermal regime and water level on recruitment and production. The model also can be used to assess the potential impact of changes in habitat quality and quantity when development reduces habitat supply or when restoration increases it.

5.5.4 Process Ecology 2: Movement and Habitat Use of Spawning Lake Whitefish in a Boreal Lake

A recent study of spawning activity of lake whitefish *Coregonus clupeaformis* in Lake 226 at the Experimental Lakes Area in northwestern Ontario demonstrates the insights into fish ecology that are obtained when GIS is used in conjunction with radio telemetry, GPS, and habitat mapping and modeling (Bégout-Anras et al. 1999; Cooley 1999). The study allowed spawning sites to be identified in Lake 226, the spatiotemporal behavior of the fish to be followed, and the associations between fish and habitat features to be discerned.

Lake 226 was subjected to an experimental lowering of lake levels, and the movements and habitat use by adult lake whitefish were followed by automated radiotelemetry. Detailed maps of substrate distributions were created (Figure 5.8A) by using field surveys. To assess the role of surface wave energy in forming the spatial distribution of depositional environments, maps of wave

(A)

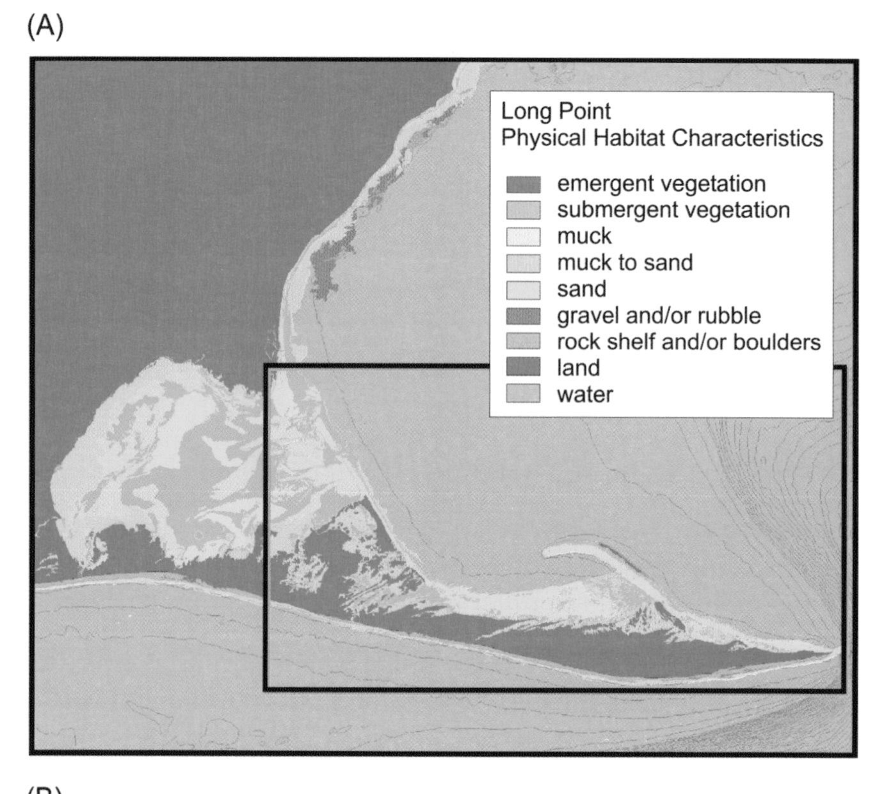

(B)

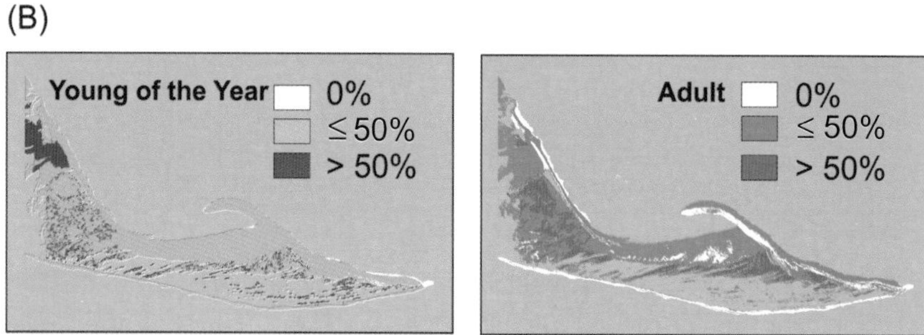

Figure 5.6 Long Point Bay, Lake Erie, nearshore physical habitat (A), as derived from airborne remote-sensing imagery and bathymetry from Canadian Hydrographic Services, provide a basis for population modeling; and physical habitat suitability maps (B) for young-of-the-year and adult northern pike. Values of 0%, ≤50%, and >50% represent locations with no suitable habitat, low-suitability habitat in the lower 50% of the cumulative weighted suitable area distribution, and high-suitability habitat in the upper 50% of the cumulative distribution, respectively.

energy were produced from custom fetch distance programs and wave theory. The relationship between substrate types and the wind-driven wave exposure was examined for the whole lake by using fetch maps, and it showed that the main nearshore habitat features could be related to the exposure regime. Cooley (1999) also was able to predict spatially the mud deposition boundary

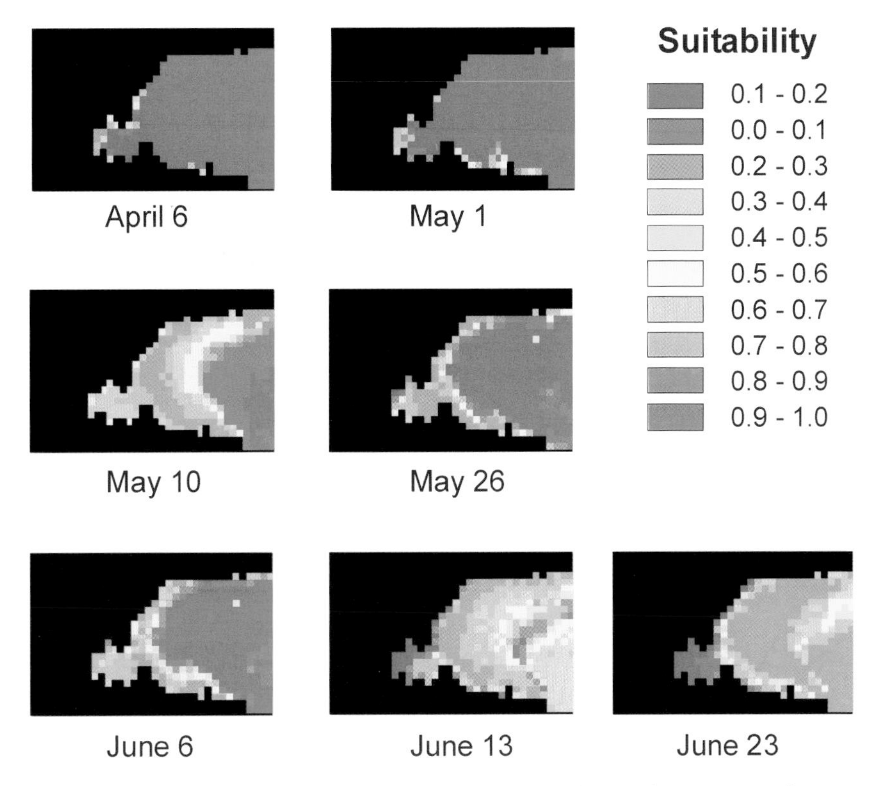

Figure 5.7 Yellow perch egg suitabilities based on bottom thermal structure changes from April 6 to June 23, 1993, in Long Point Bay, Lake Erie. Suitability is based on egg mortality rates at different temperatures, where 0 is low survival and 1 is high survival. The figure shows that a narrow window for optimal thermal habitat exists at any one location because of temporal changes in lake thermal conditions.

depth in the lake by using predicted wave energies derived from maximum fetch (Figure 5.8B). The mud deposition boundary is defined by Rowan et al. (1992) as the depth of the abrupt transition from coarse-grained noncohesive sediments of high-energy erosion environments, to fine-grained cohesive sediments of low-energy depositional environments.

Based on methods from Bégout-Anras et al. (1999), 18 fish were tagged with radio transmitters and then tracked every 100 s via a fixed array of detectors for several weeks during the fall spawning period in 1995 and 1996. Water level was 0.6 m higher in the second year. The high frequency of monitoring allowed movement rates to be computed and detailed records of habitat use to be obtained. Two primary types of tracking patterns, nearshore and offshore, were observed (Figure 5.8C). The lake whitefish found in the western basin showed nearshore movements characteristic of nearshore spawning. The lake whitefish in the eastern basin showed an offshore and repetitive pattern of increasing depth during the study

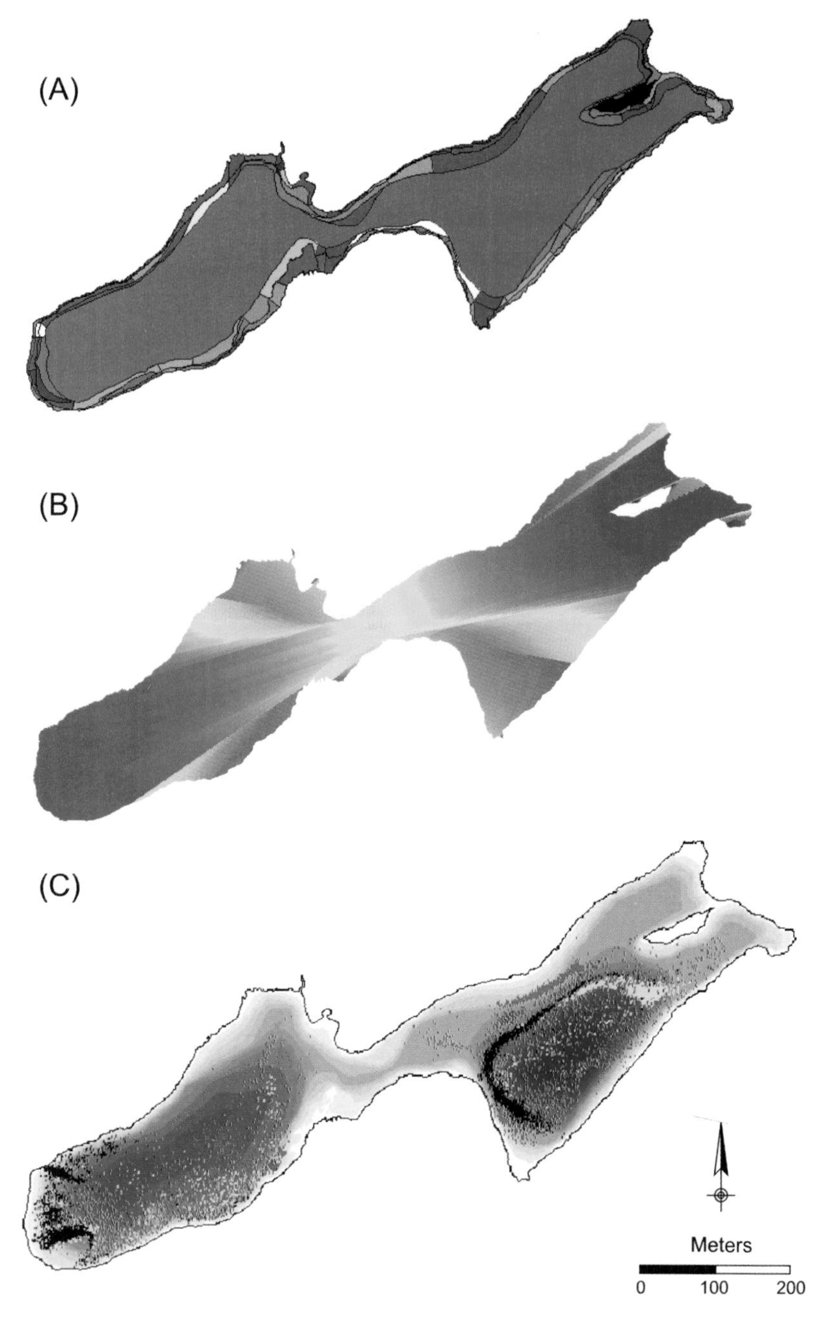

Figure 5.8 Maps of Experimental Lake 226 in northwestern Ontario showing (A) the substratum distribution (34 detailed types of material class combinations exist in the lake, including the offshore zone of deposition), (B) the maximum fetch distance distribution, and (C) the movement patterns of two lake whitefish as determined by acoustic telemetry during the spawning period, 1995. One fish was tracked in the eastern basin and the other in the western basin. (The colors of telemetry positions indicate fish positions during a 4-d period in the order of red, green, yellow, and black).

period to avoid cooler and sinking surface waters before freeze-up. The analyses showed that during spawning, the fish had a high fidelity to the boulder and cobble substrate, which often occur in combination with a detritus infill. Littoral activity was greatest in depths between 2.7 and 3.5 m, and slopes averaging 36%. Tagged fish demonstrated inshore movements to specific shorelines, repetitive shallow-water tracking patterns along the length of these shores prior to spawning, and fidelity to spawning sites among days during the inshore movement. Tagged lake whitefish also frequented other littoral sites as water depths were manipulated.

5.5.5 Fisheries Management: How Should Fish Exploitation be Managed by Using GIS to Ensure the Sustainability and Socioeconomic Benefits of Fisheries in Lakes?

As the literature review in the previous section showed, no studies were found where GIS was applied in a fisheries management context in lakes. There are examples in marine and riverine environments (Maury and Gascuel 1999; Walters et al. 1999; Rose 2000) and, hence, the absence of such studies in lakes is surprising. Section 5.3 described the aspects of fisheries management that would benefit from GIS applications. Studies with a fisheries management context are those where the enumeration data from hydroacoustic line transect surveys are analyzed in a GIS or where geospatial tools are used to derive fish abundance maps across the surface of a lake (Appenzeller 1995; Brosse et al. 1999a, 1999b). Hopefully, applications in the management of lake fisheries will develop as managers recognize the capabilities of GIS.

5.5.6 Habitat Management 1: Conservation and Protection

Essential steps in efforts to conserve the Earth's species are the identification and protection of areas supporting high levels of biodiversity (TNC 1994). The introduction of GAP analysis (Scott et al. 1993) highlighted the use of GIS for organizing and integrating spatially explicit information on the biodiversity of terrestrial landscapes. More recently, the focus of the GAP in the United States has widened to include the development of methodologies for mapping biodiversity in aquatic ecosystems (Lammert et al. 1997; NYCFWRU 1999; Morrison et al. 2001).

Most efforts to apply GIS tools to the identification of aquatic areas of high biodiversity have focused on streams (Lammert et al. 1997; NYCFWRU 1999). Projects overseen by the New York Cooperative Fish and Wildlife Research Unit and The Nature Conservancy (TNC 1994) have used information on stream and land use characteristics obtained by remote sensing, along with data on the habitat preferences of aquatic biota, to predict areas of high biodiversity. Many of these variables are not appropriate for predicting the distribution of aquatic species in lakes. Morrison et al. (2001) have proposed the use of a GIS-based approach called habitat supply analysis to identify areas of high productivity and biodiversity in lacustrine ecosystems.

Habitat supply analysis is founded on a conceptual framework that uses spatially explicit information for the habitat attributes in an area in conjunction with the habitat preferences of aquatic species to predict areas that are suitable for specific aquatic species and life stages. Observations and surveys of the distribution of aquatic biodiversity or fisheries are then used to test and calibrate the predictions. To date, this framework has only been implemented in fragmentary, pilot projects on Lake Erie (Minns and Bakelaar 1999; Minns et al. 1999b). Implementation requires intensive use of GIS to assemble the spatial maps of the various habitat attributes and to model habitat suitabilities and the potential productivity and biodiversity across the lake ecosystem. The implementation may extend up into the surrounding watersheds because many fish species require both lake and stream habitat features to complete their life cycles and persist.

5.5.7 Habitat Management 2: Local Planning Tools for Attaining Policy Goals

In Severn Sound, an International Joint Commission-designated area of concern on Georgian Bay in the Great Lakes, a fish habitat management plan has been developed to protect important fish in the littoral zone. An integral part of that plan involved the development of a classification model for nearshore habitats, where color codes are linked to a graduated scheme of development restrictions, planning, and guidelines (Minns et al. 1999a).

Development of the classification model was undertaken to provide implementers of the plan with a scientifically defensible basis for their interim fish habitat management plan. The interim plan was prepared as a guidance document for local and regional planning authorities to promote increased regard for fish habitat and the legal responsibilities where proposed developments impinge on littoral habitat.

A GIS database of littoral habitat provided the foundation for the development of a new fish habitat classification model. Most of the littoral habitat in Severn Sound between 0-m and 1.5-m depths was inventoried by the Ontario Ministries of Natural Resources and Environment over a period of several years. Field crews mapped depth, substrate, and vegetation; these data were then digitized and brought into an ArcInfo database. The fish habitat classification model has four components: (1) composite suitability index values derived for all species and life stages of the fish assemblage present in Severn Sound by using the Defensible Methods habitat suitability matrix indexing model (Minns and Nairn 1999); (2) identification of rare habitat types specific to particular thermal life-stage–trophic guilds of fish species in the littoral zone; (3) coastal wetlands identified through Ontario's provincial wetland classification system; and (4) local expert identification of important habitat areas for particular fish species and life stages (e.g., known spawning areas for walleye *Stizostedion vitreum*). Each component was implemented as a map layer in the GIS database. The classification model separated habitat units into three habitat suitability classes: low, medium, and high. Then rarity, wetland, and expert layers were used to override low and medium class assignments, raising them to the high class. The final high, medium, and low classes were then colored red, yellow, and green, respectively (Figure 5.9). The completed GIS database, including full implementation of the classification model and the associated documentation, was assembled on CD-ROMs and transferred to the Severn Sound Environmental Association for integration with the planning guide and for distribution to local planners. The combination of the color-coded maps and the development guidance assembled in the area fish habitat management plan are key elements in local efforts to restore and then conserve the Severn Sound ecosystem.

Fish habitat restoration plans are often instigated without a complete examination of all environmental factors. Improvements in physical habitats may attract fish and improve productivity in areas having greater exposure to bioaccumulating chemicals if contamination continues. This can lead to increased chemical levels in fish and increased human exposure risk as fishing improves. A current project (Morrison and Minns unpublished material) focuses on Hamilton Harbor, a highly contaminated bay on Lake Ontario and aims to predict patterns of toxic chemical accumulation in fishes in relation to differing physical habitat suitability patterns and chemical gradients. A risk assessment tool called Urban Fisheries Management model (URB-MAN) is being developed (Morrison and Minns, unpublished material).

The extent to which fish and consumers of fish bioaccumulate and are affected by toxic chemicals depends on their exposure via consumption and absorption. Bioaccumulation by fishes is influenced by their trophic position and by their use of aquatic habitats (Zlokovitz and Secor 1999). The URB-MAN builds on a food web bioaccumulation model developed for the Great Lakes (Morrison et al. 1997, 1999). At present, the URB-MAN model uses physical habitat suitability indices based on a nearshore assessment system (Minns and Nairn 1999) to predict where fish are most likely to be present and abundant, and to estimate home ranges for targeted sizes and species of fish based on an empirical model (Minns 1995). The goal is to establish a sampling grid for comparing suitability and chemical exposures. Maps of physical habitat suitability, chemical gradients in water and surface sediments, and a home range grid were overlaid in a GIS for the whole of Hamilton Harbor (Figure 5.10). Expected bioaccumulation levels were predicted with the food web model by using sediment

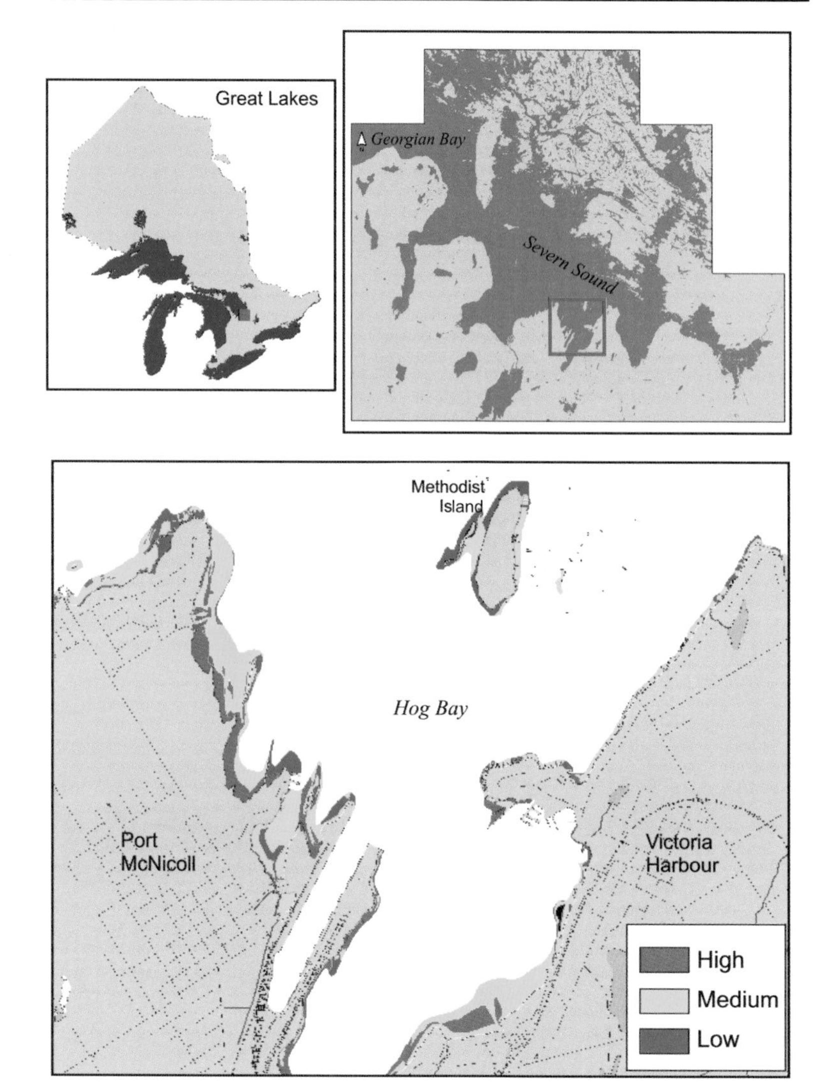

Figure 5.9 A map showing a portion of Severn Sound, Georgian Bay, where units of littoral habitat have been color coded according to the classification model developed by Minns et al. (1999a). High, medium, and low refer to the importance of a site as fish habitat.

and water concentrations as a starting point. Areas where potential bioaccumulation and habitat suitability are high pose a risk to both fish and anglers. The URB-MAN model requires fieldwork to examine habitat patch fidelity and mobility in individual fish from which average chemical exposures can be calculated and to test for spatial differences in bioaccumulation. The results of such modeling can be used to enhance plans to restore fish habitat and fisheries in heavily impacted urban ecosystems.

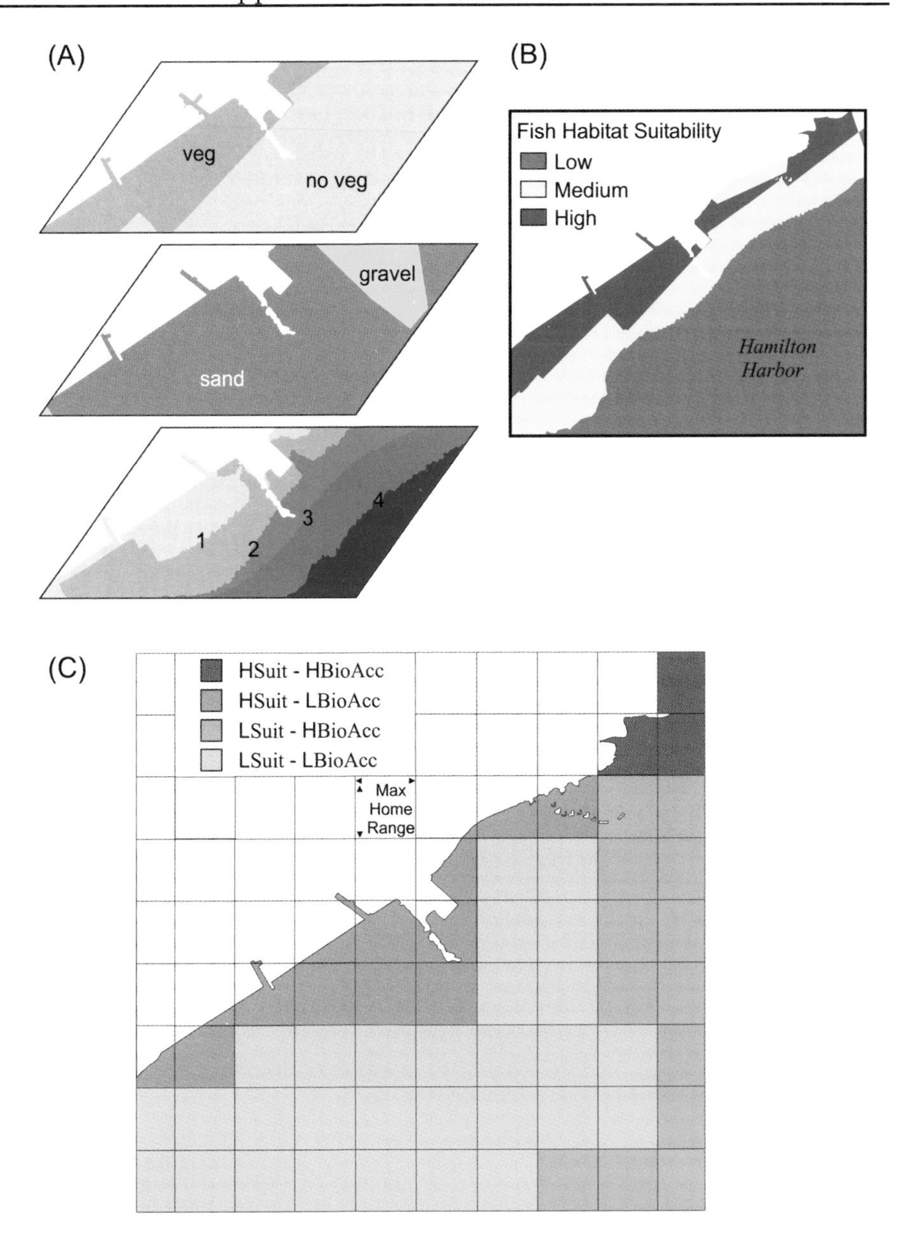

Figure 5.10 Maps showing how physical habitat and chemical exposure data can be used to assess chemical bioaccumulation and human exposure risk by using the URB-MAN model in Hamilton Harbor, Lake Ontario. (A) The physical habitat layers (vegetation, substrate, and depth) are used to form (B) a low, medium, and high fish suitability map which corresponds to fish presence and abundance. (C) Finally, four classes based on a combination of suitability and polychlorinated byphenol concentration layers randomly sampled with a grid size representing maximum home range + 20% (198.6 m^2) of a 38-cm largemouth bass are plotted to indicate the overlap between habitat suitability (Suit) and bioaccumulation potential (BioAcc). H = High; L = Low.

5.6 CONCLUSION

To conclude, we offer a summation under three headings: the state of the art, a vision of the future, and some cautionary points to ponder.

5.6.1 The State of the Art

The broad potential of GIS applications to lakes and their fish resources is apparent in the literature and case studies reviewed in the previous sections. Perhaps the potential has not been explored to the same extent as in terrestrial and other aquatic ecosystems (streams and oceans) but the results so far show considerable promise. In three of the four areas of application (determining factors, process ecology, and habitat management), there is ample evidence of the growth and sophistication of GIS use. With respect to lakes, fisheries management applications are lacking, although they are evident for other aquatic ecosystems. The three-dimensional and four-dimensional aspects of a number of significant features in lakes, such as thermal regime and macrophytes, pose complex challenges to researchers, although the conceptual frameworks have been laid out. There have been a number of efforts to integrate habitat supply accounting and the dynamic simulation of populations, laying the groundwork for increasing knowledge and better management of fisheries resources (e.g., Minns et al. 1996b, 1999b). Statistical consideration of error and uncertainty is lagging far behind, making it difficult to link lacustrine geospatial ecology in context with the rest of ecosystem science.

5.6.2 A Vision of the Future

The growing use of GIS technology in conjunction with GPS-enhanced field survey techniques will lead to both advances in the science of fish ecology in lakes and improvements in fisheries and habitat management. The establishment of coordinated geospatial data warehouses with easy, economical access (e.g., the Internet) will make it easier to develop new applications and should foster a wider use of GIS and further innovation. For lacustrine applications to increase significantly, a wider use of remote sensing (LIDAR, multibeam, satellite) will be required. Much of the necessary expertise and technology is available now to be applied. The science content of management decision making will increase as a result of the extension of information management into the spatial domain. These events also will lead to increasing integration of fisheries and habitat management planning and decision making. In turn, the linkages into wider ecosystem management concerns such as biodiversity conservation and sustainability of ecological services and human uses will expand.

5.6.3 Some Cautionary Points to Ponder

- Successful exploitation of the capabilities of GIS is contingent on the implementation of seamless, updateable geospatial information data systems cutting across jurisdictional and disciplinary boundaries. In the United States, policies of open access to basic geospatial information, especially at the federal level, are fostering the rapid development and deployment of a wide range of GIS-based applications in the renewable, natural resource sector. In Canada, an emphasis on controlled, user-pay access to information is hampering the development of applications because proprietary claims for data collected by governments are enforced.

- Sometimes the enthusiasm and claims for applications of GIS technology run ahead of the science needed to support them. The availability of GIS and mapped data does not guarantee that analyses and interpretations made with them are scientifically defensible. Maps have always had a disproportionate power to sway minds and hearts (Monmonier 1991). Moving from paper to dynamic, digital, color presentations has only increased that power. There needs to be a greater effort to look beyond the pictures to the data, concepts, and assumptions that lie behind them. One subject that is still relatively underdeveloped is the measurement of statistical uncertainty in maps.

- Conversely, there are still instances where the capabilities of the technology still lag behind the inventiveness and ingenuity of science. For example, tools enabling the close integration of multilayer three- and four-dimensional GIS mapping of lake habitats with spatially explicit dynamic models of population and communities of fish are still in their infancy.

Abler (1988) wrote, "GIS are simultaneously the telescope, the microscope, the computer, and the xerox machine of regional analysis and synthesis." More than a decade later, GIS still has much of that inflated status in the eyes of many, though it is beginning to take its proper place as a tool, albeit a powerful one. The unfortunate tendency to see GIS as the end point is disappearing; however, because modern society is visually oriented, maps continue to have an aura and a power to persuade, mislead, and distort (Monmonier 1991; Wood 1992). Each of us is guided by an internal, changeable set of "mental maps" (Gould and White 1986) and, hence, may be predisposed to accept the inferences presented to us in maps. As the capability of GIS and related technologies progress and the applications become ever more sophisticated, we must scrutinize the source data and the models and inferences used to make the maps that will guide our stewardship of natural aquatic resources.

5.7 REFERENCES

Abler, R. F. 1988. Awards, rewards, and excellence: keeping geography alive and well. Professional Geographer 40:135–140.

Appenzeller, A. R. 1995. Hydroacoustic measurement of spatial heterogeneity of European whitefish (*Coregonus lavaretus*) and perch (*Perca *fluviatilis*) in Lake Constance. Archives fur Hydrobiologica, Special Issue. Advances in Limnology 46:261–266.

Baban, S. M. J. 1999. Use of remote sensing and geographical information systems in developing lake management strategies. Hydrobiologia 395/396:211–226.

Bailey, R. G. 1988. Problems with using overlay mapping for planning and their implications for geographic information systems. Environmental Management 12:11–17.

Bégout-Anras, M. L., P. M. Cooley, R. A. Bodaly, L. Anras, and R. J. P. Fudge. 1999. Movement and habitat use by lake whitefish during spawning in a boreal lake: integrating acoustic telemetry and geographic information systems. Transactions of the American Fisheries Society 128:939–952.

Bernardo, G. J. 1997. GIS unlocks the mystery of walleye movements. GIS World 10(10):42–47.

Berry, J. K. 1987. A mathematical structure for analyzing maps. Environmental Management 11:317–325.

Beverton, R. J. H., and S. H. Holt. 1957. On the dynamics of exploited fish populations. Fisheries and Food (UK), Ministry of Agriculture, Fisheries Investigations, Series 2, 19, London.

Bonham-Carter, G. F. 1994. Geographic information systems for geoscientists: modeling with GIS. Computer methods in the geosciences, volume 13 . Pergamon Elsevier, New York.

Borgmann, U. 1987. Models of the slope of, and biomass flow up, the biomass size spectrum. Canadian Journal of Fisheries and Aquatic Sciences 44(Supplement 2):136–140.

Brandt, S. B., D. M. Mason, and E. V. Patrick. 1992. Spatially-explicit models of fish growth rate. Fisheries 17(2):23–31.

Brassard, P, and W. Morris. 1997. Resuspension and redistribution of sediments in Hamilton Harbour. Journal of Great Lakes Research 23:74–85.

Brosse, S., F. Dauba, T. Oberdorff, and S. Lek. 1999b. Influence of some topographical variables on the spatial distribution of lake fish during summer stratification. Archives fur Hydrobiologia 145:359–371.

Brosse, S., S. Lek, and F. Dauba. 1999a. Predicting fish distribution in a mesotrophic lake by hydroacoustic survey and artificial neural networks. Limnology and Oceanography 44:1293–1303.

Bruton, M. N. 1995. Have fishes had their chips? The dilemma of threatened fishes. Environmental Biology of Fishes 43:1–27.

Burrough, P. A. 1987. Principles of geographic information systems for land resources. Oxford University Press, New York.

Caddy, J. F. 1999. Fisheries management in the twenty-first century: will new paradigms apply? Reviews in Fish Biology and Fisheries 9:1–43.

Cooley, P. M. 1999. A manual for analysis and display of lake habitat and fish position information using GIS: physical and hydraulic habitat, draw down, and acoustic telemetry. Canadian Technical Report of Fisheries and Aquatic Sciences 2276.

de Vries, M.. 1996. Integrated system analysis: effects of water management on the habitat suitability for key species in the Sea of Azov. Pages A59–A70 *in* H. Leclerc, H. Capra, S. Valentin, A. Boudreault, and Y. Cote, editors. Ecohydraulics 2000—proceedings of the 2nd international symposium on habitat hydraulics, volume A. INRS-Eau, Université du Quebéc, Quebec.

Downing, J. A., C. Plant, and S. Lalonde. 1990. Fish production correlated with primary productivity, not the morphoedaphic index. Canadian Journal of Fisheries and Aquatic Sciences 47:1929–1936.

Duel, H., and W. E. M. Laane. 1996. The habitat evaluation procedure in the policy analysis of inland waters in the Netherlands: towards ecological rehabilitation. Pages A619–A630 *in* H. Leclerc, H., H. Capra, S. Valentin, A. Boudreault, and Y. Cote, editors. Ecohydraulics 2000—proceedings of the 2nd international symposium on habitat hydraulics, volume A. INRS-Eau, Université du Quebéc, Quebec.

Duel, H., B. P. M. Specken, W. D. Denneman, and C. Kwakernaak. 1995. The habitat evaluation procedure as a tool for ecological rehabilitation of wetlands in The Netherlands. Water Science and Technology 31:387–391.

Eklov, P. 1997. Effects of habitat complexity and prey abundance on the spatial and temporal distributions of perch (*Perca fluviatilis*) and pike (*Esox lucius*). Canadian Journal of Fisheries and Aquatic Sciences 54:1520–1531.

Essington, T. E., and J. F. Kitchell. 1999. New perspectives in the spatial analysis of fish distributions: a case study on the spatial distributions of largemouth bass (*Micropterus salmoides*). Canadian Journal of Fisheries and Aquatic Sciences 56(Supplement 1):52–60.

FAO (Food and Agriculture Organization). 1999. The state of world fisheries and aquaculture, 1998. FAO. Available: *www.fao.org/fi/default.asp* (September 2002).

Fee, E. J. 1980. Important factors for estimating annual integral primary production in the Experimental Lakes area. Canadian Journal of Fisheries and Aquatic Sciences 37:513–522.

Frisk, T., K. Salojarvi, and M. Virtanen. 1988. Modeling the impacts of lake regulation on whitefish stocks. Finnish Fisheries Research 9:467–475.

Gaff, H., D. L. DeAngelis, L. J. Gross, R. Salinas, and M. Shorrosh. 2000. A dynamic landscape model for fish in the Everglades and its application to restoration. Ecological Modelling 127:33–52.

Gaudreau, N., and D. Boisclair. 1998. The influence of spatial heterogeneity on the study of fish horizontal daily migration. Fisheries Research 35:65–73.

Gottgens, J. F., B. P. Swartz, R. W. Kroll, and M. Eboch. 1998. Long-term GIS-based records of habitat changes in a Lake Erie coastal marsh. Wetlands Ecology and Management 6(1):5–17.

Gould, P., and R. White. 1986. Mental maps, 2nd edition. Allen and Unwin, Winchester, Massachusetts.

Gubala, C. P., C. Branch, N. Roundy, and D. Landers. 1994. Automated global positioning system charting of environmental attributes: a limnologic case study. Science of the Total Environment 148(1):83–92.

Guénette, S., T. Lauck, and C. Clark. 1998. Marine reserves: from Beverton and Holt to the present. Reviews in Fish Biology and Fisheries 8:251–272.

Guillard, J., D. Gerdeaux, and J.-M. Chautru. 1990. The use of geostatistics for abundance estimation by echo integration in lakes: the example of Lake Annecy. Rapports et Proces-Verbaux des Reunions, Conseil International pour l'Exploration de la Mer 189:410–414.

Haining, R., and G. Arbia. 1993. Error propagation through map operations. Technometrics 35:293–305.

Haltuch, M. A., and P. A. Berkman. 1999. Lake Erie geographic information system: bathymetry, substrates, mussels (CD-ROM). National Sea Grant Library, OHSU-C-99-001/OHSU-TS-029, Narragansett, Rhode Island. Available: *http://nsgd.gso.uri.edu* (September 2002).

Herbst, D. B., and T. J. Bradley. 1993. A population model for the alkali fly at Mono Lake: depth distribution and changing habitat availability. Hydrobiologia 267:191–201.

Hinch, S. G., K. M. Somers, and N. C. Collins. 1994. Spatial autocorrelation and assessment of habitat–abundance relationships in littoral zone fish. Canadian Journal of Fisheries and Aquatic Sciences 51:701–712.

Hobbs, B. F. 1997. Bayesian methods for analysing climate change and water resources uncertainties. Environmental Management 49:53–72.

Horne, J. K., and D. C. Schneider. 1994. Analysis of scale-dependent processes with dimensionless ratios. Oikos 70:201–211.

Hrabik, T. R., and J. Magnuson. 1999. Simulated dispersal of exotic rainbow smelt (*Osmerus mordax*) in a northern Wisconsin lake district and implications for management. Canadian Journal of Fisheries and Aquatic Sciences 56(Supplement 1):35–42.

Jean, M., and A. Bouchard. 1991. Temporal changes in wetland landscapes of a section of the St. Lawrence River, Canada. Environmental Management 15:241–250.

Jennings, M. D. 2000. Gap analysis: concepts, methods, and recent results. Landscape Ecology 15:5–20.

Job, S. D., and D. R. Bellwood. 2000. Light sensitivity in larval fishes: implications for vertical zonation in the pelagic zone. Limnology and Oceanography 45:362–371.

Johannsson, O. E. 1995. Response of *Mysis relicta* population dynamics and productivity to spatial and seasonal gradients in Lake Ontario. Canadian Journal of Fisheries and Aquatic Sciences 52:1509–1522.

Johnson, C. A. 1994. Cumulative impacts to wetlands. Wetlands 14:49–55.

Jordan, C., and I. J. Enlander. 1990. The variation in the acidity of ground and surface waters in northern Ireland. International Review of Hydrobiology 75:379–401.

Kornijów, R., and B. Moss. 1998. Vertical distribution of in-benthos in relation to fish and floating-leaved macrophyte populations. Pages 227–232 *in* E. Jeppesen, Ma. Søndergaard, Mo. Søndergaard, and K. Christoffersen, editors. The structuring role of submerged macrophytes in lakes. Springer-Verlag, New York.

Koutnik, M. A., and D. K. Padilla. 1994. Predicting the spatial distribution of *Dreissena polymorpha* (zebra mussel) among inland lakes of Wisconsin: modeling with a GIS. Canadian Journal of Fisheries and Aquatic Sciences 51(5):1189–1196.

Lammert, M., J. Higgins, M. Bryer, and D. Grossman. 1997. A classification framework for freshwater communities: proceedings of The Nature Conservancy's aquatic community classification workshop. The Nature Conservancy, Arlington, Virginia.

Lamon, E. C., III, and C. A. Stow. 1999. Sources of variability in microcontaminant data for Lake Michigan salmonids: statistical models and implications for trend detection. Canadian Journal of Fisheries and Aquatic Sciences 56(Supplement 1):71–85.

Leach, J. H., L. M. Dickie, B. J. Shuter, U. Borgmann, J. Hyman, and W. Lysack. 1987. A review of methods for prediction of potential fish production with application to the Great Lakes and Lake Winnipeg. Canadian Journal of Fisheries and Aquatic Sciences 44(Supplement 2):471–485.

Legendre, P. 1993. Spatial autocorrelation: trouble or new paradigm? Ecology 74:1659–1673.

Lehmann, A., J.-M. Jaquet, and J.-B. Lachavanne. 1994. Contribution of GIS to submerged macrophyte biomass estimation and community structure modeling, Lake Geneva, Switzerland. Aquatic Botany 47(2):99–117.

Lehmann, A., J.-M. Jaquet, and J.-B. Lachavanne. 1997. A GIS approach to aquatic plant spatial heterogeneity in relation to sediment and depth gradients, Lake Geneva, Switzerland. Aquatic Botany 58(3/4):347–361.

Lehmann, A., and J.-B. Lachavanne. 1997. Geographic information systems and remote sensing in aquatic botany. Aquatic Botany 58(3/4):195–207.

Lek, S., and J. F. Guégan. 1999. Artificial neural networks as a tool in ecological modeling, an introduction. Ecological Modelling 120:65–73.

MacCall, A. D. 1990. Dynamic geography of marine fish populations. University of Washington Press, Seattle.

MacLeod, W. D., C. K. Minns, A. Mathers, and S. Mee. 1995. An evaluation of biotic indices and habitat suitability scores for classifying littoral habitats. Canadian Manuscript Report of Fisheries and Aquatic Sciences 2334.

Mason, D. M., A. Goyke, and S. B. Brandt. 1995. A spatially explicit bioenergetics measure of habitat quality for adult salmonines: comparison between Lakes Michigan and Ontario. Canadian Journal of Fisheries and Aquatic Sciences 52:1572–1583.

Maury, O., and D. Gascuel. 1999. SHADYS ("simulateur halieutique de dynamiques spatiales"), a GIS based numerical model of fisheries. Example application: the study of marine protected area. Aquatic Living Resources 12:77–88.

Meaden, G. J., and J. M. Kapetsky. 1991. Geographical information systems and remote sensing in inland fisheries and aquaculture. Food and Agricultural Organization, Technical Report 318, Rome.

Millard, E. S., E. J. Fee, D. D. Myles, and J. A. Dahl. 1999. Comparison of phytoplankton photosynthesis methodology in Lakes Erie, Ontario, the Bay of Quinte, and the northwest Ontario Lakes size series. Pages 441–468 *in* M. Munawar, T. Edsall, and I. F. Munawar, editors. State of Lake Erie (SOLE)—past, present and future. Backhuys Publishers, Leiden, The Netherlands.

Millard, E. S., D. D. Myles, O. E. Johannsson, and K. M. Ralph. 1996. Phytoplankton photosynthesis at two index stations in Lake Ontario 1987–1992: assessment of the long-term response to phosphorus control. Canadian Journal of Fisheries and Aquatic Sciences 53(5):1112–1124.

Minns, C. K. 1995. Allometry of home range size in lake and river fishes. Canadian Journal of Fisheries and Aquatic Sciences 52(7):1499–1508.

Minns, C. K., and C. N. Bakelaar. 1999. A method for quantifying the supply of suitable habitat for fish stocks in Lake Erie. Pages 481–496 *in* M. Munawar, T. Edsall, and I. F. Munawar, editors. State of Lake Erie—past, present and future. Backhuys Publishers, Leiden, The Netherlands.

Minns, C. K., C. N. Bakelaar, J. E. Moore, R. W. Dermott, and R. Green. 1996c. Measuring differences between overlapping but unpaired spatial surveys using a geographic information system. Environmental Monitoring and Assessment 43:237–253.

Minns, C. K., P. C. E. Brunette, M. Stoneman, K. Sherman, R. Craig, C. B. Portt, and R. G. Randall. 1999a. Development of a fish habitat classification model for littoral areas of Severn Sound, Georgian Bay, a Great Lakes' area of concern. Canadian Manuscript Report of Fisheries and Aquatic Sciences 2490.

Minns, C. K., V. W. Cairns, R. G. Randall, and J. E. Moore. 1993a. UET: a tool for fish habitat management? Pages 236–245 *in* J. Kozlowski and G. Hill, editors. Towards planning for sustainable development. Queensland University Press, Brisbane, Australia.

Minns, C. K., V. W. Cairns, R. G. Randall, and J. E. Moore. 1993b. UET: A practical application in Canadian lakes. Pages 246–267 *in* J. Kozlowski and G. Hill, editors. Towards planning for sustainable development. Queensland University Press, Brisbane, Australia.

Minns, C. K., S. E. Doka, C. N. Bakelaar, P. C. E. Brunette, and W. M. Schertzer. 1999b. Identifying habitats essential for pike, *Esox lucius* L., in the Long Point region of Lake Erie: a suitable supply approach. Pages 363–382 *in* L. Benaka, editor. Fish habitat: essential fish habitat and rehabilitation. American Fisheries Society, Symposium 22, Bethesda, Maryland.

Minns, C. K., J. R. M. Kelso, and R. G. Randall. 1996a. Detecting the response of fish to habitat alterations in freshwater ecosystems. Canadian Journal of Fisheries and Aquatic Sciences 53(Supplement 1):403–414.

Minns, C. K., and J. E. Moore. 1992. Predicting the impact of climate change on the spatial pattern of freshwater fish yield capability in eastern Canada. Climatic Change 22:327–346.

Minns, C. K., and J. E. Moore. 1995. Factors limiting the distributions of Ontario's freshwater fishes: the role of climate and other variables, and the potential impacts of climate change. Canadian Special Publication of Fisheries and Aquatic Sciences 121:137–160.

Minns, C. K., J. E. Moore, D. W. Schindler, and M. L. Jones. 1990. Assessing the potential extent of damage to inland lakes in eastern Canada due to acidic deposition: III. Predicted impacts on species richness in seven groups of aquatic biota. Canadian Journal of Fisheries and Aquatic Sciences 47:821–830.

Minns, C. K., and R. B. Nairn. 1999. Defensible methods: applications of a procedure for assessing developments affecting littoral fish habitat on the lower Great Lakes. Pages 15–35 *in* T. P. Murphy and M. Munawar, editors. Aquatic restoration in Canada. Backhuys Publishers, Leiden, The Netherlands.

Minns, C. K., R. G. Randall, J. E. Moore, and V. W. Cairns. 1996b. A model simulating the impact of habitat supply limits on northern pike, *Esox lucius*, in Hamilton Harbour, Lake Ontario. Canadian Journal of Fisheries and Aquatic Sciences 53(Supplement 1):20–34.

Monmonier, M. 1991. How to lie with maps. University of Chicago Press, Chicago.

Morrison, H. A., F. A. P. C. Gobas, R. Lazar, D. M. Whittle, and G. D. Haffner. 1997. Development and verification of a benthic/pelagic food web bioaccumulation model for PCB congenors in western Lake Erie. Environmental Science and Technology 31:3267–3273.

Morrison, H. A., C. K. Minns, and J. F. Koonce. 2001. A methodology for identifying and classifying aquatic biodiversity investment areas: application in the Great Lakes Basin. Aquatic Ecosystem Health and Management 4:1–12.

Morrison, H. A., D. M. Whittle, C. D. Metcalfe, and A. J. Niimi. 1999. Application of a food web bioaccumulation model for the prediction of PCB, dioxin, and furan congener concentrations in Lake Ontario aquatic biota. Canadian Journal of Fisheries and Aquatic Sciences 56:1389–1400.

Narumalani, S., J. R. Jensen, J. D. Althausen, S. Burkhalter, and H. E. Mackey, Jr. 1997. Aquatic macrophyte modeling using GIS and logistic multiple regression. Photogrammetric Engineering and Remote Sensing 63:41–49.

NYCFWRU (New York Cooperative Fish and Wildlife Research Unit). 1999. Application of GAP analysis to aquatic biodiversity conservation. Available: *www.dnr.cornell.edu/hydro2/aquagap.htm* (September 2002).

Odum, E. P. 1971. Fundamentals of ecology, 3rd edition. Saunders Company, Philadelphia, Pennsylvania.

Omernik, J. M., C. M. Rohm, R. A. Lillie, and N. Mesner. 1991. Usefulness of natural regions for lake management: analysis of variation among lakes in northwestern Wisconsin, USA. Environmental Management 15:281–292.

Pauly, D. 1980. On the relationships between natural mortality, growth parameters and mean environmental temperature in 175 fish stocks. Journal du Conseil International pour l'Exploration de la Mer 39:175–192.

Polfeldt, T. 1999. On the quality of contour maps. Environmetrics 10:785–790.

Pulliam, H. R., and B. J. Danielson. 1991. Sources, sinks, and habitat selection: a landscape perspective on population dynamics. American Naturalist 137:S50–S66.

Rejwan, C., N. C. Collins, J. Brunner, B. J. Shuter, and M. S. Rigdway. 1999. Tree regression analysis on the nesting habitat of smallmouth bass. Ecology 80:341–348.

Remillard, M. M., and R. A. Welch. 1993a. GIS technologies for aquatic macrophyte studies: modeling applications. Landscape Ecology 8:163–175.

Remillard, M. M., and R. A. Welch. 1993b. GIS technologies for aquatic macrophyte studies: I. Database development and changes in the aquatic environment. Landscape Ecology 7:151–162.

Richardson, J. R., and E. Hamouda. 1995. GIS modeling of hydroperiod, vegetation, and soil nutrient relationships in the Lake Okeechokee marsh ecosystem. Ergebnisse der Limnologie 45:95–115.

Ricker, W. E. 1975. Computation and interpretation of biological statistics of fish populations. Fisheries Research Board of Canada Bulletin 191.

Rose, K. A. 2000. Why are quantitative relationships between environmental quality and fish populations so elusive? Ecological Applications 10:367–385.

Rowan, D. J., J. Kalff, and J. B. Rasmussen. 1992. Estimating the mud deposition boundary depth in lakes from wave theory. Canadian Journal of Fisheries and Aquatic Sciences 49:2490–2497.

Rudstam, L. G., and J. J. Magnuson. 1985. Predicting the vertical distribution of fish populations: analysis of cisco, *Coregonus artedii*, and yellow perch, *Perca flavescens*. Canadian Journal of Fisheries and Aquatic Sciences 42:1178–1188.

Scheffer, M., J. M. Baveco, D. L. DeAngelis, K. A. Rose, and E. H. van Nes. 1995. Super-individuals a simple solution for modeling large populations on an individual basis. Ecological Modelling 80:161–170.

Scheffer, M., M. R. de Redelijkheid, and F. Noppert. 1992. Distribution and dynamics of submerged vegetation in a chain of shallow eutrophic lakes. Aquatic Botany 42:199–216.

Schernewski, G., V. Podsetchine, M. Asshoff, D. Garbe-Schonberg, and T. Huttula. 2000. Spatial ecological structures in littoral zones and small lakes: examples and future prospects of flow models as research tools. Archiv fuer Hydrobiologie Special Issue 55:227–241.

Schertzer, W. M., and A. M. Sawchuk. 1990. Thermal structure of the lower Great Lakes in a warm year: implications for the occurrence of hypolimnion anoxia. Transactions of the American Fisheries Society 119:195–209.

Schmieder, K. 1997. Littoral zone— GIS of Lake Constance: a useful tool in lake monitoring and autoecological studies with submersed macrophytes. Aquatic Botany 58(3/4):333–346.

Scott, J. M., F. Davis, B. Csuti, R. Noss, B. Butterfield, C. Groves, H. Anderson, S. Caicco, F. D'Erchia, T. Edward, Jr, J. Ulliman, and R. G. Wright.

1993. GAP analysis: a geographic approach to protection of biological diversity. Wildlife Monographs 123:1–41.

Selgeby, J. H. 1982. Decline of lake herring (*Coregonus artedii*) in Lake Superior: analysis of the Wisconsin herring fishery, 1936–78. Canadian Journal of Fisheries and Aquatic Sciences 39(4):554–563.

Steedman, R. J., and H. A Regier. 1987. Ecosystem science for the Great Lakes: perspectives on degradative and rehabilitative transformations. Canadian Journal of Fisheries and Aquatic Sciences 44(Supplement 2):95–103.

Stephenson, T. D. 1990. Fish reproductive utilization of coastal marshes of Lake Ontario near Toronto. Journal of Great Lakes Research 16:71–81.

Terrell, J. W., editor. 1984. Proceedings of a workshop on fish habitat suitability index models. U. S. Fish and Wildlife Service Biological Report 85(6).

Thierfelder, T. 1998. The morphology of landscape elements as predictors of water quality in glacial/boreal lakes. Journal of Hydrology (Amsterdam) 207(3/4):189–203.

Tischendorf, L. 1997. Modeling individual movements in heterogeneous landscapes: potentials of a new approach. Ecological Modelling 103:33–42.

TNC (The Nature Conservancy). 1994. The conservation of biological diversity in the Great Lakes ecosystem: issues and opportunities. The Nature Conservancy, Chicago.

Tudge, C. 1990. Underwater, out of mind. New Scientist November 3 1990:40–45.

Tyler, J. A., and K. A. Rose, 1994. Individual variability and spatial heterogeneity in fish population models. Reviews in Fish Biology and Fisheries 4:91–123.

Ullman, D., J. Brown, P. Cornillon, and M. Timothy. 1998. Surface temperature fronts in the Great Lakes. Journal of Great Lakes Research 24:753–775.

Walters, C., D. Pauly, and V. Christensen. 1999. Ecospace: prediction of mesoscale spatial patterns in trophic relationships of exploited ecosystems, with emphasis on the impact of marine protected areas. Ecosystems 2:539–554.

Wang, J., and Y. Xie. 1994. Application of geographical information systems to toxic chemical mapping in Lake Erie. Environmental Technology 15(8):701–714.

Williams, D. C., and J. G. Lyons. 1997. Historical aerial photographs and a geographic information system (GIS) to determine effects of long-term water level fluctuations on wetlands along the St. Mary's River, Michigan, USA. Aquatic Botany 58(3/4):363–378.

Williams, J. E., and R. R. Miller. 1990. Conservation status of the North American fish fauna in fresh water. Journal of Fish Biology(Supplement A):79–85.

Willis, T. V., and J. J. Magnuson. 2000. Patterns of fish species composition across the interface between streams and lakes. Canadian Journal of Fisheries and Aquatic Sciences 57:1042–1052.

Wood, D. 1992. The power of maps. The Guilford Press, New York.

Zlokovitz, E. R., and D. H. Secor. 1999. Effect of habitat use on PCB body burden in Hudson River striped bass (*Morone saxatilis*). Canadian Journal of Fisheries and Aquatic Sciences 56:86–93.

Chapter 6

Geographic Information Systems Applications in Aquaculture

James McDaid Kapetsky

6.1 INTRODUCTION

6.1.1 An Overview of Aquaculture

Aquaculture commands increasing attention worldwide. It continues to be the world's fastest growing food production sector (FAO Fisheries Department 2001). From 1970–1999, global output of cultured plants and animals increased twelvefold, attaining nearly 43 million metric tons in 1999 (FAO Fisheries Department 2003). In comparison, output from marine and inland capture fisheries together was about 94 million metric tons in 1999.

Mariculture accounts for 50% of aquaculture output, and freshwater culture is about 45%. Both are increasing at a rapid rate. In comparison, brackish-water culture makes up about 5% of output, and it is increasing at a relatively modest rate. Aquaculture production is greatly skewed among continental areas with Asia accounting for 91%. In terms of value, finfish account for 55%, crustaceans for 18%, mollusks for 16%, and aquatic plants for 10% (FAO Fisheries Department 2001).

Aquaculture is complex. It is a vast array of water-based, controlled production that can be viewed as a matrix of categories within criteria (Table 6.1). In consideration of its complexity, economic importance, and recent rapid growth, it should not be surprising that aquafarming faces many development and management.

The fundamental issue in aquaculture is to ensure its sustainability. For any expanse of area at any location, or at any level of detail, sustainability can be addressed with two basic criteria in mind: (1) suitability *for* aquaculture, and (2) consequences *of* aquaculture.

In turn, suitability can be measured in terms of the needs of the culture system that is used and the requirements of the organism to be cultured. Suitability also takes into account the broad "ecology" of aquaculture that includes the politico-administrative and socioeconomic climates, as well as the physical, chemical, and biological environments. Likewise, the consequences of aquaculture can be viewed from the same perspectives. Sustainability of aquaculture encompasses many other related issues that have spatial elements, and it is these

Table 6.1 A broad matrix of aquaculture.

Criteria	Categories
Organisms	Plants, animals
Systems	Open, closed
Intensity	Extensive, semi-intensive, intensive
Location	Inland, coastal, offshore
Products	Food for human consumption, chemicals (from plants), recreation (some culture-based fisheries), aquarium plants and animals

that open the door for a broad spectrum of applications of geographic information systems (GIS).

6.1.2 Chapter Objectives

The objectives of this chapter, from a global perspective, are to (1) characterize GIS applications in aquaculture in the context of focus, geography, environment, and organisms; (2) provide examples of the use of GIS in aquaculture in the framework of development and management issues affecting the sustainability aquaculture; and (3) consider the future direction of GIS applications in terms of unrealized opportunities for applications and constraints.

6.2 CHARACTER OF GIS APPLICATIONS IN AQUACULTURE

6.2.1 Sources of Information

The following sources were used to build a database of applications and characteristics of GIS uses in aquaculture: (1) *Aquatic Sciences and Fisheries Abstracts* (ASFA), from 1990 through October 2001, and *Oceanic Abstracts,* from 1985 through October 2001, by using "GIS" as the sole keyword; (2) material in the author's own collection of print and electronic papers and reports; and (3) Internet searches by using various keywords incorporating GIS, aquaculture, mariculture, and fish and shellfish farming.

Altogether, 127 examples of applications were assembled. Of these, 78 are in the form of complete works, 42 are abstracts, and 7 are references only. Most of the applications were distributed among journals (45), proceedings from symposia (35), and technical reports (30). Additionally, 8 appeared in magazines, and 6 in newsletters, while 2 applications were taken from Web pages, and 1 appeared in a book.

At first glance, it may appear that the Internet was not an important source of information on applications; however, this was not the case. Internet searches led to individuals and institutions from which material was obtained in print or as files that could be cited. Web pages featuring recent symposia or pages that previewed upcoming meetings were among the best sources.

Regarding the completeness of coverage of GIS applications in aquaculture, there is little doubt that there are many more applications of GIS in aquaculture than it has been possible to bring to light in this chapter. This is partly because GIS applications in aquaculture carried out in important aquafarming countries such as Japan and China are not finding their way into the media and languages that are the most common sources of information, namely journals and the Internet, in English. Another reason is that a significant part (24%) of the applications identified herein appeared in technical reports, most of which are not disseminated widely. Another significant part (28%) was made up of papers presented at symposia. Sometimes, proceedings are not published or the papers are distributed in print or on computer disks that are not easily accessible. Another limitation is that some GIS applications in aquaculture may not appear in a journal that is abstracted by ASFA or other abstracting services. Finally, the applications that appear here tend to be from sources where GIS was the end in itself. No doubt there are many interesting and essential applications of GIS in aquaculture that did not come to light because GIS was the means, not the end.

6.2.2 Character of GIS Applications

The purpose of this section is to provide a brief overview of the character of GIS applications according to focus, geography, the main environments in which GIS was applied, the organisms involved, and the applied research or operational nature of the use of GIS.

6.2.2.1 ***Focus.*** As would be expected from the topic at hand, nearly all of the material is focused fairly narrowly on aquaculture development or management. However, for completeness, the range of GIS applications that are important to aquaculture development and management, including GIS in culture-based fisheries and in integrated management that includes aquaculture, also have been included.

6.2.2.2 ***Geography.*** Geographic information systems applications in aquaculture could be identified within 23 countries. The distribution among continents of applications identified within specific countries was North America, 48; Asia, 20; Africa, 4; Europe, 13; Latin America, 11; and Oceania, 2. Five applications covered regions consisting of a number of countries in the Caribbean, southern Africa, Oceania, and South Asia. Ten applications were directed at entire continents, which included Africa, Latin America, and Asia. The remainder had no specific geographic association or were global.

6.2.2.3 ***Environments.*** The great majority of applications, 82, were aimed at coastal aquaculture, while 32 were for inland areas. Only 2 applications were offshore. The remainder were comprehensive or directed at combinations of environments.

6.2.2.4 ***Organisms.*** Geographic information system applications by target organisms have been listed by Aguilar-Manjarrez (1996). In this review, seaweeds were represented in only one study. Among the animals, shellfishes, and among shellfishes, oysters were the most frequent targets of GIS studies. Second most frequent were penaeid shrimps. This goes along with the fact that nearly two-thirds of the applications were in coastal waters and involved economically valuable shrimps. In inland waters, catfishes (Clariidae), carps, and tilapias were the main targets.

6.3 APPLICATIONS OF GIS IN AQUACULTURE

6.3.1 Framework for the Classification of GIS Applications in Aquaculture

Four major categories of GIS uses in aquaculture can be identified (Table 6.2): (1) GIS training and promotion, (2) GIS aimed at aquaculture development, (3) GIS for aquaculture practice and management, and (4) GIS for multisectoral planning that includes aquaculture.

Geographic information systems for aquaculture development encompass a broad variety of applications. Development, as used here, pertains to GIS applications that precede the establishment of aquaculture and involve planning. These can be seen as relating to the environment for aquaculture development and as anticipating the environmental consequences of aquaculture. In both cases, "environment" is used in the broadest sense: not only the physical, bio-

Table 6.2 Categories of GIS applications in aquaculture.

Category	Specific application
GIS training and promotion	GIS training
	Promotion of GIS
GIS aimed at development	Suitability of the site
	Strategic planning for aquaculture development
	Anticipating the consequences of aquaculture
	Web-based aquaculture information systems
	Marketing
GIS in culture practice and management	Inventory and monitoring of aquaculture and the environment
	Environmental impacts of aquaculture
	Restoration of aquaculture habitats
Multisectoral planning that includes aquaculture	Management of aquaculture together with fisheries
	Planning for aquaculture among other uses of land and water

logical, and chemical environments but also the economic, political, and social environments. Thus, the environment for aquaculture includes not only the suitability of sites for the culture system and organism(s), but also the synoptic planning for aquaculture over large areas, that is, strategic assessments, at ecosystem, state, national, and continental levels of organization. Similarly, anticipation of the environmental consequences of aquaculture is broadly considered and includes both impacts on the ecosystem and on society. The GIS-assisted Internet-based aquaculture information systems are set apart because of their present novelty and their likely increasing importance to aquaculture development.

A GIS for aquaculture practice and management includes inventory and monitoring of aquaculture and the environment, environmental impacts of aquaculture, and restoration of habitat for aquaculture. Finally, GIS for multisectoral planning that includes aquaculture is viewed within two subcategories: management of aquaculture together with fisheries and planning for aquaculture among other uses of land and water. Geographic information system planning is set apart because of its fundamental importance in stimulating the use of GIS in aquaculture.

In the rest of this section, the objective is to provide the reader with the broadest possible spectrum of GIS applications following the outline in Table 6.2. A brief overview of each use of GIS, or similar uses in groups, is provided. Applications are addressed according to purpose, environment, and location, as well as by culture system and cultured organism, where relevant.

6.3.2 GIS Training and Promotion

6.3.2.1 ***Training.*** Geographic information systems applied to aquaculture, in so far as the citable literature is concerned, goes back only as far as 1985. Mooneyhan (1985) developed an exercise for FAO (Food and Agriculture Organization of the United Nations) training courses that estimated inland aquaculture potential in south Florida. The exercise took into account mangrove vegetation and proximity to roads, water bodies, and settlements (Figure 6.1). The exercise, based on Landsat (East Lansing, Michigan) Multispectral Scanner (MS) data, is indicative of the fact that GIS for coastal and inland aquaculture was, and continues to be, firmly rooted in satellite remote sensing, which is an indispensable source of up-to-date land and underwater cover, land and water use, and water quality information.

Early training in GIS applications in aquaculture, combined with inland fisheries, and including remote sensing, took several forms. One form was regional, continent-wide (FAO 1989, 1991, 1992), or global 2-week to 3-week workshops (e.g., Kapetsky 1989a) aimed at fisheries and aquaculture personnel from developing countries working at technical levels. Another variant was shorter and aimed at decision makers from developing countries (e.g., Kapetsky 1991). The objectives were to raise awareness of GIS and remote-sensing appli-

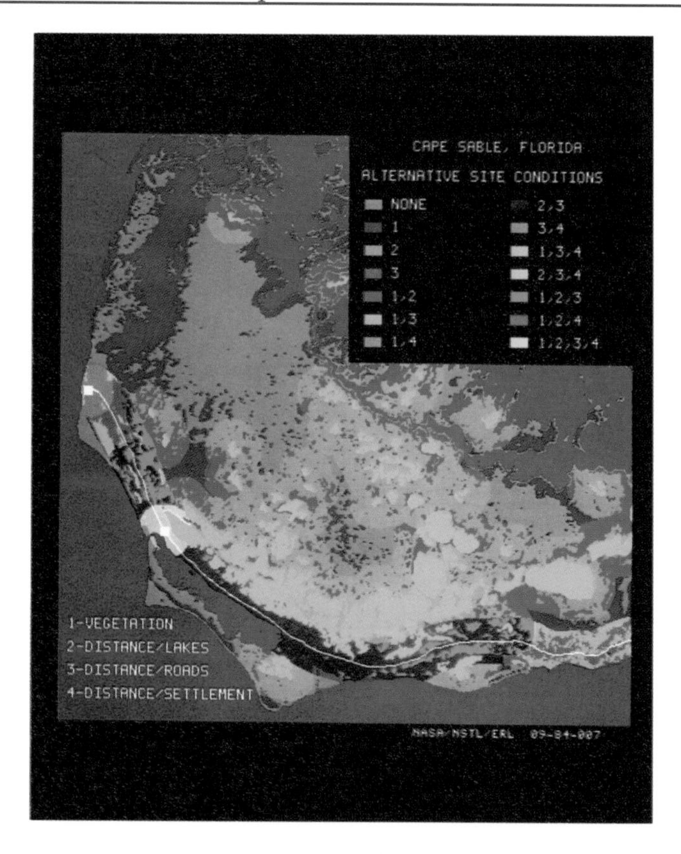

Figure 6.1 Aquaculture siting simulation (from Mooneyhan 1985). The lower left numbers designate where (1) siting only within red mangrove *Rhizophora mangle* and black mangrove *Avicennia germinans* was allowed, and (2) optimum requirements were based on a maximum distance of 1 km from a tidal water lake with an area greater than 4 ha and (3) the same distance from a settlement and (4) a roadway. The numbers in the upper right are combinations of the site conditions meeting vegetation and distance requirements 1 through 4. The areas that meet all requirements are yellow and labeled "1, 2, 3, 4."

cations, while at the same time familiarizing trainees with technical and administrative constraints, as well as benefits of GIS and remote sensing. Underlying these objectives was the promotion of national-level planning for fisheries management and aquaculture development.

Another form was self-training manifest in the production and wide dissemination in English and Spanish of a manual on GIS in inland fisheries and aquaculture (Meaden and Kapetsky 1991). Currently, an Internet-based search will reveal several pages of formal GIS training opportunities in natural resources offered by software vendors, universities, and government agencies. Self-training opportunities with accompanying software also are readily available.

Still another direction is university-level training in GIS in the fisheries sector. The Institute of Aquaculture, University of Stirling,

United Kingdom (Institute of Aquaculture 2003), is unique in emphasizing graduate training in GIS and remote-sensing applications in aquaculture. In North America, a survey of GIS-related research conducted at U.S. universities with fisheries graduate programs was made by Fisher and Toepfer (1998).

6.3.2.2 ***Promotion of GIS.*** A valuable part of the literature on GIS in aquaculture and an important stimulus for its expanding use are promotional articles in newsletters and technical magazines (e.g., Kapetsky 1993a, 1993b, 1993c; Aguilar-Manjarrez and Ross 1993, 1995a; Kapetsky et al. 1996; Klotz-Shiran 1999) and journals (Nath et al. 2000). Newsletters and magazines have been important in the dissemination of GIS in a timely way because they usually summarize information in difficult-to-access technical reports and because they often preview results of ongoing studies and of upcoming activities. Nath et al. (2000) viewed GIS applications in aquaculture from the viewpoint of constraints, and they reviewed GIS terminology and methodology, presented case studies, and considered future trends.

Finally, it could be said that papers presented at symposia are, in part, promotional (e.g., Kapetsky and Travaglia 1995; Preston et al. 1997; Salam and Ross 1999; Kapetsky 2000). In this regard Meaden's (2001) comprehensive review of GIS applications in the fishery sector, including aquaculture, will be an important promotional work for GIS applications in the fishery sector for some time to come.

6.3.3 GIS Aimed at Aquaculture Development

6.3.3.1 ***Suitability of the Site.*** Assessing the suitability of a site is one of the most critical steps in the development of aquaculture. It also is one of the most frequent applications of GIS in aquaculture. With one exception, farming of channel catfish *Ictalurus punctatus* in Louisiana (Kapetsky et al. 1988), examples come entirely from coastal aquaculture. As a group, they pertain mainly to siting of farming of penaeid shrimp *Penaeus* spp., fish culture in cages, and intertidal and subtidal shellfish farming.

One of the earliest examples of the comprehensive application of remote-sensing–assisted GIS to coastal aquaculture planning, and typical of the other early studies, was in the Gulf of Nicoya, Costa Rica (Kapetsky et al. 1987a, 1987b). The objective of the study was to assess the capability of GIS and satellite remote sensing to rapidly provide synoptic information for aquaculture development. Three kinds of aquaculture opportunities were evaluated in terms of locations and areal expanses: (1) culture of mollusks in intertidal and subtidal areas, suspended culture of mollusks, and cage culture of fishes; (2) extensive culture of fish and shrimp in existing solar salt ponds; and (3) semi-intensive farming of the whiteleg shrimp *Penaeus vannamei* along the gulf shoreline outside of mangroves. Criteria for evaluation included salinity, transportation infrastructure, and water quality for all culture types; bathymetry, shelter from winds and waves, and security for intertidal, subtidal, and suspended cultures; locations of existing salt ponds and availability of shrimp postlarvae for extensive shrimp farming; locations of shrimp farms, soil suitability, distance from freshwater and saltwater, and land uses for semi-intensive shrimp farming. Many uses were made of Landsat Thematic Mapper data (30-m resolution) to support the GIS analyses. These included inventory of salt ponds with the idea that some could be used for mullet culture during the rainy season, the location of shrimp farms to compare actual and predicted site suitability, and identification of mangroves as unsuitable areas. Water quality of inflowing streams was inferred by drainage basin from the relative distributions of land cover types. Land cover types also were used to infer the relative costs of land acquisition and land preparation for conversion to shrimp ponds.

In Malaysia, Hanafi et al. (1991) selected sites for ponds for the giant tiger prawn *Penaeus monodon* and the banana prawn *P. merguiensis*, and for other unspecified mariculture. In Indonesia, Wibowo et al. (1994) inventoried three coastal districts for salt pans, paddy fields, salt marshes, and ponds that could be adapted to, or improved, for penaeid shrimp farming. In eastern Sri Lanka, the National Aquatic Resources Agency developed a GIS for coastal aquaculture and selected pond sites for giant tiger prawn and brine shrimp *Artemia* sp. farming (Meaden 1999). Siting for farming of giant tiger prawn and mud crab *Scylla serrata* in Bangladesh was carried out by Salam and Ross (1999, 2000). The potential for shrimp farming in general, and specifically for the whiteleg shrimp, in the Baia de Sepetiba, Brazil, was assessed by Scott and de N. Vianna (2001).

In other coastal aquaculture, siting of oyster farms in France was evaluated by Durand et al. (1994a, 1994b). Siting of mangrove oyster *Crassostrea rhizophorae* and Mexihao mussel *Perna perna* culture in Brazil was reported by Scott (1998), Scott and Ross (1998), and Scott (2001). In Maine, Congleton et al. (1993) used C++ programming to create a template for temporal and spatial simulation of shellfish culture, and Congleton et al. (1999) dealt broadly with site selection for culture of juvenile soft-shell clam *Mya arenaria*. Parker et al. (1998) combined site characteristics in a GIS with biological data to predict and map growth rates at potential seeding sites for juvenile shellfish, also in Maine. Beveridge et al. (1994) broadly addressed potential for coastal aquaculture, whereas Ross et al. (1993) specifically addressed salmonid cage siting in Scotland.

In comparison with the early study by Kapetsky et al. (1987a) that used noncommercial software (Earth Resources Laboratory Applications Software, National Aeronautics and Space Administration, Stennis Space Center, Mississippi) and very simple models, recent GIS-assisted site selection studies are quite sophisticated. They use many more criteria in a broader range of factors and constraints in order to evaluate suitability of the site for the culture systems and cultured organisms, weighting, and submodels to handle the criteria, and can take into account interactions among several kinds of cultures (e.g., Salam and Ross 1999, 2000; Figure 6.2).

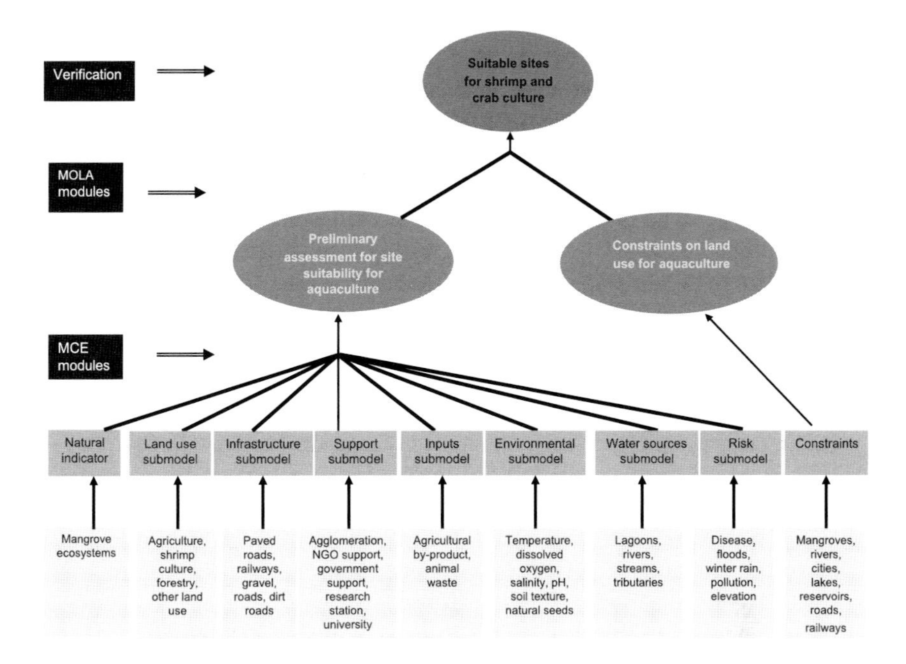

Figure 6.2 A hierarchical modeling scheme with multicriteria evaluation (MCE) and multiobjective land allocation (MOLA) to evaluate suitability of locations for giant tiger prawn and mud crab in southwestern Bangladesh (modified from Manjarrez and Ross 1995). Potential sites for brackish water shrimp and crab farming were based on 35 physical, environmental, and socioeconomic criteria that were organized into submodels and constraints (land and water unavailable for aquaculture).

Several sources relatively rich in data are needed for land-based aquaculture site selection. This is because the same data that are required for inland aquaculture site selection are generated to satisfy many uses in, for example, agriculture, forestry, animal husbandry, hydrology, and demography. This contrasts with the difficulty of characterizing the intertidal environment, particularly currents, in relation to suitability for several types of mariculture siting including shellfish seeding and grow out as described by Congleton et al. (1999). They describe a holistic intertidal GIS that includes bottom types, elevations, currents and eddies, water temperature, and infauna and epifauna population densities that can be displayed and queried for site selection. Additionally, impacts of nearshore land use could be studied and regulatory decisions could be taken.

6.3.3.2 ***Strategic Planning for Aquaculture Development.*** Examples of long-range, large-area planning for aquaculture that use GIS are numerous. These investigations show the potential for aquaculture development at various levels of suitability in areal quantities, sometimes in terms of expected production, and usually according to administrative boundaries rather than by ecosystems. Often, more than one species is the focus and more than one kind of culture system is included. Those that have dealt with coastal aquaculture at a state level include Johore (Kapetsky 1989b) and Sabah (Anonymous 1996) in Malaysia. The former dealt with the potential for giant tiger prawn and banana prawn in ponds, and sea bass *Lates* spp. and grouper *Epinephelus* spp. in floating cages. The latter included giant tiger prawn farming in ponds, fish in floating cages, and sites for culture of seaweed *Eucheuma* spp. Those at the national level include a study of the further development and management of mariculture in the Republic of Korea (Kapetsky and Ataman 1991), and a GIS-based decision support system for the development and management of Atlantic salmon *Salmo salar* farming in eastern Canada (Keizer 1994).

Inland aquaculture potential at the state level has included catfish and crawfish *Procambrus clarki* (also known as red swamp crawfish *Procambarus clarkii*) in Louisiana (Kapetsky et al. 1990). Inland aquaculture planning at the national level has been carried out for culture-based fisheries development in small reservoirs in Zimbabwe (Chimowa and Nugent 1989) for farming of Nile tilapia *Oreochromis niloticus* and North African catfish *Clarias gariepinus* in Ghana (Kapetsky and MacPherson 1990; Kapetsky et al. 1991) and for the culture of unspecified carps in Pakistan (Ali et al. 1991). Models for freshwater aquaculture development in Vietnam were developed by Tu and Demaine (1997).

Assessments of inland aquaculture potential at the continental level were made for Africa by Kapetsky (1994a, 1994b) and Aguilar-Manjarrez and Nath (1998) and for Latin America by Kapetsky and Nath (1997). Regional studies of aquaculture potential have been carried out for southern Africa by Kapetsky (1994c) and for Caribbean inland aquaculture by Kapetsky and Chakalall (1998). The purpose of all these studies was to stimulate planning for aquaculture development at the national level and to provide indicative estimates of national aquaculture potential that were comparable across continents or regions. Estimating inland aquaculture potential at the continental level initially required using relatively few criteria. Four were used to estimate the potential for small-scale fish farming in ponds: water loss, the potential for farm gate sales, soil and terrain suitability for ponds, and availability of agriculture byproducts as feed or fertilizer inputs. A fifth criterion was added in order to estimate the potential for commercial fish farming: urban market potential. These criteria were weighted in different ways on the basis of expert advice in order to develop small-scale and commercial fish farming models. Numbers of crops per year for up to three widely cultured species including the Nile tilapia, the North African catfish, and the common carp *Cyprinus carpio,* as well as two lesser-known indigenous species, were predicted based on monthly climatic variables. By varying feeding levels and harvest sizes, small-scale and commercial-level outputs were simulated (Figure 6.3).

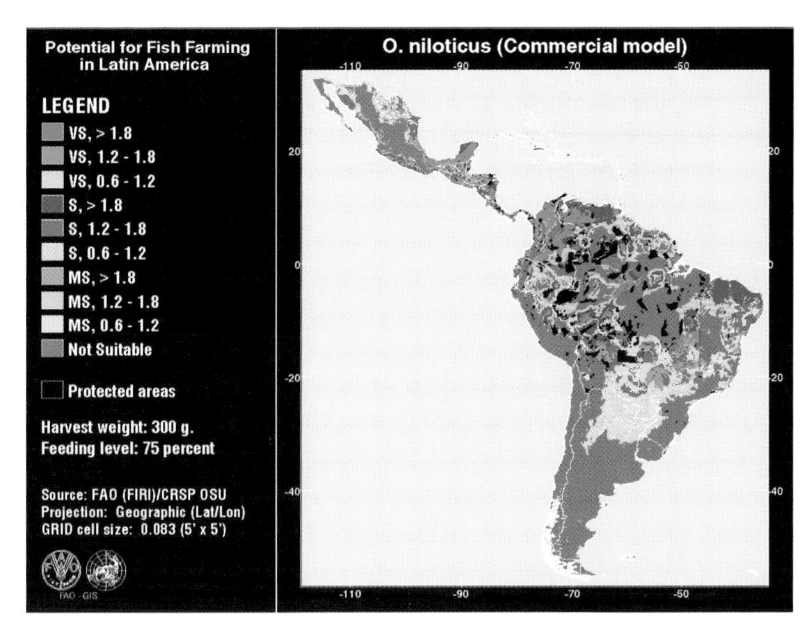

Figure 6.3 The potential for commercial farming of Nile tilapia in Latin America. The legend shows various levels of suitability for culture in ponds based on the criteria listed in the text (VS = very suitable, S = suitable, MS = moderately suitable). The adjacent numbers are the crops per year that can be obtained by feeding at 75% satiation and harvesting the fish at 300 g (from Figure 3.62, Kapetsky and Nath 1997).

Combining the small-scale and commercial models with the simulations of fish production provided overall suitability ratings for each five-arc-minute grid cell (9.3 km × 9.30 km at the equator) in Latin America and the Caribbean and each three-arc-minute grid cell (5.6 km × 5.6 km) in Africa. Potential was reported as the absolute and relative amounts of surface area in each country meeting the criteria with various levels of suitability.

The reassessment of the aquaculture potential in Africa by Aguilar-Manjarrez and Nath (1998) illustrates the rapid evolution in the availability and quality of data and sophistication of approach over a relatively short period of time since the earlier study by Kapetsky (1994a). For example, for the later investigation, resolution was increased significantly and remotely sensed land-cover data and fish growth models were incorporated in the GIS.

Moving to global assessments, Kapetsky (1998) describes a GIS-assisted approach to address fishery enhancements, including aquaculture-based fisheries, which increasingly are being practiced to expand benefits from inland fisheries both in developed and developing countries. This study dealt with the availability of inland water resources and the effects of climate and the availability of inputs on enhancement prospects.

6.3.3.3 ***Anticipating the Consequences of Aquaculture.*** Planning for aquaculture development usually is seen as consisting mainly of determining the suitability of a site, or of a larger areal expanse, for the organism to be cultured and the culture system to be used. However, more and more, part of the predevelopment planning process has to include anticipating the social, economic, and political consequences of aquaculture. Despite the apparent need for such studies, relatively few came to light from among the sources used herein.

In looking at the potential for fish farming and mussel culture in Yaldad Bay, Chile, Kriege and Muslow (1990) were careful to specify that candidate sites should be assessed to locate benthic assemblages on parts of the seafloor with a high potential to metabolize and mineralize labile detritus. Their approach included developing community measures (number of species and density of macroinfauna) that were combined with sediment organic content to generate a "benthos" layer. In turn, this layer was used to infer the short-term ability of various parts of the bay to assimilate, or recycle, organic waste products. Taking this approach further, Pérez et al. (2002) note that validated spatial models are considered the most cost-effective tools for predicting environmental impacts of marine fish cages. They use an improved version of an existing predictive particulate waste dispersion model. Incorporation of GIS into the prediction of cage effects has several advantages for spatial modeling, including fast image generation and manipulation, flexibility to run alternative scenarios, statistical analysis of the image, and color outputs that help visual interpretation of results.

The potential effects of hard clam aquaculture on submerged aquatic vegetation were evaluated in the Chesapeake Bay (Center for Coastal Resources Management 1999). Such effects are ecologically and economically important because the submerged aquatic vegetation is the habitat for fish and crab resources that constitute important fisheries in the bay (Figure 6.4).

Halvorson (1997) described a broad approach to public policy issues in aquaculture development of sea scallop *Plactopecten magellanicus*. The principal issues were sea scallop culture technologies appropriate for Massachusetts; siting criteria, including consideration of user conflicts, potential environmental impacts, regulatory restraints, economic feasibility, public education, and developing a better knowledge base for sea scallop biology and aquaculture technology. Similarly, Chenon et al. (1992) evaluated pearl oyster culture in relation to other water uses in a Pacific island state. There, the extremely fast development of pearl culture activities in atoll lagoons can result in a "gold rush" that leads to numerous conflicts in the lagoon spatial allocations. Resolution of the conflicts through better administrative and technical management of activities was based on remotely sensed data and spatial analyses via GIS.

Finally, the development of the use of native species in aquaculture in relation to concerns over the loss of biodiversity was addressed by Ross and Beveridge (1995). In this regard, the role foreseen for GIS is strategic planning and facility location.

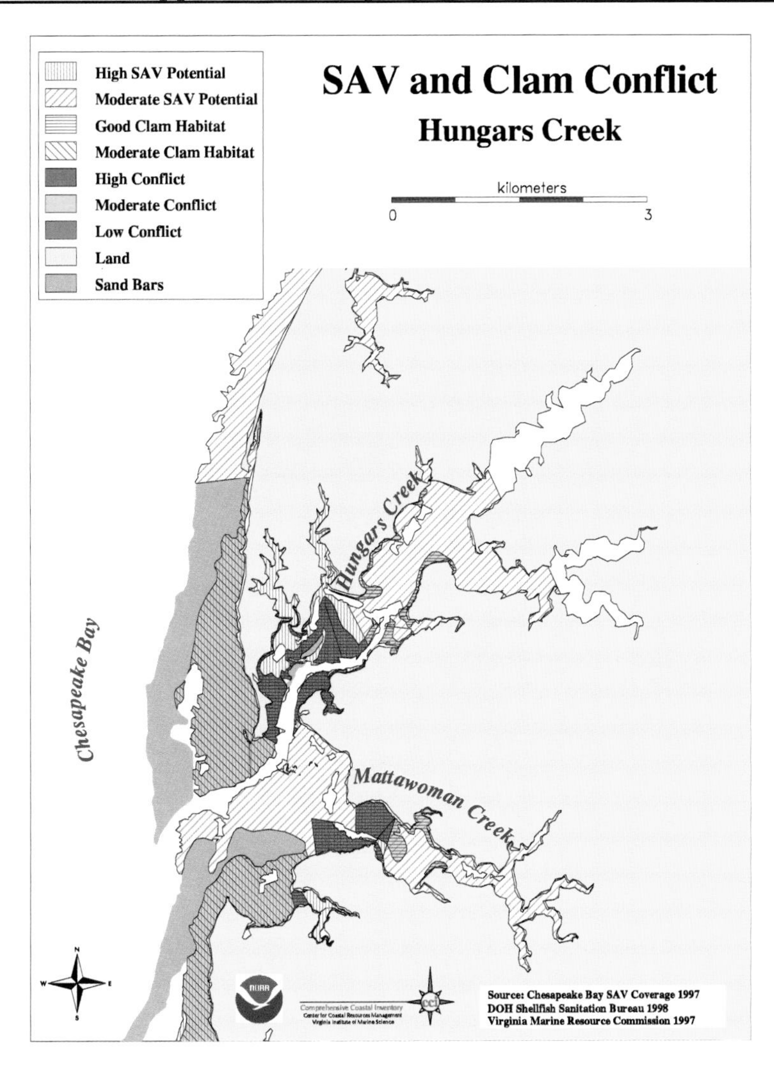

Figure 6.4 Submerged aquatic vegetation (SAV) and clam aquaculture conflicts in Hungars Creek, Virginia (from Center for Coastal Resources Management 1999). Nets and covered trays protect the clams from predators such as crabs and sting rays *Dasyatis* spp. Conflicts occur by their presence, if they kill existing SAV and exclude the growth of SAV into the area on which they are placed.

6.3.3.4 ***Web-Based Aquaculture Information Systems.*** The Aquaculture Geographic Information System (AquaGIS 2003) is a Web-based comprehensive system to collect, manage, and distribute aquaculture information that currently covers Newfoundland, Canada. This GIS-based system integrates data from multiple government departments with the main focus to aid in the development of the aquaculture industry.

The Aquaculture Information Project that led to AquaGIS commenced in April 1997. The goals were to build an information system that would help coordinate administration for aquaculture licensing and industry growth. There are currently over 20 agencies involved with the approval process for an aquaculture license. Each agency maintains its own information about the industry and its growth. Therefore, a system to help maintain and share information was spawned. Since much of the information related to aquaculture is geospatial, the use of GIS became part of the solution. The system was designed to be easy to access, low cost for users, and low cost to maintain, and to provide the most current information maintained by each agency. Its purpose is to serve regional economic, financial, and environmental planning activities, and it is to be used both by the aquaculture industry, as well as by the government. Information can be added so that it can become part of a broader integrated fisheries and coastal zone resource management system. The AquaGIS features interactive reporting, Web-based (HTML and Java) map viewers, automated updates and dynamic linking of various agency databases, and a security model for various levels of system access.

The capabilities of AquaGIS include (1) display of aquaculture sites (Figure 6.5) on marine charts, relative to the adjacent shoreline and other geographic objects such as sewage outfalls, wharves, and navigation routes; (2) display of information specific to an aquaculture site, such as ownership, species, and size; (3) support to the aquaculture licensing process by providing uniform baseline data to all users; (4) a catalog or database tool to assist users in finding information available in the system; (5) reports pertaining to the aquaculture industry such as species farmed, site layout, and other licensing issues; and (6) reports on the state of the industry, such as numbers of species farmed and increases over previous years.

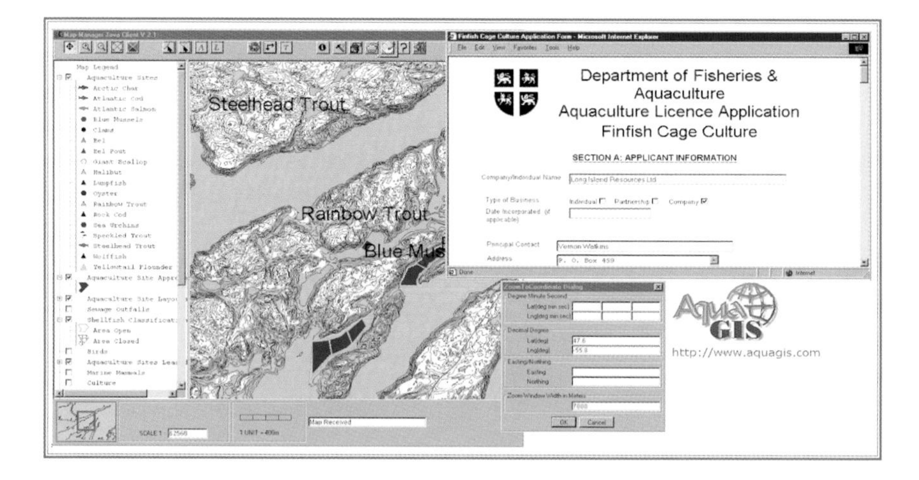

Figure 6.5 AquaGIS Aquaculture Geographical Information System. A few of the capabilities of AquaGIS are illustrated, including the display of aquaculture sites on marine charts relative to the adjacent shoreline and other geographic objects, and information specific to an aquaculture site, such as ownership, species cultured, and size of installation.

The perceived benefits to users include (1) aid in the planning of future aquaculture development by identifying opportunities and constraints for specific geographic areas, (2) a single, easy-to-use information tool that will assist decision making by industry and government, (3) assistance in conflict prevention and resolution among aquafarmers and other resource users, (4) reduction in the licensing approval time for proposed aquaculture sites, and (5) easy access to aquaculture information by interfacing with existing computer systems.

As technologies continue to improve and as the need for improved administration of aquaculture increases, both for the development and management of aquaculture, AquaGIS-like solutions are bound to become more and more common. In this light, it is interesting to know the constraints that were overcome to create the system and to ensure its effective use. They include the following (T. Fradsham, Canadian Coast Guard, personal communication):

- monitoring and controlling the scope of the project so that the common needs of all participants are met first and specialized interests are met second, as time and resources allow;

- integration of new technologies with existing policies and procedures of the departments concerned; training and technical support were required;

- dealing with varying degrees of data quality, accuracy, and completeness: much time was needed to evaluate data sets, identify information that is pertinent to the aquaculture industry, and to connect it to the system;

- cost of system access: if cost of the information was very high, many clients would not use AquaGIS. The AquaGIS was developed as a client-server architecture. Changes are made on the server only, and client changes are automatically reflected, thus keeping technical support costs down; and

- making effective use of the system: training courses were not as effective as demonstrations and one-on-one training due to differing needs and skill levels of the clients. Demonstrations showed the efficiency of the system and individual training allowed it to be used in daily work.

The AquaGIS is currently being transferred to the Newfoundland Department of Fisheries and Aquaculture in order to move it from the developmental to operational phase, where it will continue to be developed under a federal-provincial partnership (M. Warren, Newfoundland Department of Fisheries and Aquaculture, personal communication).

6.3.3.5 ***Marketing.*** Geographic information systems applications addressing the marketing of aquafarmed products appear to be scarce, although numerous Web pages describe applications of GIS to market other products. Aquaculture is a competitive industry and, in order, to be profitable, product markets need to be targeted. The use of census data to conduct an initial assessment was described by Kohler (1996). As a demonstration, she used the live tilapia market in urban areas in the United States.

6.3.4 GIS in Aquaculture Practice and Management

In addition to development planning, culture practice and management are among the other broad domains of GIS applications in aquaculture. Culture practice and management have been considered in several subcategories that include inventory and monitoring of the culture environment, environmental impacts of aquaculture, and restoration of aquaculture habitats.

6.3.4.1 ***Inventory and Monitoring of Aquaculture and the Environment.*** There is only one application of GIS for in-pond management. Based on periodic observations, Porter and Rice (2001) used GIS to generate a surface dissolved oxygen gradient map for each site. The resulting slope map was combined with a pond base map and surrounding land-use map for generating a surface oxygen distribution map for each time interval. These maps showed distinct oxygen gradients and patterns in response to environmental factors throughout the night. Using GIS to integrate aeration and wind data, the water quality manager can quickly evaluate the placement and effectiveness of mechanical aerators and intervene rapidly when required. Computerized data collection can facilitate production of real-time maps for interactive water quality management. By integrating aeration and wind data, the water quality manager can evaluate placement and effectiveness of mechanical aerators.

Many of the studies that apply GIS for inventory and monitoring in support of management pertain to oyster culture along the East and Gulf coasts of the United States. Among these, Jefferson et al. (1991) point out that past efforts to manage fish and shellfish resources have been based primarily on catch and effort statistics, but they show that management decisions can be enhanced by resource management tools that incorporate the spatial distribution and abundance of a resource. They developed data layers to characterize the distribution and size of oyster reefs at Murrells Inlet, South Carolina, and to document spatial recruitment patterns. In addition, the distribution of oyster populations and oyster reef recruitment were related to hydrography and adjacent land-use patterns.

In Atlantic Canada on the east coast of Prince Edward Island, a pilot GIS study was conducted by Legault (1992) to (1) quantify the shellfish habitat affected by the closure of shellfish beds contaminated by coliform bacteria, (2) estimate the market lost or jeopardized as a result of the closures, and (3) identify potential point sources of pollution to determine their relationship to closure zones. The study was aimed at providing timely data to managers and visually representing data for decision makers. In another area suffering from the same problem, Stevens et al. (1997) recount how a nonpartisan association of federal, state, and local government agencies, as well as nonprofit environmental organizations and private agencies evolved as an initiative to unconditionally reopen Greenwich Bay's (Rhode Island) highly productive shellfish beds. The beds had been closed due to dangerously high levels of fecal coliform bacterial pollution. The role of GIS, along with remote sensing, was to monitor the sources of pollution.

Ellis et al. (1993) used hydroacoustics to inventory natural and man-made reefs for eastern oyster *Crassostrea virginica* and GIS to ascertain trends in kinds and areal expanses of the reefs in Galveston Bay, Texas. Similarly, near Charleston, South Carolina, Boyd et al. (1996) compared intertidal oyster surveys from the mid-1980s and 1890–1891 to the most recent oyster survey in 1995. By using GIS, the comparison of the three surveys illustrated the dynamic nature of the resource over an extended period. In the same area, W. D. Anderson and G. M. Yianopoulos (South Carolina Department of Natural Resources, Marine Resources Center, abstract from Aquaculture 2001: Book of Abstracts, 2001) describe how GIS, global positioning systems (GPS), laser rangefinders, and digital photography were used for intertidal oyster resource assessments. The resulting maps show commercial harvest areas and are disseminated to recreational oyster harvesters and made available on the South Carolina Department of Natural Resources Web page.

Smith and Jordan (1993) found that integration of data input and analysis programs with a personal computer–based commercial GIS system showed promise in improving monitoring of oyster disease and population parameters in the Chesapeake Bay. Taking this approach further, Bushek et al. (1998) and White et al. (1998) used a GIS with geostatistic kriging to compare spatial patterns of disease in Dermo *Perkinsus marinus* with patterns of land use, anthropogenic activity, and estuarine hydrography in two similar South Carolina estuaries—one urbanized, one pristine. The results indicated that land-use patterns might affect the distribution of the disease. Along similar lines, and in the same area, Kelsey et al. (2003) used GIS overlays and statistical analysis to evaluate the effects of human encroachment on estuarine surface water quality and oyster health. Their results showed that fecal coliform bacteria pollution was associated with urbanization and that closure of shellfish harvesting waters may be the most significant, quantifiable impact from urbanization.

In Louisiana's Barataria and Terrebonne estuaries, four subtidal oyster resource zones were identified in relation to salinity and survival in dry and wet years (Melancon et al. 1994, 1997). Louisiana's wetland loss is being mitigated by diverting relatively large quantities of freshwater from the Mississippi River into the estuaries in an effort to restore historical salinity patterns and to enhance coastal restoration. In effect, the diversions change the spatial distribution of wet and dry years and, thereby, shift oyster production. State and federal resource managers are using the GIS as an environmental assessment tool for oyster beds that will be impacted. Related to the same activity, C. A. Wilson and H. H. Roberts (Coastal Fisheries Institute, Louisiana State University, abstract from Aquaculture 2001: Book of Abstracts, 2001) have used digital, high-resolution, acoustic instruments to build a baseline of information on oyster habitats. Side-scan sonar and a sub-bottom profiler were used to characterize habitats. Verification showed that numerically indexed acoustic reflectance intensities correlated closely with surface shell and oyster reef density.

Smith et al. (1994) integrated hydroacoustical data and GIS to test the assumption that severe sedimentation has been a critical factor in making large expanses of the Maryland portion of Chesapeake Bay nonproductive for oyster harvest. In a related study, Smith and Greenhawk (1996) conducted a three-dimensional analysis by using GIS to compare oyster habitat loss from the turn of the century until the mid 1970s due to sedimentation. In a follow-up study, Smith et al. (1997) documented sediment accumulation over historic oyster bars. There was an apparent correspondence of presently buried basement reef to historic, turn-of-the-century charted oyster bars. The effects of sedimentation on oyster shell plantings for rehabilitation purposes also were demonstrated.

In order to support interest in expansion of the oyster industry in Mobile Bay, Alabama, a decision support system was developed by Rodgers and Rouse (1997) using GIS. The system simulates the effects of management and culture practices on the oyster resources of the bay. It incorporates spatial as well as qualitative effects of management alternatives, so that the manager will be better able to predict where various alternative management practices should be attempted.

In France, Populus et al. (1997) and Loubersac et al. (1997) described a GIS to address a number of administrative and environmental management problems associated with the growing area of the valuable Pacific oyster *Crassostrea gigas*, the Bassin de Marennes-Oléron. One activity was to bring 22,000 oyster leases and their attributes into the GIS in relation to their management groups. Another was to assess sedimentation in relation to the use of off-bottom structures upon which oysters are grown to decide on dredging operations that would lead to better flushing of natural sediments and animal wastes. Yet another activity was to use GIS to establish a relationship between environmental parameters and oyster mortality in summer (Goulletquer et al. 1998). According to Le Moine et al. (1998), these preliminary experiments demonstrated the technical feasibility of simultaneously monitoring environmental parameters, depth, and GPS positions, and of visualizing all the overlaid information by using a mapping system. But these experiments also showed the need for more appropriate sampling strategies, considering the highly variable characteristics of the environment within an estuarine ecosystem. In the same bay area, Soletchnik et al. (1999) used GIS to compare growth, sexual maturation, survival rates, and environmental variables at 15 sites of on-bottom and off-bottom oyster culture.

Stolyarenko (1995) described an approach to shellfish surveying using spatial regression that permits rapid data analysis and real-time mapping aboard a research vessel. The method allows adaptive sampling by using currently available information during a survey as a means of improving the sampling scheme.

In anticipating the environmental changes that affect coastal aquaculture, Loubersac et al. (1999) assessed the quality of water from coastal rivers by combining numerical water quality models with a GIS to analyze spatial relationships of the areas impacted.

Looking much more broadly at GIS for shellfish culture, the Shellfish Information Management System (SIMS) is a cross-boundary, multi-administration information system that is the management counterpart of AquaGIS. The SIMS is currently under development and is described in some detail by Kracker (1999). Several agencies, including the U.S. National Ocean Service Center for Coastal Environmental Health and Biomolecular Research in cooperation with the states of New Jersey, South Carolina, Florida, and Washington, as well as the U.S. Food and Drug Administration (FDA), U.S. Environmental Protection Agency (EPA), and the Interstate Shellfish Sanitation Conference, are developing the SIMS. The SIMS is intended to (1) provide a system for input and update of shellfish growing area monitoring, survey, and classification data; (2) aid the states in quickly producing growing area classification maps from the data; and (3) establish a database of current classifications and other shellfish information nationwide useful to state, federal, and industry users. Beyond uses at the state level, SIMS information and GIS functionality should prove useful to various coastal monitoring and assessment elements of the National Oceanic and Atmospheric Administration (NOAA). The SIMS may serve as an information resource to the FDA on shellfish safety program issues and actions. Further, data from SIMS may be overlaid with other EPA and state water quality monitoring and land-use management data to enhance knowledge of coastal water measurements and perhaps streamline some monitoring efforts.

6.3.4.2 ***Environmental Impacts of Aquaculture.*** There are few GIS applications that deal with the environmental impacts of aquaculture. Aquaculture was introduced to the coastal areas of Taiwan in the early 1970s as an alternative to traditional agricultural production. Shrimp (family Penaeidae), eels (family Anguilidae), tilapias *Oreochromis* spp., carp (family Cyprinidae), and milkfish *Chanos chanos* are raised in brackish water and freshwater ponds. The impacts of aquaculture have included subsidence due to the use of groundwater. Other environmental problems resulting from the expansion of aquaculture include coastal plain erosion and groundwater pollution. A GIS and spatial statistics were used by Tsai et al. (1997) and Tsai et al. (2001) to describe and predict the spatial dynamics and temporal characteristics of aquaculture in two areas that accounted for about a half of aquaculture production in Taiwan. As part of a broad study to assess the impact of tropical shrimp culture on the environment, mainly in Indonesia and Vietnam, Fuchs et al. (1998) used remote sensing and geographic databases to correlate ecological indicators with average production of shrimp farms.

Hassen and Prou (2001) used GIS to predict concentrations of inorganic nitrogen and phosphorus loading of salt marshes from fish and shellfish pond outfalls along the mid-Atlantic French coast. The fish and oysters under intensive culture were, respectively, European sea bass *Dicentrarchus labrax* (also known as European bass *Morone labrax*) and Pacific oyster, and under extensive culture were European eel *Anguilla anguilla*, mullet *Mugil* spp. and gilthead bream *Sparus aurata*.

6.3.4.3 ***Restoration of Aquaculture Habitats.*** All of the GIS applications dealing with restoration pertain to oysters. Jordan et al. (1995) reported that the GIS developed to manage and interpret data from Maryland's Chesapeake Bay oyster monitoring and management programs (see work of Smith and collaborators cited above) proved especially useful in supporting the information needs of the state's Oyster Recovery Action Plan. For example, the GIS has provided managers, scientists, and policy makers with clear, graphical portrayals of oyster habitat, population and disease status, salinity gradients, and management history, with a minimum of effort. As new experimental management efforts develop, the GIS is being used to maintain a standard, geographically precise database for documenting and tracking the performance of the management efforts.

White et al. (1999, 2000) used a GIS-based holistic approach to study bacterial loading that caused oyster bed closures adjacent to a small watershed in North Carolina. They combined in-column water quality, storm-water quality, storm and stream flow measures, dye studies, and historical aerial photography to detect hydrologic modification. They also made door-to-door surveys to determine the status of septic systems, the numbers of pets, and the presence of wildlife. The ultimate aim was to assess techniques that can be used to mitigate the effects of land-use change on shellfish closures. Finally, Grizzle and Castagna (2000) studied the distribution and abundance of natural, intertidal oyster reefs at various spatial scales on Florida's east coast to gain insight on the design of constructed and restored reefs. Water movements were found to be of major importance to reef development and maintenance at all spatial scales.

6.3.5 Multisectoral Planning that Includes Aquaculture

Future development of aquaculture will follow a much smoother path if the requirements and consequences of its development become more widely known and, in particular, if aquaculture is integrated into broader development and management plans.

6.3.5.1 ***Management of Aquaculture Together with Fisheries.*** There are a number of ways that aquaculture and fisheries interact. Perhaps the most obvious interactions are competition for space and markets. An example of a resolution of spatial conflicts between a fishery and shellfish culture comes from the Indian River lagoon on the eastern coast of central Florida where hard clams *Mercenaria* spp. support an important commercial fishery (Arnold et al. 1996). In response to the extreme variability in landings and income that characterizes the hard clam fishery in Florida, many local residents have become involved in the culture of hard clams. As the industry continues to develop and expand, conflicts have arisen. Original conflicts focused on the allocation of submerged bottomland between aquaculture operations and fishing for naturally occurring clam populations. However, those conflicts have expanded to involve upland property owners, environmentalists, and natural resource managers. Further, the geographic scope of conflict has expanded concomitant with the expansion of hard clam aquaculture to other regions of Florida. Arnold and Norris (1998) and Arnold et al. (2000) have developed refined GIS technology for conflict resolution in the hard clam aquaculture industry. Site-specific information on water depth, water quality, upland land use, site accessibility, transportation corridors, distribution and abundance of clams and submerged aquatic vegetation, and other parameters are used to determine the suitability of a site for hard clam culture. Geographic information systems have been used to allocate hard clam leases at several sites by using a pro-active approach. This approach maximizes the probability that the lease-site applicant will be successful in obtaining a lease and in realizing an economically successful operation, while complaints from competing user groups are contemporaneously resolved.

On a broader scale, Rubec et al. (1998) described the Florida Marine Resources Geographic Information System that has been created for research and management of fisheries and aquaculture coastwide. It relates fishery and aquaculture species to habitat and anthropogenic effects on ecosystems.

Although it seems logical to implement GIS for fisheries and aquaculture together because of common approaches and ample opportunities to share data and analyses (e.g., Kapetsky 2001), in practice it is difficult to do so because of divergent goals of individuals and the technical units involved. One approach showed where GIS results were beneficial across the administrative and scientific disciplines for fisheries management and aquaculture development in the Lower Mekong Basin (Kapetsky 1999). These included dynamic inventory of water bodies and aquatic habitats and the environmental health of subbasins.

6.3.5.2 ***Planning for Aquaculture among Other Uses of Land and Water.*** The basic issue in aquaculture is its sustainability. As has been shown above, it is necessary to anticipate the environment for aquaculture before it is implemented. After aquaculture is developed, it is necessary to cope with environmental change in order to ensure sustainability. Anticipating and mitigating environmental change can be facilitated if planning for aquaculture is integrated with planning in other sectors.

In planning for aquaculture, it is necessary to keep in mind that, in much of the world, aquaculture is extensive and is carried out as an activity supplemental to farming. Even at the commercial level, aquaculture may be side by side with agriculture under the same ownership and management. Therefore, there is ample justification for land-use planning for aquaculture development in the context of general planning for agriculture. There is an obvious role for GIS in such planning (Gordon and Kapetsky 1991) and opportunity for cooperation on the use of data layers of mutual interest. Planning for further coastal aquaculture development while excluding major agricultural land use and mangrove reforestation areas has been reported by Paw et al. (1992, 1994) in the Lingayen Gulf Area, Philippines, and Chanthaburi Province, Thailand, by Tookwinas and Leeruksakiat (1999).

In contrast to Asia where, in some places, coastal aquaculture may be the main activity, increasing urbanization of coastal areas in the United States is an important concern. This is a particularly acute problem in areas that are dependent on tourism, as well as on traditional fishing and shellfish harvesting activities. Porter et al. (1996) describe a six-year study of the impacts of urbanization on small, high-salinity estuaries in South Carolina. The project used geographic information processing to integrate data and scientific expertise for the identification, assessment, and modeling of relationships within coastal estuaries and impacts associated with urbanization and natural activities. This goal was achieved through the continued development and use of the multi-agency Coastal Geographic Information System.

In Sinaloa State, Mexico, models for land-based aquaculture were developed by Aguilar-Manjarrez and Ross (1995b). Their study is a good example, not only of planning for aquaculture in itself, but also for the consideration of agriculture as another possible production activity competing for the same land and water. Aguilar-Manjarrez (1996), in the same area, dealt with forestry, livestock rearing, urban development, agriculture, and aquaculture as competing land uses.

6.4 FUTURE DIRECTIONS OF GIS APPLICATIONS IN AQUACULTURE

It seems certain that aquaculture will continue its current upward development trend for some time to come; however, future development, like current practice, is likely to be patchy. Numbers and kinds of GIS applications in aquaculture are on an increasing trend but, considering the role that GIS might play in the development and management of aquaculture, they are remarkably few. Consequently, a basic question is, "What can be done to further GIS applications in aquaculture?" It is assumed that the use of GIS in aquaculture planning and management can significantly contribute to the sustainability of aquaculture while at the same time play an important role in providing the information needed to minimize the ecological, social, and economic outcomes that are negative. Along with this idea is the proviso that the results of the GIS analyses must be objectively interpreted and disseminated. In this light, future directions of GIS applications in aquaculture are addressed from the perspective of unrealized opportunities. These include promotion and training, opportunities for cross-administrative, multidisciplinary GIS applications, and gaps that GIS could fill in some environments and in many geographical areas. Finally, some constraints imposed by data quality and quantity are considered.

6.4.1 Promotion and Training

Increasing the use of GIS in aquaculture has to begin with promotion and training. Many professionals at working technical levels already have some concept of GIS and its applications. But one of the weak links in the promotion of the wider use of GIS in aquaculture (and fisheries) is that mid- to higher-level management personnel in government and academic institutions are not aware of the benefits. Awareness raising and training at that level are necessary in order to ensure that GIS is effectively used, whether by an agency's GIS unit, through cooperation with other agencies, or by contracting to the private sector. It is up to the technical people to communicate to their supervisors the potential benefits of GIS and the need for familiarization at the managerial level and for applications training at technical levels. Going along with these activities are realistic appraisals of implementation costs, personnel requirements, and other commitments for various alternatives of GIS applications.

6.4.2 Expansion of GIS Applications across Administrative Boundaries

Reduction of risk, or said differently, improved reliability in the prediction of expected outcomes of aquaculture development, is a fundamental task in which GIS can play a much more prominent role than at present. The broader the scope of such studies, extending as far as the data permit into present and future competing uses of land and water, and to social, economic, administrative, and political consequences of development, the better the reliability of the results is likely to be.

It follows that the broader the scope, the more disciplines that have to be involved. Thus, interdisciplinary, cooperative studies that cross administrative and political boundaries at all levels are a way to achieve a comprehensive and widely usable body of information on aquaculture potential and potential problems. One example of such a system in aquaculture that already is operational is AquaGIS. Although the system presently caters to government entities and benefits the commercial side of aquaculture indirectly by facilitating and accelerating the licensing process (T. Fradsham, Canadian Coast Guard, personal communication), it does have the capacity for expansion in several directions that could take it more directly into the commercial side of the aquaculture industry and to a larger geographic area.

It is the smallest administrative unit that may be most in need of reliable, objective advice on the consequences of aquaculture development, and in the worst position to acquire it, especially in developing countries. Present applications tend to be strategic assessments at relatively coarse resolutions, or small-area studies at fine resolutions. However, because of the ready availability of high-resolution data and increasing computing power to deal with it, it is increasingly within the capability of GIS to accommodate the small areas, while at the same time serving the information needs of the larger areas and administrations.

6.4.3 Other Opportunities for the Expansion of the Use of GIS in Aquaculture

As this survey shows, there are few applications of GIS for management and development of inland culture-based fisheries and there were no examples of GIS for sea ranching. A common weakness of both of these kinds of aquaculture-based fisheries is a lack of information by which to evaluate the performance of stocking activities (FAO Inland Water Resources and Aquaculture Service 1999a, 1999b). Analyzing biological characteristics, quantities, and locations of recoveries in relation to the environment; optimizing hatchery locations and stocking sites in relation to a favorable environment for grow out or in relation to access by recreational fishermen; and siting artificial structures for attraction or to facilitate capture are but a few of the potential GIS applications.

Although the environmental impact of aquaculture is a prominent issue, mainly in developed countries, I found surprisingly few GIS applications in this important aspect of aquaculture sustainability. Visualization of the aesthetic impact of aquaculture is but one of the many concerns for which no examples were found.

The survey also shows that there has been almost no attention to the use of GIS for assessing the markets for aquacultured products. Surrogates for potential farm gate sales and urban market demand are included in several strategic assessments of inland fish farming potential, as is the transportation of fish to urban markets over road systems (Kapetsky et al. 1991; Kapetsky and Nath 1997; Aguilar-Majarrez and Nath 1998), but there was only one study dedicated to marketing and none on the distribution of cultured fish.

The GIS applications in aquaculture are unevenly distributed among countries and continents. As pointed out in Section 6.1, most of the world's aquaculture activity is in Asia, but only about 16% of the GIS applications cover that region. When it is considered that some of the applications are conceived and carried out away from the target countries, regions, and continents for which they are intended, except for verification, the data are even more skewed. This polarization is not entirely along the lines of developed versus developing countries. Although the largest number of GIS applications in aquaculture is in the United States (42), only 13 of 50 states are represented!

Thus, it appears there is ample opportunity to expand the geographic range of GIS applications in aquaculture.

Regarding environments, offshore applications of GIS in aquaculture are in short supply. One reason is that the commercial companies that are involved in offshore aquaculture development consider their GIS-assisted development approaches as proprietary, so they are not found in the literature. However, as part of the NOAA National Marine Aquaculture Initiative for Fiscal Year 2000, requests for five proposals for research, developmental, and programmatic activities were advertised, and one was titled, "Demonstration of the use of geographic information system-based use-mapping of federal and/or state waters useful to the potential siting of marine aquaculture projects."

Another problem for development offshore is lack of data with the spatial and temporal resolution to be effective for siting. In this regard, the Ocean Planning Information System (OPIS; OPIS 2003) may prove to be helpful. The OPIS is a prototype online, regional, marine GIS covering the landward coastal zone boundaries of North Carolina, South Carolina, Georgia, Florida, out to the U.S. Exclusive Economic Zone. The project "provides marine resource managers with timely and convenient access to downloadable data, cutting-edge online mapping functionality, and guidance on how to use geographic information systems (GIS) in a meaningful way with respect to ocean management." Another seemingly rich source of environmental data for coastal and offshore aquaculture in North America is that already assembled, or being compiled, for determinations of essential fish habitat by the National Marine Fisheries Service (Benaka 1999).

In studies of GIS applications in aquaculture, data availability is most likely to be perceived as more limiting than any other factor. For example, in looking at the role of agrometeorological data in inland fisheries and aquaculture from a GIS perspective, Kapetsky (2000) concluded that the fundamental requirements for aquaculture are climate data sets that provide measures of central tendency and of extremes for the prediction of future conditions at a relatively high resolution of 1 km^2 for strategic studies. For these kinds of data, the extremes are as important as the means in order to predict the economic consequences of the "worst case" and "best case" production outcomes and to simulate tests of the durability of the culture systems. Continuous data sets with which to simulate the effects of cyclic phenomena also would be useful. But, not only do these data have to be available, they have to be GIS-ready, and this seems too costly at present.

Continuing on the theme of data, the increasingly important role of data capture technologies in support of GIS analyses has to be emphasized both for modeling and field verification. In the realm of remote sensing, satellite and airborne sensors as sources of data for use in GIS for aquaculture increasingly are becoming more varied, useful, and affordable. There has been an evolution from Landsat MS, with 80-m resolution (e.g., Mooneyhan 1985), to Landsat Thematic Mapper, with 30-m resolution and more bands (e.g., Kapetsky et al. 1987a), to Satellite Pour l'Observation de la Terre, with choices of resolutions and frequencies of coverage (e.g., Populus and Lantieri 1991). Meanwhile, 1-km resolution NOAA weather satellite data were adapted to estimates of land cover that made an important contribution to the strategic assessment of aquaculture potential by Aguilar-Manjarrez and Nath (1998). At present, synthetic aperture radar shows promise for certain inventory and monitoring in aquaculture, such as shrimp farming (Travaglia et al. 1999). Also, in the realm of remote sensing, digital acoustic technologies are being adapted for surveys and monitoring (e.g., Wilson and Roberts, abstract from Aquaculture 2001: Book of Abstracts, 2001), and Anderson and Yianopoulos (abstract from Aquaculture 2001: Book of Abstracts, 2001) used laser rangefinders in support of oyster resource assessments.

Regarding location, GPS has become the indispensable tool of a host of GIS-based studies, and its importance in efficiently and precisely locating data for use in GIS cannot be overstated; however, its use in GIS-assisted studies of aquaculture is relatively recent. For example, Aguilar-Manjarrez (1996) pioneered the use of GPS to verify models for the planning and management of coastal aquaculture in Mexico. As predicted by Nath et al. (2000), mobile data collection devices will only increase in their capabilities and user friendliness.

Finally, lack of good quality data may be perceived as limiting when it is not. One aspect of this issue is the extent to which investigators are willing to use surrogate data. Another aspect is that, by and large, GIS practitioners are imaginative in identifying new applications for GIS in aquaculture. Imagination, together with the available technology, tends to outrun the data that are available to address the new and innovative tasks.

6.5 CONCLUSIONS

There is an important, but still largely unrealized, role for GIS in the development and management of aquaculture, but there is a trend for an increase in such applications. Geographical gaps are evident in the patchiness of applications within large countries and among countries, regions, and continents. The dearth of applications is particularly noticeable in relation to the areas where the value and production from aquaculture are high.

Among the various categories of applications, the use of GIS for site selection, for strategic assessments of inland aquaculture potential, and for the inventory and management of some kinds of cultures is relatively well developed. One exception is that there are few examples of incorporating growth and production models in site selection and in strategic assessments of potential. In contrast, there are unrealized opportunities for the use of GIS in several kinds of applications in: monitoring within culture systems, offshore environments, anticipating the consequences of aquaculture, estimating the environmental impacts of aquaculture, culture-based fisheries and sea ranching, socioeconomics, the distribution and marketing of aquacultured products, and multiadministration, multidisciplinary planning and management. Web-based aquaculture information systems appear to be useful tools to spur development, to improve management and to promote the use of GIS in aquaculture.

Even though quantity and quality of data are an impediment to GIS applications in aquaculture, the situation continues to improve. Use of GIS in aquaculture can expand, if technical personnel take it upon themselves to ensure that mid- and upper-level managers are made aware of the potential benefits of GIS. In this regard, it is necessary to draw attention to those aspects of GIS that, according to M. Beveridge (University of Stirling, personal communication) are the key to sustainable aquaculture: engendering debate and identifying issues that, prior to GIS were addressed in a much less defined manner.

6.6 REFERENCES

Aguilar-Manjarrez, J. 1996. Development and evaluation of GIS-based models for planning and management of coastal aquaculture: a case study in Sinaloa, Mexico. Doctoral dissertation. Institute of Aquaculture, University of Stirling, Stirling, Scotland.

Aguilar-Manjarrez, J., and S. S. Nath. 1998. A strategic reassessment of fish farming potential in Africa. FAO-CIFA, Technical Paper 32, Rome.

Aguilar-Manjarrez, J., and L. G. Ross. 1993. Aquaculture development and GIS. Mapping Awareness and GIS Europe 7(4):49–52.

Aguilar-Manjarrez, J., and L. G. Ross. 1995a. GIS enhances aquaculture development. GIS World 8(3):52–56.

Aguilar-Manjarrez, J., and L. G. Ross. 1995b. Geographical information systems (GIS): environmental models for aquaculture development in Sinaloa State, Mexico. Aquaculture International 3:103–115.

Ali, C. Q, L. G. Ross, and M. C. M Beveridge. 1991. Microcomputer spreadsheets for the implementation of geographic information systems in aquaculture: a case study on carp in Pakistan. Aquaculture 92:199–205.

Anonymous. 1996. Masterplan for aquaculture development in Sabah. Department of Fisheries, Sabah, and Network of Aquaculture Centres in Asia-Pacific (NACA), Final Report MAL/93/013, part 1: main report, part 2: Sabah aquaculture maps, Bangkok, Thailand.

AquaGIS. 2003. Newfoundland and Labrador aquaculture information and mapping. Government of Newfoundland and Labrador Fisheries and Aquaculture. Available: *http://gis.gov.nf.ca/aquagis/* (September 2003).

Arnold, W. S., and H. A. Norris. 1998. Integrated resource management using GIS: shellfish aquaculture in Florida. Journal of Shellfish Research 17:318.

Arnold, W. S., H. A Norris, and M. E. Berrigan. 1996. Lease site considerations for hard clam aquaculture in Florida. Journal of Shellfish Research 15:478–479.

Arnold, W. S., M. W. White, H. A. Norris, and M. E. Berrigan. 2000. Hard clam (*Mercenaria* spp.) aquaculture in Florida, USA: geographic information system applications to lease site selection. Aquaculture Engineering 23(1–3):203–231.

Benaka, L. R., editor. 1999. Fish habitat: essential fish habitat and rehabilitation. American Fisheries Society, Symposium 22, Bethesda, Maryland.
Beveridge, M. C. M., L. G. Ross, and A. Q. M. Mendoza. 1994. Geographic information Systems (GIS) for coastal aquaculture site selection and planning. Pages 26–47 *in* K. Koop, editor. Ecology of marine aquaculture. International Foundation for Science, Stockholm, Sweden.
Boyd, J. G., W. D. Anderson, and G. M Yianopoulos. 1996. Using real time data with a PC-based GIS for shellfish management. Journal of Shellfish Research 15:521.
Bushek, D., D. White, D. Porter, and D. Edwards. 1998. Land-use patterns, hydrodynamics and the spatial pattern of Dermo disease in two South Carolina estuaries. Journal of Shellfish Research 17:320.
Center for Coastal Resources Management. 1999. Shallow water resource use conflicts: clam aquaculture and submerged aquatic vegetation. Center for Coastal Resources Management, Virginia Institute of Marine Science, Gloucester Point.
Chenon, F., H. Varet, L. Loubersac, S. Grand, and A. Hauti. 1992. SIGMA POE RAVA, a GIS of the fisheries and aquaculture territorial department. A tool for a better monitoring of public marine ownerships and pearl oyster culture. Pages 561–570 *in* W. Bour and L. Loubersac, editors. Remote sensing and insular environments in the Pacfic: integrated approaches. Pix 'Iles 90, Noumea, New Caledonia, France.
Chimowa, M., and C. Nugent. 1989. An initial analysis of the numbers, distribution and size of Zimbabwe's small dams. FAO/UNDP support for rural aquaculture extension project. FAO Technical Report FAO/UNDP/ZIM/88/02.
Congleton, W. R., Jr., B. F. Beal, B. R Pearce, and R. C. Bayer. 1993. Application of object oriented programming and remote sensing to assess intertidal sites for aquaculture. Bulletin of the Aquaculture Association of Canada 93(4):115–116.
Congleton, W. R., Jr., B. R. Pearce, M. R. Parker, and B. F. Beal. 1999. Mariculture siting: a GIS description of intertidal areas. Ecological Modelling 116:63–75.
Durand, H., B. Guillaumont, and S. Labbe. 1994a. Maquette d'un SIG littoral en vue de la recherche de sites ostreicoles en eau profonde. Gutlar, French Research Institute for Exploitation of the Sea, Groupe Sillage, Brest, France.
Durand, H. B., B. Guillaumont, R. Loarer, L. Loubersac, J. Prou, and M. Heral. 1994b. An example of GIS potentiality for coastal zone management: preselection of submerged oyster culture areas near Marennes Oléron (France). European Association of Remote Sensing Laboratories workshop on remote sensing and GIS for coastal zone management. European Association of Remote Sensing Laboratories, Delft, The Netherlands.
Ellis, M. S., J. Song, and E. N. Powell. 1993. Status and trends analysis of oyster reef habitat in Galveston Bay, Texas. Journal of Shellfish Research 12:154.
FAO (Food and Agriculture Organization of the United Nations). 1989. Report of the FAO Asian Region workshop on geographical information systems applications in aquaculture. FAO, Fisheries Report 414, Rome.
FAO (Food and Agriculture Organization of the United Nations). 1991. Report of the FAO Africa Region workshop on applications of geographical information systems and remote sensing in aquaculture and inland fisheries in cooperation with UNEP/GRID. FAO, Fisheries Report 451, Rome.

FAO (Food and Agriculture Organization of the United Nations). 1992. Workshop on applications of geographical information systems (GIS) and remote sensing applied to aquaculture and inland fisheries in Africa. FAO, Fisheries Report 466, Rome.

FAO Fisheries Department. 2001. Aquaculture and inland fisheries: fact sheet. FAO Aquaculture Newsletter 27:18–27.

FAO Fisheries Department. 2003. FAO fisheries data, information and statistics service. FAO. Available: *www.fao.org/fi/statist/statist.asp* (September 2003).

FAO Inland Water Resources and Aquaculture Service. 1999a. Review of the state of world fishery resources: inland fisheries. FAO Fisheries Circular 942.

FAO Inland Water Resources and Aquaculture Service.1999b. Global characterization of inland fishery enhancements and associated environmental impacts. FAO Fisheries Circular 945.

Fisher, W. L., and C. S. Toepfer. 1998. Recent trends in geographic information systems education and fisheries research applications in U.S. universities. Fisheries 23(5):10–13.

Fuchs, J., J. L. M. Martin, and J. Populus. 1998. Assessment of tropical shrimp aquaculture impact on the environment in tropical countries, using hydrobiology, ecology and remote sensing as helping tools for diagnosis. French Research Institute for Exploitation of the Sea, Final report of European Contract TS3-CT 94-00284, Issy-les-Moulineaux, France.

Gordon, C., and J. M. Kapetsky. 1991. Land use planning for aquaculture: a West African case study. Pages 109–121 *in* FAO world soil resources report 68. FAO, Rome.

Goulletquer, P., P. Soletchnik, O. Le Moine, D. Razet, P. Geairon, N. Faury, and S. Taillade. 1998. Summer mortality of the Pacific cupped oyster *Crassostrea gigas* in the Bay of Marennes-Oleron (France). International Council for the Exploration of the Sea, Theme session on population biology, Copenhagen.

Grizzle, R., and M. Castagna. 2000. Natural intertidal oyster reefs in Florida: can they teach us anything about constructed/restored reefs? Journal of Shellfish Research 19:609.

Halvorson, H. O. 1997. Addressing public policy issues on scallop aquaculture in Massachusetts. Journal of Shellfish Research 16:287.

Hanafi, H. H., I. A. Hassan, M. M. Y. Nafiah, and L. D. Rajamanickam. 1991. Present status of aquaculture practices and potential areas for their development in South Johore, Malaysia. Pages 65–73 *in* L. M. Chou, T. E. Chua, P. E. Lim, P.K. Paw, G. T. Silvestre, M. J. Valencia, A. T. White, and P. K. Wong, editors. Towards an integrated mangement of tropical coastal resources. National University of Singapore, National Science and Technology Board, Singapore, and International Centre for Living Aquatic Resources Management, ICLARM, Conference Proceedings 22, Manila.

Hassen, M. B., and J. Prou. 2001. A GIS assessment of aquacultural nonpoint source pollution in the Fier d'Ars Bay. Ecological Applications 11:800–814.

Institute of Aquaculture. 2003. Geographical information systems and applied physiology. Institute of Aquaculture, University of Stirling. Available: *www.aqua.stir.ac.uk/GISAP/index.htm* (September 2003).

Jefferson, W. H., W. K. Michener, D. A. Karinshak, W. Anderson, and D. E. Porter. 1991. Developing data layers for estuarine resource management. Pages 331–342 *in* Proceedings of GIS/LIS '91, volume 1. Inforum, Atlanta.

Jordan, S. J., K. Greenhawk, and G. F. Smith. 1995. Maryland oyster geographical information system: management and scientific applications. Journal of Shellfish Research 14:269.

Kapetsky, J. M. 1989a. Base-line environmental information, remote sensing and geographical information systems for mariculture development. Pages 225–264 *in* Report of the 12th UN/FAO international training course in cooperation with the government of Italy. Contribution of remote sensing to marine fisheries. FAO, Remote Sensing Center Series 49, Rome.

Kapetsky, J. M. 1989b. A geographical information system for aquaculture development in Johor State. FAO Technical Cooperation Program Project, Land and Water Use Planning for Aquaculture Development, Field Document TCP/MAL/6754, Rome.

Kapetsky, J. M. 1991. Remote sensing and GIS for decision-making in fisheries and aquaculture: applications in Africa. Economic Commission for Africa, Report of the regional workshop on remote sensing and geographic information sytems for African decision-makers, ECA/NRD/CRSU/RWRSGIS/6, Addis Abba, Ethiopia.

Kapetsky, J. M. 1993a. Geographical information systems in aquaculture. FAO Aquaculture Newsletter 3:11–14.

Kapetsky, J. M. 1993b. Aquaculture and geographical information systems. INFOFISH International 1993(4):40–43.

Kapetsky, J. M. 1993c. Geographical information systems for aquaculture in Africa. Aquaculture for local community development program, Harare, Zimbabwe. ALCOM News 9:4–5.

Kapetsky, J. M. 1994a. A strategic assessment of warm-water fish farming potential in Africa. FAO-CIFA Technical Paper 27, Rome.

Kapetsky, J. M. 1994b. The potential for warm water fish farming in Africa: a strategic assessment from a continental viewpoint. Pages 67–73 *in* P. Kestemont, J. Muir, F. Sevila, and P. Wilmot, editors. Measures for success. Meteorology and instrumentation in aquaculture management. Institut de recherche pour l'ingénierie de l'agriculture et de l'environnement Editions, Bordeaux, France.

Kapetsky, J. M. 1994c. The potential for warm water fish farming in the SADC countries. FAO Aquaculture Newsletter 6:2–6.

Kapetsky, J. M. 1995. A first look at the potential contribution of warm water fish farming to food security in Africa. Pages 547–572 *in* J.-J. Symoens and J. C. Micha, editors. Proceedings of the seminar on the management of integrated freshwater agro-piscicultural ecosystems in tropical areas. Technical Centre for Agricultural and Rural Cooperation, Wageningen, The Netherlands.

Kapetsky, J. M. 1998. Geography and constraints on inland fishery enhancements. FAO Fisheries Technical Paper 374:37–63.

Kapetsky, J. M. 1999. The development and training requirements for geographic information systems (GIS) and remote sensing (RS) applications in the Lower Mekong Basin (basin-wide) in relation to inland fisheries (including aquaculture). Assessment of Mekong fisheries: fish migrations and spawning and the impact of water management project (AMFP). AMFP Technical Report 4/99, Vientiane, Lao PDR.

Kapetsky, J. M. 2000. Present applications and future needs of meteorological and climatological data in inland fisheries and aquaculture. Journal of Agricultural and Forest Meteorology 103:109–117.

Kapetsky, J. M. 2001. Recent applications of GIS in inland fisheries. Pages 339–359 *in* T. Nishida, P. J. Kailola, and C. E. Hollingworth, editors. Proceedings of the first international symposium on GIS in fishery science. Fishery GIS Research Group, Saitama, Japan.

Kapetsky, J. M., and E. Ataman. 1991. Report of a mission to the National Fisheries Research and Development Agency, Republic of Korea, on an information base for the orderly development of mariculture and a regional project for training on mariculture development and management. FAO/UNDP regional seafarming development and demonstration project, RAS/90/002, Working paper SF/WP/91/6, Bangkok, Thailand.

Kapetsky, J. M., and B. Chakalall. 1998. A strategic assessment of the potential for freshwater fish farming in the Caribbean Island states. FAO, Comision de Pesca Continental para America Latina, Technical Paper 10 Supplement, Rome.

Kapetsky, J. M., J. M. Hill, and L. D. Worthy. 1988. A geographical information system for catfish farming development. Aquaculture 68:311–320.

Kapetsky, J. M., J. M. Hill, L. D. Worthy, and D. L. Evans. 1990. Assessing potential for aquaculture development with a geographic information system. Journal of the World Aquaculture Society 21:241–249.

Kapetsky, J. M., and N. MacPherson. 1990. Use of a geographic information system (GIS) to select priority for fish farm development. Technical Assistance and Investment Framework for Aquaculture in Ghana, FAO Technical Cooperation Program Project, Field document FI:TCP/GHA/0051, Field working paper 9, Rome.

Kapetsky, J. M., L. McGregor, and H. Nanne E. 1987a. A geographical information system to assess opportunities for aquaculture development: a FAO-UNDP/GRID cooperative study in Costa Rica. Pages 519–535 *in* GIS '87, volume 2. Second annual international conference, exhibits and workshops on geographical information systems. American Society for Photogrammetry and Remote Sensing, Falls Church, Virginia.

Kapetsky, J. M., L. McGregor, and H. Nanne E. 1987b. A geographical information system and satellite remote sensing to plan for aquaculture development: a FAO-UNDP/GRID cooperative study in Costa Rica. FAO Fisheries Technical Paper 287.

Kapetsky, J. M., and S. S. Nath. 1997. A strategic assessment of the potential for freshwater farming in Latin America. FAO, Comision de Pesca Continental para America Latina, Technical Paper 10, Rome.

Kapetsky, J. M., S. S. Nath, and J. P. Bolte. 1996. Inland fish farming potential in Latin America: an overview of the study and a preview of the results. FAO Aquaculture Newsletter 13:18–21.

Kapetsky, J. M., and C. Travaglia. 1995. Geographical information systems and remote sensing: An overview of their present and potential applications in aquaculture. Pages 187–208 *in* K. Nambiar and T. Singh, editors. Aquaculture towards the 21st century. INFOFISH, Kuala Lumpur, Malaysia.

Kapetsky, J. M., U. N. Wijkstrom, N. MacPherson, M. M. J. Vincke, E. Ataman, and F. Caponera. 1991. Where are the best opportunities for fish farming in Ghana? The Ghana aquaculture geographical information system as a decision-making tool. Technical Assistance and Investment Framework for Aquaculture in Ghana, FAO Technical Cooperation Program Project, Field document FI:TCP/GHA/0051, Field Technical Paper 5, Rome.

Keizer, P. D. 1994. An integrated approach to aquaculture site selection and management. Pages 53–60 *in* A. Ervik, P. K Hansen, and V. Wennevik, editors. Proceedings of the Canada–Norway workshop on environmental impacts of aquaculture. Havforskningsinstituttet, Bergen, Norway.

Kelsey, H. E., G. Scott, D. E. Porter, B. Thompson, and L. Webster. 2003. Using multiple antibiotic resistance and land use characteristics to determine sources of fecal coliform bacterial pollution. International Journal of Environmental Monitoring and Assessment 81:337–348.

Klotz-Shiran, I. 1999. A GIS fishing expedition. GEOEurope 8(7):32.

Kohler, S. T. 1996. Using census data and geographic information systems to identify target markets for aquaculture products in the USA. World Aquaculture 27(4):23–35.

Kracker, L. 1999. Shellfish information management system. NOAA National Ocean Service Center for Coastal Environmental Health and Biomolecular Research. Available: *www.chbr.noaa.gov/presentations/Sims/index.htm* (September 2003).

Krieger, Y., and S. Mulsow. 1990. GIS application in marine benthic resource management. Pages 671–683 *in* GIS for the 1990s. Proceedings of the national conference. Canadian Institute of Surveying and Mapping, Ottawa.

Legault, J. A. 1992. Using a geographic information system to evaluate the effects of shellfish closures on shellfish leases, aquaculture and habitat availability. Canadian Technical Report of Fisheries and Aquatic Sciences 1882E.

Le Moine, O., D. Razet, P. Soletchnik, P. Goulletquer, N. Faury, P. Geairon, and S. Robert. 1998. Mortalités estivales de l'huître creuse *Crassostrea gigas* dans le bassin de Marennes-Oléron: développement d'une méthodologie de cartographie spatiale des paramètres hydrologiques à l'aide d'un Système d'Information Géographique (S. I. G.). Pages 167–168 *in* Aquaculture Europe '98—aquaculture and water: fish culture, shellfish culture and water usage. European Aquaculture Society, Special Publication 26, Ostende, Belgium.

Loubersac, L., J. Populus, C. Durand, J. Prou, M. Kerdreux, and O. Le Moine. 1997. System d'information a references spatiale et gestion d'un espace de production ostreicole: le cas du Bassin de Marennes Oleron (France). Bordomer 97. Amenagement et Protection de L'environnement Littoral. Actes du Colloque 2:186–197.

Loubersac, L., J. Salomon, M. Breton, C. Durand, and C. Gaudineau. 1999. Perspectives offertes par la communication entre un modele hydrodynamique et un SIG pou l'aide au diagnostic environnmental. Caracterisation de la dynamique et de la qualite de masses d'eaux cotieres. CoastGIS'99. Geomatics and coastal environment. Actes de Colloques 25:173–185.

Meaden, G. J. 1999. Zoning for coastal aquaculture. Revitalization and acceleration of aquaculture development. FAO Technical Cooperation Program Project, Field document TCP/SRL/6715, Bangkok, Thailand.

Meaden, G. J. 2001. GIS in fisheries science: foundations for a new millennium. Pages 3–30 *in* T. Nishida, P. J. Kailola, and C. E. Hollingworth, editors. Proceedings of the first international symposium on GIS in fishery science. Fishery GIS Research Group, Saitama, Japan.

Meaden, G. J., and J. M. Kapetsky. 1991. Geographical information systems in inland fisheries and aquaculture. FAO Fisheries Technical Paper 318.

Melancon, E. J., T. M. Soniat, V. Cheramie, J. Barras, R. Dugas, and M. Lagarde. 1997. Oyster resource zones based on wet and dry estuarine cycles and its implication to coastal restoration efforts in Louisiana, U.S.A. Journal of Shellfish Research 16:272.

Melancon, E. J., Jr., T. M Soniat, V. J. Cheramie, and R. J. Dugas. 1994. Barataria and Terrebonne estuaries' oyster resource mapping project. Journal of Shellfish Research 13:304.

Mooneyhan, W. 1985. Determining aquaculture development potential via remote sensing and spatial modeling. Pages 217–247 *in* Applications of remote sensing to aquaculture and inland fisheries. Report of the ninth UN/FAO international training course in co-operation with the government of Italy. FAO Remote Sensing Center, Series 27, Rome.

Nath, S. S., J. Bolte, L. G. Ross, and J. Aguilar-Manjarrez. 2000. Applications of geographical information systems (GIS) for spatial decision support in aquaculture. Aquaculture Engineering 23(1–3):233–278.

OPIS (Ocean Planning Information System). 2003. Ocean Planning Information System homepage. National Oceanic and Atmospheric Administration. Available: *www.csc.noaa.gov/opis/* (September 2003).

Parker, M. R., B. F. Beal, W. R. Congleton, Jr., B. R. Pierce, and L. Morin. 1998. Utilization of GIS and GPS for shellfish growout site selection. Journal of Shellfish Research 17:1491–1495.

Paw, J. N., D. A. D. Diamante, N. A. Robles, and T. E. Chua. 1992. Site selection for brackishwater aquaculture development and mangrove reforestation in Lingayen Gulf, Philippines using geographic information systems. International Center for Living Aquatic Resource Management, Contribution 784, Manila.

Paw, J. N, F. Domingo, Z. N. Alojado, and J. Guiang. 1994. Land resource assessment of brackishwater aquaculture development in Lingayen Gulf area, Philippines. International Center for Aquatic Resource Management and International Development, Technical report on the geographic information systems application for coastal area management and planning, Lingayen Gulf area, Philippines, part 3, Manila.

Pérez, O. M., L. G. Ross, T. C. Telfer, and M. C. M Beveridge. 2002. Geographical information systems (GIS) as a simple tool for modelling waste distribution under marine fish cages. Estuarine Coastal and Shelf Science 54:761–768.

Populus, J., and D. Lantieri. 1991. Use of high resolution satellite data for coastal fisheries. 1. Pilot study in the Philippines. 2. General review. FAO Remote Sensing Center, Series 58, High Resolution Satellite Data Series 5, Rome.

Populus, J., L. Loubersac, J. Prou, M. Kerdreux, and O. Le Moine. 1997 Geomatics for the management of oyster culture leases and production. *In* D. Green and G. Massie, editors. CoastGIS '97. Proceedings of the second international symposium on GIS and computer mapping for coastal zone management. University of Aberdeen, Aberdeen, Scotland, UK.

Porter, D. E., W. K. Michener, T. Siewicki, D. Edwards, and C. Corbett. 1996. Geographic information processing assessment of the impacts of urbanization on localized coastal estuaries: a multidisciplinary approach. Belle W. Baruch Library in Marine Science 20:355–387.

Porter, M. D., and D. Rice. 2001. Application of geographic resource analysis support system (GRASS) in analyses of dissolved oxygen declination in commercial catfish ponds. Pages 295–301 *in* T. Nishida, P. J. Kailola, and C. E. Hollingworth, editors. Proceedings of the first international symposium on GIS in fishery science. Fishery GIS Research Group, Saitama, Japan.

Preston, N., I. Macleod, P. Rothlisberg, and B. Long. 1997. Environmentally sustainable aquaculture production—an Australian perspective. Pages 471–477 *in* D. A. Hancock, D. C. Smith, A. Grant, and J. P. Beumer, editors. Developing and sustaining world fisheries resources. The state of science and management: 2nd World Fisheries Congress. CSIRO Publishing, Melbourne, Australia.

Rodgers, L. J., and D. B. Rouse. 1997. A GIS based decision support system for oyster management in Mobile Bay, Alabama. Journal of Shellfish Research 16:337.

Ross, L. G., and M. C. M. Beveridge. 1995. Is a better strategy necessary for development of native species for aquaculture? A Mexican case study. Aquaculture Research 26:539–547.

Ross, L. G., E. A. Q. Mendoza, and M. C. M. Beveridge. 1993. The application of geographical information systems to site selection for coastal aquaculture: an example based on salmonid cage culture. Aquaculture 112:165–178.

Rubec, P. J., J. D. Christensen, W. S. Arnold, H. Norris, P. Steele, and M. E. Monaco. 1998. GIS and modeling: coupling habitats to Florida fisheries. Journal of Shellfish Research 17:1451–1457.

Salam, M. A., and L. G. Ross. 1999. GIS modelling for aquaculture in Southwestern Bangladesh: comparative production scenarios for brackish and freshwater shrimp and fish. Pages 141–145 *in* Proceedings of geosolutions in a coastal world: integrating our world. Adams Media, Vancouver.

Salam, M. A., and L. G. Ross. 2000. Optimizing sites selection for development of shrimp (*Penaeus monodon*) and mud crab (*Scylla serrata*) culture in southwestern Bangladesh. Proceeding of the GIS 2000 Conference, GeoWorld-Adams Business Media, Toronto.

Scott, P. C. 1998. Considerações sobre o uso da Baía de Sepetiba, RJ para Maricultura apoiadas num Sistema de Informação Geográfica (SIG). Instituto de Ciências Biológicas e Ambientais, Universidade Santa Úrsula, Report 98.001, Rio de Janeiro, Brazil.

Scott, P. C. 2001. GIS-based environmental modelling for management of coastal aquaculture and natural resources: a case study in Baia de Sepetiba, Brazil. Pages 261–274 *in* J. Coimbra, editor. Modern aquaculture in the coastal zone-lessons and opportunities. NATO Science Series: Series A: Life Sciences, Amsterdam.

Scott, P. C., and L. F. de N. Vianna. 2001. Determinacao de areas potenciaia para o desenvolvimento da carcinicultura em sistema de informacao geografica. Panorama da Aquicultura 11(63): 42–49.

Scott, P. C., and L. G. Ross. 1998. O potencial da mitilicultura na Baía de Sepetiba. Panorama da Aquicultura 8(49):13–19.

Smith, G. F., and K. N. Greenhawk. 1996. Morphological differentiation of the fringing and patch oyster reef types in Chesapeake Bay: a comparative evaluation. Journal of Shellfish Research 15:522.

Smith, G. F., K. N. Greenhawk, and M. L Homer. 1997. Chesapeake Bay oyster reef - An examination of resource loss due to sedimentation. Journal of Shellfish Research 16:275.

Smith, G. F., and S. J. Jordan. 1993. Utilization of a geographical information system (GIS) for the timely monitoring of oyster population and disease parameters in Maryland's Chesapeake Bay. Journal of Shellfish Research 12:130.

Smith, G. F., S. J. Jordan, and K. N. Greenhawk. 1994. An oyster management information system: integrating biological, physical, and geographical dimensions. Journal of Shellfish Research 13:284.

Soletchnik, P., O. Le Moine, N. Faury, D. Razet, P. Geairon, and P. Goulletquer. 1999. Summer mortality of the oyster in the Bay Marennes-Oleron: Spatial variability of environnement and biology using a geographical information system (GIS). Aquatic Living Resources Resources Vivantes Aquatique 12:131–143.

Stevens, J., M. Pilaro, and D. Geagan. 1997. The Greenwich Bay initiative: A case study of shellfish habitat restoration through land use planning. Journal of Shellfish Research 16:275.

Stolyarenko, D. A. 1995. Methodology of shellfish surveys based on a microcomputer geographic information system. ICES Marine Science Symposium 199:259–266.

Tookwinas, S., and P. Leeruksakiat, 1999. Application of geographic information system (GIS) technique for shrimp farm and mangrove forest development in Chanthaburi Province, Thailand. Thai Marine Fisheries Research Bulletin 7:1–16.

Travaglia, C., J. M. Kapetsky, and G. Profeti. 1999. Inventory and monitoring of shrimp ponds in Sri Lanka by ERS SAR data. FAO Environmental and Natural Resources Service, Sustainable Development Department, FAO environment and natural resources working paper 1, Rome.

Tsai, B.-W., K.-T. Chang, and Y.-T. Chang. 1997. Spatial analysis in GIS-the land use changes in the coastal area of Yunlin County, Taiwan. Journal of Geographical Science 23:1–12.

Tsai, B.-W., K.-T. Chang, Y.-T. Chang, and J. M. Chu. 2001. Measuring spatial association. The case of aquacultural land use in Yunlin County, Taiwan. Modeling change in spatial and temporal characteristics of aquaculture in Taiwan. Journal of Geographical Science 29:1–36.

Tu, T. N., and H. Demaine. 1997. Potentials for different models for freshwater aquaculture development in the Red River Delta (Vietnam) using GIS analysis. Naga, the ICLARM Quarterly 19(1):29–32.

White, D. L., D. Bushek, D. E. Porter, and D. Edwards. 1998. Geographic information systems (GIS) and kriging: analysis of the spatial and temporal distributions of the oyster pathogen *Perkinsus marinus* in a developed and an undeveloped estuary. Journal of Shellfish Research 17:1473–1476.

White, N. M., D. E. Line, J. D. Potts, W. Kirby-Smith, B. Doll, and W. F. Hunt. 2000. Jump Run shellfish restoration project. Journal of Shellfish Research 19:473–476.

White, N. W., L. E. Danielson, and M. V. Holmes. 1999. Development of hydrologic modification indicators to support watershed-based restoration of shellfish resources impacted by fecal coliform contamination. Journal of Shellfish Research 18:730.

Wibowo, A., W. Hastuti, J. Populus, and A. Karsidi. 1994. The applications of remote sensing and GIS technology for monitoring and site selection of aquaculture ponds on coastal areas. ASEAN-EEC Aquaculture Development and Coordination Program (AADCP), Working paper 44, Bangkok, Thailand.

Chapter 7

Geographic Information Systems Applications in Coastal Marine Fisheries

TIMOTHY A. BATTISTA AND MARK E. MONACO

7.1 INTRODUCTION

The world's marine resources are under multiple stresses from both human activities and natural environmental perturbations. The primary factors contributing to habitat loss and impacts on living marine resources include urbanization, fishing activities, increase in water temperatures, and nonpoint-source pollution (Haddad et al. 1996). As pressures on marine resources continue to increase, it has become evident that data required to make informed management decisions, bounded by policy aspects, are either lacking or not easily accessed. The collection, compilation, and synthesis of complex data sets are necessary steps in providing timely and useful information to environmental managers. Information technologies provide a suite of capabilities that can be used to organize, analyze, and disseminate the data and the information to the management community. The use of geographic information system (GIS) technology is a powerful tool to organize, visualize, and conduct assessments of geographic environmental data in support of the management of the world's estuarine, coastal, and marine habitats, and associated living marine resources (Parker 1988; Haddad and Michener 1991).

A GIS is a computer system that stores and links nongraphic attributes or geographically referenced data with graphic map features to enable a broad range of information processing and display operations, as well as map production, analysis, and modeling (Antenucci et al. 1991). Geographic information systems often are perceived to be products (e.g., digital maps); however, it is important to recognize that not every GIS has the same functionality. A GIS is not simply technology, nor merely hardware and software, but rather it represents an organization of integrated data sets (Dangermond 1991). Geographic information systems are evolving into a primary tool for addressing coastal and marine resource use management (Haddad and Michener 1991), but it presently lacks adequate exposure or use in fisheries management (Isaak and Hubert 1997).

Many issues or problems in coastal marine environments can be addressed by using a GIS. These include: simulating potential impacts to living marine resources from various management alternatives; defining boundaries of marine protected areas based on biological, economic, and political considerations; and defining important fishery habitats in space and time. The Environmental Systems Research Institute (1990) grouped the types of questions a GIS can address into four general types. First, what are the attributes associated with a specific location (e.g., depth zone)? Second, given a set of criteria, which areas meet those criteria (e.g., display all coral reef hard bottom areas)? Third, what spatial patterns exist (e.g., display all grouper locations that occur in the fore reef zone)? Fourth, a GIS can model databases to answer "what if" questions (e.g., if estuarine salinity increases by 5 ppt, how will seatrout habitat be affected?). In many cases, more than one type of question is addressed in a given study.

A GIS enables users to increase the accuracy and speed of a response to a suite of questions, thus allowing managers to evaluate multiple management scenarios and conduct more informed discussions (Isaak and Hubert 1997). In addition, data integration and visualization enables spatial statistical analyses in the GIS (Arlinghaus 1996). Spatial statistics do not assume independence of observations. Instead, they account for the autocorrelation often present in spatial databases (Griffith 1996). This may lead to estimations with better statistical properties and more accurate modeling efforts, often improving results drawn from overlays of spatial data (Isaak and Hubert 1997).

This chapter provides a brief overview of GIS use in the coastal marine environment, with an emphasis on nekton spatial ecology and its relevance to fishery and habitat management. The chapter highlights three examples that explicitly use GIS to enable research and assessments to directly support management decisions that minimize environmental resource use conflicts. The first case study depicts the integration of biological and physical databases by using GIS technology to formulate management strategies. The study addresses the potential biological impacts from human-induced changes to freshwater inflow on the biota found in Apalachicola Bay, Florida. The second case study provides an example of how GIS can be used to define species habitat affinities based on their use of specific habitats (Monaco et al. 1998). The study focuses on habitat affinities to predict the density of nekton by habitat type in Galveston Bay, Texas. This research was formulated to aid in defining the spatial distribution of essential fish habitats (EFH; Clark et al. 1999; Rubec et al. 2001). The final case study describes the use of GIS technology to develop digital benthic habitat maps for coral reef ecosystems.

In addition, GIS has been used to define the spatial and temporal use patterns of living marine resources found across reef zones and habitats (Kendall et al. 2000). This work uses GIS as a cornerstone technology to define potential biologically relevant boundaries of marine protected areas by integration of spatial analyses and visualization of habitat and species distribution data. This includes, but is not limited to, analyses similar to those used by the gap analysis program (GAP; Scott et al. 1993), geopositioning of boundaries, and organizing and displaying habitat classes in digital georeferenced map products.

7.2 TECHNIQUES AND CHALLENGES OF USING GIS IN COASTAL MARINE ENVIRONMENTS

Historically, spatial information has been underutilized in resource management because managers have lacked the capability to explore fully the spatial and temporal relationships between species distributions and environmental gradients across large spatial scales. One approach to resolving this issue has been to develop custom coastal marine environment applications that use mathematical modeling methodologies to explain the relationship between marine habitats and species distributions. For instance, habitat suitability index (HSI) modeling was an approach first developed by the U.S. Fish and Wildlife Service to predict optimum habitats for terrestrial and aquatic species. A GIS can be used to enhance the HSI methodology to provide more quantitative and spatially derived habitat models. The HSI technique provides the means to estimate the spatial distributions of species in relation to habitat across multiple environmental gradients by using GIS spatial overlay capabilities (Figure 7.1). The National Oceanic and Atmospheric Administration (NOAA) has conducted HSI modeling in a variety of estuaries (i.e., Casco, Sheepscot, Pensacola, Tampa, and Chesapeake bays) for both invertebrate and pelagic species (Battista 1998; Christensen et al. 1998; Brown et al. 2000), and other management agencies such as the Florida Department of Environmental Protection have implemented and refined the use of HSI for delineating important fishery habitat (Rubec et al. 1998, 2001).

While the development of GIS methodologies (e.g., HSI modeling) is important, ultimately, the transfer of these capabilities to the resource management community is of greater need and is equally as challenging. Coastal managers have expressed increasing interest in GIS tools for conducting a variety of coastal assessments, ranging from impact analysis of habitat modifications to the biological and physical characterizations of estuarine and marine ecosystems, to regional human population assessments in coastal watersheds. To address this need, NOAA has developed an ArcView extension (i.e., habitat suitability modeling [HSM]), which provides resource managers with specific GIS functionality to perform coastal resource assessment, analysis, and modeling (Gill et al. 2001).

Habitat suitability modeling is a custom ArcView extension that uses the Spatial Analyst grid processing capabilities; HSM uses theoretical or empirical model parameters defined outside the application to generate maps of habitat suitability for selected fishery species and to predict the effects of environmental change. For example, HSM can predict the changes in species distribution resulting from dredging or changes in freshwater inflow (NOAA 1999b).

Habitat suitability modeling predicts the suitability of an area for a species based on the species' habitat requirements and the area's environmental characteristics; HSM assigns HSIs to environmental measurements (e.g., 10 ppt salinity) in the data layers (e.g., fall salinity grid), based on the species habitat affinities and requirements. By using a user-selected (e.g., geometric mean, weighted mean, or regres-

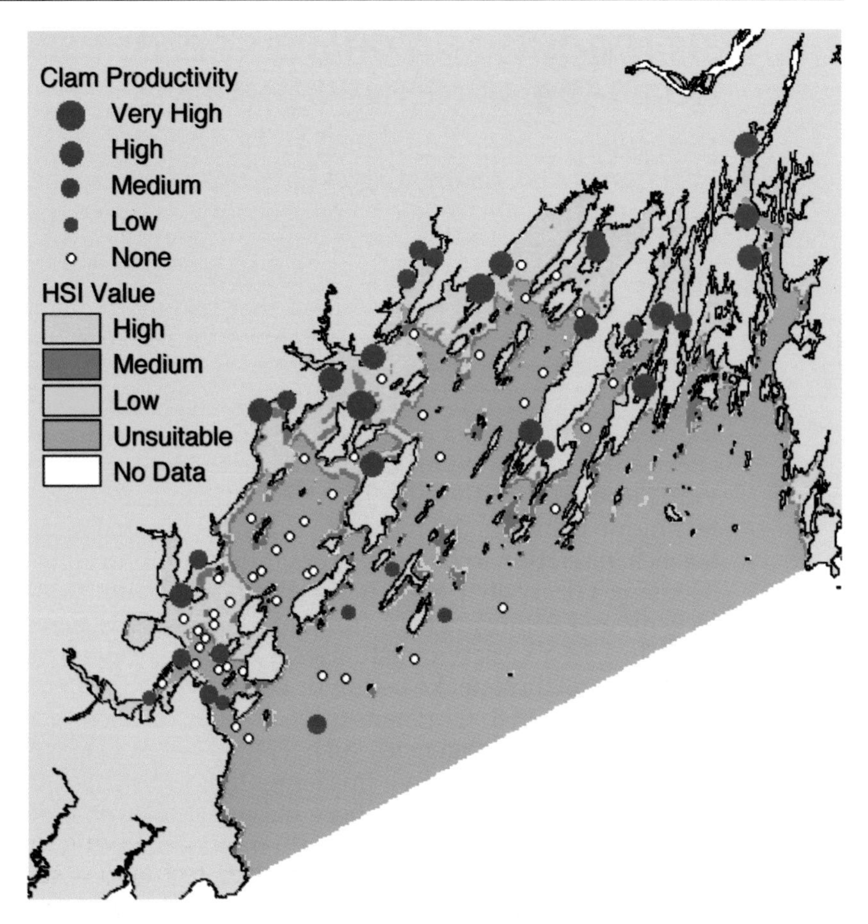

Figure 7.1 Map of habitat suitability index (HSI) classes for adult and juvenile life stages of HSI model for softshell clam *Mya arenaria* (also known as softshell) in Casco Bay, Maine. The model was run under summer conditions by using temperature, salinity, substrate, and depth as environmental layers and depicts clam productivity for validation purposes. The model was validated by statistically comparing catch data (depicted here as clam productivity) versus predicted habitat model results.

sion equation) or user-programmed algorithm, HSM combines the weighted layers to calculate the overall habitat suitability for the species and the area of interest, and generates predictive maps of the relative habitat suitability for the species. To examine the accuracy of the model, the user can import and overlay actual survey or field data (e.g., fish trawl survey data) and generate maps and statistical summaries of the predicted and actual species distributions (Figure 7.1).

In addition to developing coastal marine modeling methodologies and applications, users also are challenged with choosing the most appropriate interpolation technique to best represent spatial patterns in coastal marine environmental data. Interpolation is the process of estimating values from point observations to create a continuous representation of the feature. A variety of standard interpolation procedures are available to create grid surfaces; however, the resulting surface may not always accurately represent the true spatial behavior of the data, depending upon the mathematical formula of the interpolation procedure. Investigators using GIS must be aware that the validity of each interpolation technique is dependent upon the type of data being interpolated and the ability of the survey data to adequately capture the temporal and spatial variability of the feature of interest (Laurini and Thompson 1992). For instance, if a bathymetric surface were created from depth soundings taken in an area with high topographic relief, it would be necessary to choose an interpolation technique that could simulate the dramatic changes in depth. Furthermore, if the survey data do not adequately capture the true behavior of the bathymetric features (i.e., pinnacles, ridgelines, and trenches), then these distinguishing features will not be represented in the resulting continuous surface.

The GIS user must understand the spatial properties of the data prior to selecting the interpolation technique and conducting the surface generation process. For instance, users can ask detailed questions about the data sets in order to determine the most appropriate interpolation technique. These questions include: (1) What do the data points represent or what kind of data are they? (2) Is it important to retain the value of the original survey points in the derived surface? (3) What is the distribution of survey points? (4) Is interpolation processing speed a factor? For instance, interpolation techniques such as nearest neighbor analysis and triangulated irregular network are most appropriately applied to elevation data. If preserving the exact value of the data point is not important, then techniques such as inverse distance weighting and kriging are more suitable. If data sample points are highly clustered, then either nearest neighbor, triangulated irregular network, or kriging would produce the most reasonable surfaces. Or if speed of processing is a factor, then the user should interpolate with triangulated irregular network or inverse distance weighting techniques, since these are faster than other techniques such as kriging.

Although this section highlights a few of the challenges users must address to conduct GIS analysis of coastal marine environments, there are a multitude of additional factors, such as spatial scale, data autocorrelation, and temporal dimensionality, that also may be relevant (Isaaks and Srivastava 1989; Wiens 1989; Levin 1992). A more detailed discussion of case specific technical issues, methodology, and GIS functionality is described in the next section.

7.3 APPLICATIONS IN COASTAL MARINE ENVIRONMENTS

The following three case studies in estuarine, marine, and coral reef ecosystems are intended to introduce innovative and program-level implementations of GIS designed to support management decisions. The examples emphasize techniques and approaches used to map and model the distribution of fish and invertebrate species to help define the ecological function of marine habitats in space and time.

7.3.1 Case Study: Ecological Impacts of Alteration to Freshwater Inflow

A study was conducted through NOAA's Biogeography Program (Monaco and Christensen 1997) for the U.S. Army Corps of Engineers, Mobile District, to assess the potential changes in distribution and abundance of the eastern oyster *Crassostrea virginica* in response to altered freshwater inflow conditions in Apalachicola Bay, Florida (Wilber 1992; Christensen et al. 1998). A GIS was used to integrate a suite of biological and physical models to demonstrate potential oyster biological responses (mortality and growth) as a consequence of altered hydrodynamic and ecological conditions. A series of digital maps was produced by using a GIS that showed the altered physical regimes that would develop in the bay under various seasonal freshwater inflow and demand scenarios (i.e., drought, low-moderate, and flood conditions) being considered by management agencies (Livingston 1997). Alteration in the timing and magnitude of freshwater inflow would contribute to regions of reduced salinity variability in oyster-rich portions of the bay, thereby generating conditions conducive for the proliferation of oyster drills *Thais haemastoma*, a major marine predator. The GIS models predicted that prolonged and stable salinities would likely be spatially coincident with the most productive oyster bars in Apalachicola Bay (Livingston et al. 2000), which would in turn promote a significantly elevated rate of oyster mortality due to predation by oyster drills (NOAA 1999a).

The Apalachicola Bay modeling effort demonstrates the utility of creating a statistically robust, spatially explicit resource management predictive tool. The success of this effort was largely a function of the ability to use a GIS system to integrate and display the results from separate biological and physical modeling components (Figure 7.2). Hydrodynamic conditions in the bay during ambient and reduced inflow scenarios were predicted from estimates of freshwater inflow and other factors in a GIS. Biological components were included by conducting stepwise multiple regression to identify statistically significant and biologically relevant environmental correlates from the hydrodynamic model output (i.e., surface and bottom temperature, salinity, current velocity, residual velocity, residual flow direction, salinity deviation, temperature deviation, salinity stratification, tidal elevation, temperature minima and maxima, and salinity minima and maxima) and in situ oyster population data collected from Apalachicola Bay (Livingston et al. 2000). A multivariate, nonlinear oyster mortality regression model then was developed by using the variables identified by stepwise multiple regression. A GIS was used to depict oyster distribution and abundance during ambient and reduced freshwater inflow by relating oyster survival from the mortality model with environmental conditions derived from the hydrodynamic model (Christensen et al. 1998).

The Apalachicola Bay GIS model predicted that significant oyster mortality will likely occur in locations where salinities become elevated and stable beyond ambient conditions (Figure 7.2).

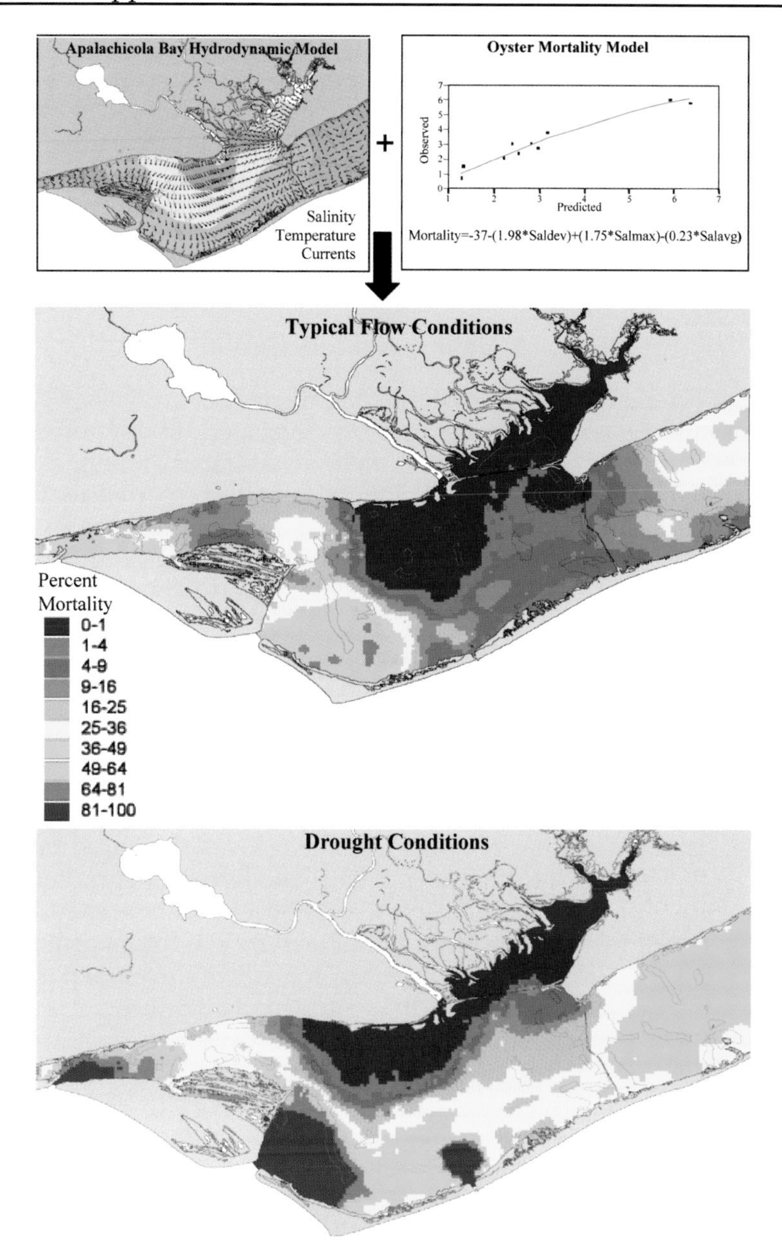

Figure 7.2 Conceptual model of hydrologic and biological model integration in Apalachicola Bay, Florida. The hydrodynamic model output depicts a temporal snapshot of surface current flow magnitude and direction (arrows), and surface salinity (blue to red, increasing). The resulting model outputs indicate the potential eastern oyster mortality for typical flow year and drought year conditions in Apalachicola Bay, Florida. Saldev = salinity deviation, Salmax = salinity maxima, and Salavg = average salinity.

This response to salinity and its underlying structure is consistent with various investigators' findings, suggesting that oyster drill is a stenohaline marine predator whose populations are limited by the physiological bounds of osmoregulatory stress (Galtsoff 1964; MacKenzie 1977). Results from the Apalachicola Bay study showed that GIS models can be used to conduct ecological forecasting of potential anthropogenic alterations and, thereby, provide a unique and useful predictive tool for marine resource management. This work is being extended to include GIS analyses of the potential impact of estuarine salinity changes on fisheries managed by the U.S. Government (NOAA 1996).

7.3.2 Case Study: Modeling Nekton Habitat Use

The reauthorization of the Magnuson-Stevens Fisheries Conservation and Management Act in 1996 mandated the identification, management, and conservation of fishery habitat as an integral component of federal fishery management. Initial efforts to identify EFH areas were based on limited information of species presence or absence (NOS 1997) or on qualitative models (Christensen et al. 1997) to establish ecological linkages between habitat and fishery production. A study conducted in Galveston Bay, Texas, investigated a more robust method of defining EFH than other methods used by the Gulf of Mexico Fishery Management Council to date. The GIS-based approach examined species habitat use patterns through spatial depiction of statistically derived ecological responses.

Considerable variation in structural habitat exists in northern Gulf of Mexico estuaries, including intertidal marsh, submerged aquatic vegetation, oyster reef, mangroves, tidal mudflats, and subtidal bay bottom. Numerous physical and structural gradients exist within each of these habitat types, and these gradients may affect the functional role or importance of a habitat type for a particular species. In this analysis, an innovative approach was developed to construct predictive models in a GIS that compare habitat use patterns and interactions with density-independent processes (e.g., salinity and temperature) in Galveston Bay. Sixteen years of nekton density data collected from three distinct structural habitat types (marsh edge, submerged aquatic vegetation, and submerged nonvegetated bottom) were analyzed to identify potential EFH for the federally managed species brown shrimp *Farfantepanaeus aztecus* (also known as *Penaeus aztecus*), white shrimp *Litopenaeus setiferus* (also known as *Penaeus setiferus*), and pinfish *Lagodon rhomboides* (Minello 1999). Multivariate models were developed for individual species, examining density patterns among estuarine habitats and responses to stochastic processes. Model results were then combined with a GIS to provide a spatial mosaic of potential EFHs. This approach provided a tool for fishery managers to compare habitat requirements of different species and to examine the relationships between habitat use and temporal, spatial, and ecological factors.

Density values (number/m^2) were calculated for all fish and decapod crustaceans at all sample locations, and a natural log transformation was used to correct the heteroscedacity in the data. Once transformed, a forward stepwise multiple regression (generalized linear model) procedure was used to identify statistically significant predictors of nekton density (i.e., season, habitat type, and salinity zone). The data were averaged by these variables to develop a mean log-density matrix for all possible season, habitat, and salinity combinations. The resultant matrices were then used to develop spatially explicit multivariate models depicting predicted maximum habitat use. Density estimates were classified into five equal percentiles based on their resultant distribution and subsequently mapped by using a GIS (Figure 7.3; Clark et al. 1999).

The performance of the spatially explicit biological models was assessed by comparing predicted species densities (model) in Galveston Bay with observed densities (catch) from nearby Texas estuaries with similar habitat (i.e., Aransas, Matagorda, and San Antonio bays). Statistical validation of the results was conducted by using linear regression to test differences in observed versus predicted model results. The validation procedures confirmed similar species habitat utilization patterns among Texas estuaries. Successful transfer of the empirical model to nearby and similar systems indicated that, when necessary, the GIS-based multivariate biological model could be applied to estuaries lacking sufficient empirical catch data (Clark et al. 1999). Furthermore, the spatially explicit GIS models that used species density data provided a more robust, resolved, and improved approach over current methods used to delineate EFH.

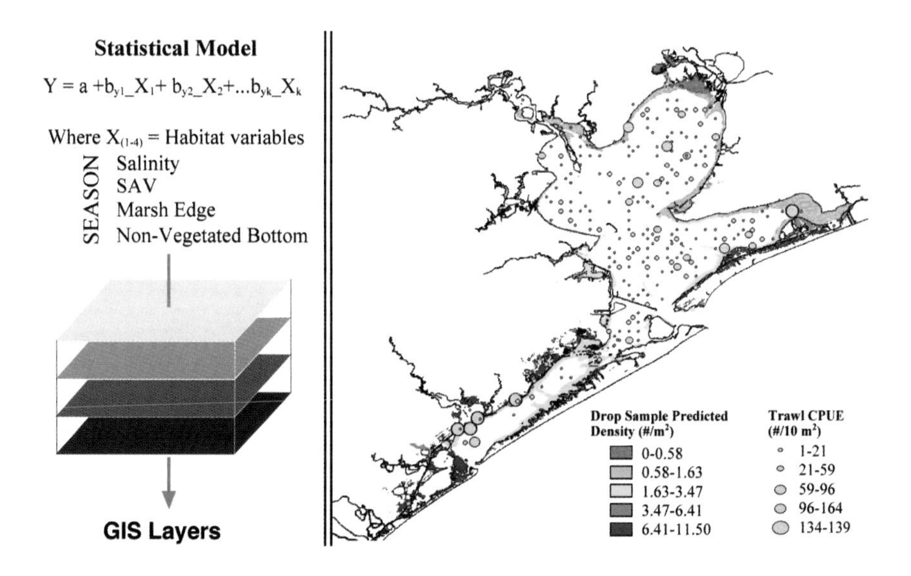

Figure 7.3 Conceptual model of nekton habitat use in Galveston Bay, Texas, based on salinity, submerged aquatic vegetation (SAV), marsh edge, and nonvegetated bottom used to define essential fish habitat. Model results depict the spatial distribution of predicted brown shrimp density from National Marine Fisheries Service drop sampler and observed otter trawl catch per unit effort (CPUE).

7.3.3 Case Study: Benthic Habitat Mapping

Accurate habitat maps are critically important for resource managers to make informed decisions about the protection and use of coastal areas (Maniere et al. 1986). In December 1998, NOAA's National Ocean Service acquired color aerial photography for the nearshore waters of Puerto Rico and the U.S. Virgin Islands at altitudes of 7,300 m, 4,500 m, and 3,000 m (1:48,000, 1:30,000, and 1:20,000 scales, respectively). These images were used to create digital maps of the region's marine resources, including coral reefs, seagrass beds, mangrove forests, and other important benthic habitats (Armstrong 1981; Ibrahim and Hashim 1990; Maniere et al. 1994; Chauvaud et al. 1998). In addition, visual fish sampling was conducted, and the resultant data were organized by reef zone location (e.g., backreef) and habitat (e.g., seagrass) combination in concordance with the habitat classes delineated from interpretation of aerial photography (NOAA 2000a). This effort supports the U.S. Caribbean Fishery Management Council in defining EFHs of managed reef fishes and invertebrates (NOAA 1998). In addition, the integrated habitat and species distribution maps support defining ecologically relevant boundaries for potential MPAs.

The primary objective of the project was to create digital georeferenced products by using digital scans of the aerial photography to conduct the classification procedures rather than hard copy photographs. Photointerpretation then could be conducted on digital images in a GIS environment by using a custom digitizing interface. Once the aerial photography was acquired, the first step involved scanning the photographs at high resolution to create a digital copy of each image (Figure 7.4). A scan resolution of 25 m was selected because it provided a manageable file size of 64 MB for a 23-cm × 23-cm photograph and a pixel size of approximately 2 m, which provided sufficient resolution to discriminate the benthic features of interest. The photointerpreter also made use of hard copy photographs and photographic diapositives to complement the digital imagery in areas where features were difficult to distinguish.

Soft copy photogrammetry then was then used to rectify geometrically and to georeference the digital imagery (Figure 7.4). The mission was flown with an airborne kinematic global positioning system, which provides very accurate vertical and horizontal control along the flight lines at the time of image capture (Curry and Schuckman 1993). This capability allows postprocessing of the imagery to correct for location and orientation of the airplane at the time of imaging (Thorpe 1993).

Once the imagery was corrected geometrically, the digital photographs were mosaicked together by using common tie points between overlapping photographs and were edge-matched to produce a seamless data set of only the "usable" segments of each photograph (Figure 7.4; Afek and Brand 1998). Typically, the usable area of any individual photograph is reduced to less than half due to interference from sun glint, cloud cover, or wave action. It was possible to eliminate these

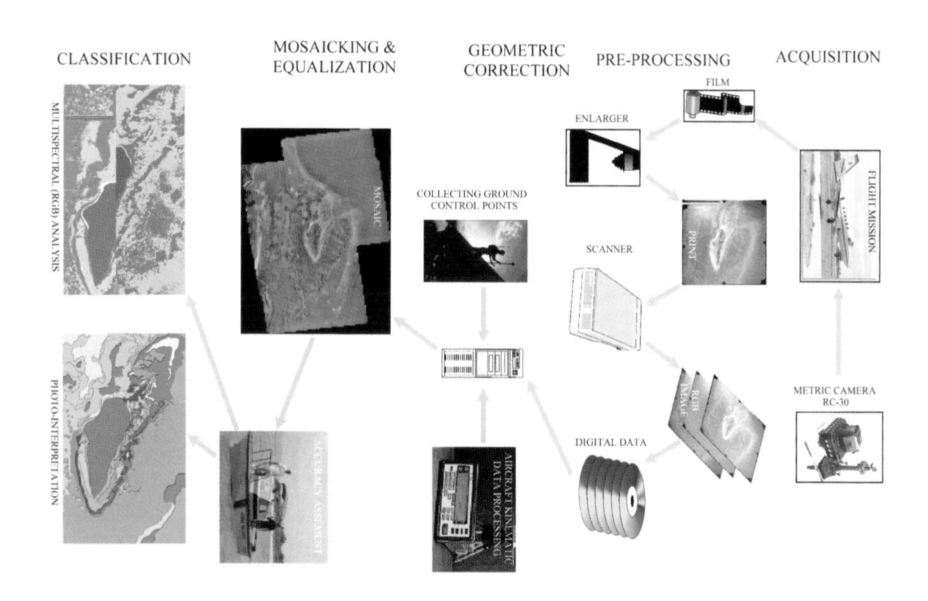

Figure 7.4 Steps to create benthic habitat classification maps by using aerial photography for the nearshore waters of the U.S. Virgin Islands and Puerto Rico.

sections since the flight lines were acquired with 60% side-lap and 30% end-lap between photographs, which ensures that every pixel is captured in at least two photographs. The images included in the mosaic then can be adjusted radiometrically to balance ambient light conditions at the time of exposure, thereby providing a smooth color tone transition between the scenes (Afek and Brand 1998).

Unlike hard copy photographs, digital imagery provides the advantage of georeferencing to a common local coordinate system (Projection UTM, Datum WGS84, spheroid units meters) (Fowler 1999). This allows resource managers and researchers to overlay additional GIS data layers (e.g., infrastructure, such as roads, drainage areas, effluent sites; areas of particular concern, such as marine parks and fishery reserves; and other data sets, such as ship grounding sites or fishery catch data) on the georeferenced imagery (Rulli and Shah 1998). This enabled the photointerpreter to work with a data set consisting of the most useful photosegments to delineate benthic habitat directly from the georeferenced digital photography.

Once the georeferenced mosaic was complete, photointerpretation was conducted by using the custom built ArcView "Habitat Digitizer" extension created by NOAA (2000b). By using a digitizing tablet or a computer mouse, the benthic habitat map is created by drawing habitat polygons on the digital photomosaic as it appears on the computer monitor with custom or preexisting hierarchical classification schemes. Simple pull-down menus are used to quickly attribute each new polygon with the appropriate habitat type. Photointerpreters can adjust the zoom in questionable parts of the digital imagery just as they would if they were varying magnification with hard copy photographs. The result is a georeferenced draft of the habitat map. After field validation of this draft, the GIS interface allows for simple adjustments of any polygon shape or attributes to produce the final map product. This approach offers the same advantages of traditional photointerpretation techniques but with four important improvements: a georeferenced mosaic is provided to users as part of the process, line work on benthic habitat maps in the GIS is aligned perfectly with georeferenced imagery, geolocational errors are reduced by eliminating the need to transcribe line work from a Mylar sheet into a GIS, and there is the ability to overlay additional GIS data sets (i.e., bathymetry) to aid in photointerpretation of benthic features (Figure 7.4; Logan 1993; Barrette et al. 2000).

Spectral analysis also was conducted on the digital photomosaics, to augment visual photointerpretation. The technique, also known as multispectral analysis or RGB analysis, is based on statistical analysis of the red, green, and blue spectral values recorded for each pixel in a scanned photograph (Pasqualini et al. 1997). Supervised classification was conducted by manually selecting a subset of pixels in the mosaic that is representative of each habitat type of interest (e.g., land, unconsolidated sediment, coral, or submerged vegetation). The remaining pixels in the image were assigned to one of the habitat types based on their statistical similarity to each of the preselected pixel (habitat) types. Due to the limited spectral resolution of digital photographs, this technique cannot discriminate as many habitat types as visual photointerpretation (6 habitat types are discriminated in multispectral analysis versus 27 in visual techniques). However, the technique does have the advantages of rapid habitat classification and much greater spatial detail (i.e., minimum mapping unit) of mapped features since each pixel is assigned a habitat type (rather than large polygons as is done in visual interpretation) (Shepard et al. 1995).

The resulting digital habitat maps provided spatial geographies to depict patterns of habitat utilization through ontogenetic life stages of a given species within the seascape. Figure 7.5 shows a single species model (blue tang *Acanthurus coeruleus*) that represents the probability of encountering the animal over a range of benthic habitats in Buck Island National Park, U.S. Virgin Islands.

7.4 CONCLUSION

7.4.1 Current State of GIS in Coastal Marine Environments

The use of GIS by scientists and resource managers for spatial data analysis of coastal environments has rapidly proliferated in recent years. This growth is evidence of both the success and acceptance of GIS as a valuable tool for analyzing complex and diverse coastal information (e.g., geological, geomorphological, ecological, hydrological, meteorological). The rapid rise in GIS usage in both aquatic and terrestrial environments is a function of multiple factors, including

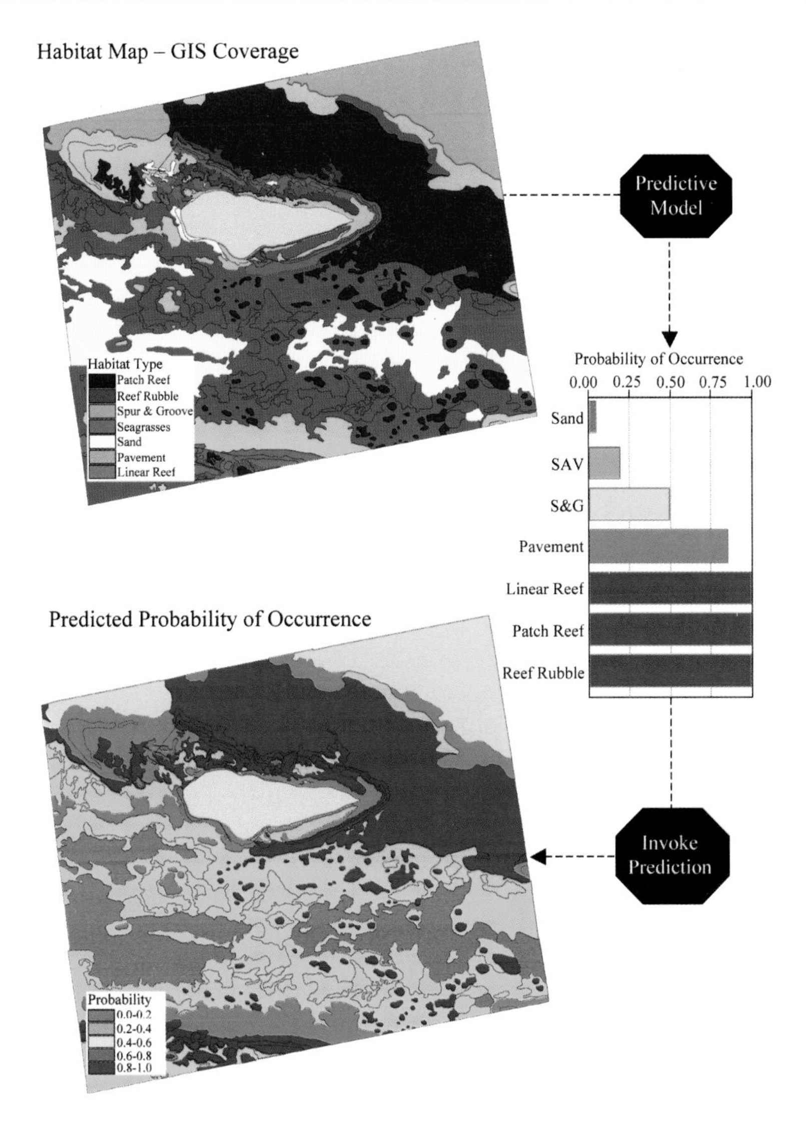

Figure 7.5 Integrating benthic habitat maps and in situ biological observations to predict species' (blue tang) habitat utilization patterns at Buck Island Reef National Monument, U.S. Virgin Islands. A logistic regression model was used to predict the probability of occurrence of blue tang in various habitat types found at Buck Island Reef National Marine Sanctuary. Habitat types were mapped by using aerial photography as indicated in Figure 7.4. SAV = submerged aquatic vegetation; S&G = spur and groove.

tremendous advances in computer processing speed, emphasis on desktop GIS software capability, the advent of Web-enabled GIS applications, and increased recognition of GIS as an interdisciplinary problem-solving tool. Radical improvements and changing market strategies by software vendors in information technologies has greatly benefited GIS users and has enhanced the software utility. Some of the changes include a transition to more inexpensive, user-friendly, and readily customizable desktop systems. The transition to object-oriented programming (e.g., Visual Basic) and a more versatile graphical user interface, reflects a major paradigm shift from the former command line, server-based, and technically challenging GIS.

The end effect of these changes is that contemporary GIS provide robust capabilities to even the most novice users and have, therefore, stimulated more widespread use of GIS technology by fishery scientists. Likewise, there has been greater acceptance and realization by fishery scientists of the potential uses of GIS (Isaak and Hubert 1997). Whereas GIS software was formerly perceived to be purely a mapping capability by coastal resource managers, many professionals increasingly recognize the importance and utility of a GIS for exploring the spatial component of coastal data. State and federal resource programs continue to invest in personnel training and infrastructure because management has embraced GIS as the means to analyze the highly dimensional, spatially explicit, and multidisciplinary nature of coastal data.

Despite the increased computational capability of current systems, the quantity and resolution of data available for analysis often limits coastal GIS applications. The dynamic nature of marine systems and the limited temporal and spatial dimensionality of survey data generally preclude robust quantitative analytical capabilities and undermine the statistical confidence of spatial interpolation techniques. However, the rapid development of Internet data portals has made it easier for researchers to find coastal and marine databases that were developed for other projects but are applicable to other studies. The advent of Internet and Intranet GIS capabilities has made remote viewing, mapping, querying, and serving of spatial data by users more readily available (Zeiler 1999). Coastal GIS users and developers have benefited greatly from the combination of distributed GIS systems and spatial geography network portals, which has provided greater access to federal, state, and private sector data sources.

7.4.2 Future of GIS in Coastal Marine Environments

The continued advancement of coastal GIS will be determined by the synthesis of three interdependent factors: software enhancement, data access, and user-derived statistical, spectral, and spatial data exploratory techniques. The simultaneous development of these components will ensure that GIS continues to enable researchers to readily visualize complex spatial relationships among species, assemblages, habitats, and physical regimes of marine environments. Although the inherent data limitations of coastal environments will continue to constrain our ability to make full use of GIS capabilities, reductions in data collection cost should alleviate some of these difficulties.

The development of more cost-effective and resolved remote-sensing technologies is being pursued aggressively to enhance the collection of synoptic data for large coastal and island areas (Mumby et al. 1999). Satellite and airborne remote-sensing imagery offers the advantages over field surveys of increased frequency of coverage, extensive coverage at low cost, archival data, and fast results (Sabins 1996). For instance, several new sensors provide dramatically improved spatial and spectral resolution (i.e., Ikonos, SeaWiFS, and MODIS) that have been instrumental in identifying coastal habitats and dynamic marine processes (Holden and LeDrew 1998). In addition, NOAA presently is conducting fixed-wing hyperspectral imaging in the U.S. Hawaiian Islands (NOAA 2002) to provide highly resolved spectral libraries for rapid and accurate delineation of submerged ecosystems.

Conditions for using remote-sensing data will continue to improve as the number of users and demand for data increases. This will stimulate an increase in the number of imaging platforms, better ranges of sensors, decreased sensor price, better algorithms for rapid processing of the data, and, ultimately, reduced data costs. In addition to optical remote sensing, future developments in acoustical sonar data collection will enhance coastal GIS capabilities. For instance, side-scan sonar provides the ability to conduct seafloor mapping based on the isonification of physical properties (i.e., hardness and roughness) of habitat features (Sotheran et al. 1997; Auster et al. 1998; Barnhardt et al. 1998). Rapid, highly detailed, in situ mapping of seafloor topography, bathymetry, and habitat classification will continue to improve as the ability to process complex and voluminous swath beam (fixed hull and towed transducer arrays) spatial data sets increases (Mayer et al. 1999).

Fishery scientists have a history of using visualization tools to analyze data in three and four dimensions, but these capabilities continue to evolve within GIS systems. Three-dimensional GIS systems offer the unique capability of explaining patterns of highly dimensional spatial and temporal variables. The use of inexpensive and commonly available tools such as virtual reality modeling will facilitate the creation of three-dimensional visualizations. The ability to analyze these data is dependent upon the ability of databases to optimize the storage and retrieval of independent thematic layers. Scalable object-oriented databases are currently being devised that will provide users with intelligent data querying capabilities. Relational databases are being designed to decipher the intrinsic behaviors of data and to automate the identification and retrieval of relevant data sources. For instance, present relational database management systems provide the dramatic improvement of being able to simultaneously store various types of imagery and the full informational content of GIS polygon, point, and line features (Oracle 1999). The operability of intelligent databases and three-dimensional visualization is dependent on the development of recognized data standards. The participation of federal, state, and private sector users as stakeholders will aid in the development of "Open GIS" standards so as to provide the infrastructure for evolving data portals and data transfer capabilities. File type conformity will create conditions more conducive for the distribution, acquisition, and integration of data by users into their GIS applications.

In summary, the use of GIS in coastal and marine environments can provide scientists, managers, and the public with a powerful tool to address complex spatial issues for natural resource management. While coastal GIS will continue to meet the requirements of most users as a marine resource mapping and inventory tool, the evolution of GIS, coupled with multivariate data analysis and hypothesis testing, will prove to be most valuable to scientists and resource managers. Further efforts must continue to integrate and expand the use of mathematical modeling, spatial depiction of bioenergetic models, predator–prey interaction, and ecological forecasting through predictive modeling capabilities (Christensen et al. 1997, 1998; Livingston et al. 2000). However, the coastal and marine community must continue to invest in GIS technology and, most importantly, commit to training staff and students in the use of GIS and related information technologies (e.g., image analysis). These investments will facilitate the use of GIS technology that will support development of research strategies and better management decisions to protect our coastal and marine natural resources.

7.5 REFERENCES

Afek, Y., and A. Brand. 1998. Mosaicking of orthorectified aerial images. Photogrammetric Engineering and Remote Sensing 64:115–125.

Antenucci, C. J., K. Brown, P. L. Croswell, M. J. Kevany, and H. Archer. 1991. Geographic information systems: a guide to the technology. Von Nostrand Reinhold, New York.

Arlinghaus, S. L. 1996. Practical handbook of spatial statistics. CRC Press, New York.

Armstrong, R. A. 1981. Changes in a Puerto Rican coral reef from 1936–1979 using aerial photoanalysis. Pages 309–316 in Proceedings of the fourth international coral reef symposium. Marine Sciences Center, University of the Phillippines, Quezon City.

Auster, P. J., C. Michalopoulos, R. Robertson, P. C. Valentine, K. Joy, and V. Cross. 1998. Use of acoustic methods for classification and monitoring of seafloor habitat complexity: description of approaches. Pages 186–197 *in* N. W. P. Munro and J. H. M. Willison, editors. Linking protected areas with working landscapes, conserving biodiversity. Science and Management of Protected Areas Association, Wolfville, Nova Scotia.

Barnhardt, W. A., J. T. Kelley, S. M. Dickson, and D. F. Belknap. 1998. Mapping the Gulf of Maine with side-scan sonar: a new bottom-type classification for complex seafloors. Journal of Coastal Research 14:646–659.

Barrette, J., P. August, and F. Golet. 2000. Accuracy assessment of wetland boundary delineation using aerial photography and digital orthophotography. Photogrammetric Engineering and Remote Sensing 66:409–416.

Battista, T. A. 1998. Habitat suitability index model for the eastern oyster, *Crassostrea virginica*, in the Chesapeake Bay: a geographic information system approach. Master's thesis. University of Maryland, College Park.

Brown, S. K., K. R. Buja, S. H. Jury, M. E. Monaco, and A. Banner. 2000. Habitat suitability index models for eight fish and invertebrate species in Casco and Sheepscot bays, Maine. North American Journal of Fisheries Management 20:408–435.

Chauvaud, S., C. Bouchon, and R. Maniere. 1998. Remote sensing techniques adapted to high resolution mapping of tropical marine ecosystems (coral reefs, seagrass beds, and mangrove). International Journal of Remote Sensing 19:3625–3639.

Christensen, J. D., T. A. Battista, M. E. Monaco, and C. J. Klein. 1997. Habitat suitability modeling and GIS technology to support habitat management: Pensacola Bay, Florida. National Oceanic and Atmospheric Administration, National Ocean Service, Strategic Environmental Assessments Division, Silver Spring, Maryland.

Christensen, J. D., M. E. Monaco, T. A. Battista, C. J. Klein, R. L. Livingston, G. Woodsum, B. Galeprin, and W. Huang. 1998. Potential impacts of freshwater inflow on Apalachicola Bay, Florida oyster (*Crassostrea virginica*) populations: coupling hydrologic and biological models. National Oceanic and Atmospheric Administration, National Ocean Service, Strategic Environmental Assessments Division, Silver Spring, Maryland.

Clark, R. D., T. J. Minello, J. D. Christensen, P. A. Caldwell, M. E. Monaco, and G. A. Mathews. 1999. Modeling nekton habitat selection in Galveston Bay, Texas: an approach to define essential fish habitat (EFH). National Oceanic and Atmospheric Administration, National Ocean Service, Center for Coastal Monitoring and Assessments, Silver Spring, Maryland.

Curry, S., and K. Schuckman. 1993. Practical considerations for the use of airborne GPS for photogrammetry. Photogrammetric Engineering and Remote Sensing 59:1611–1617.

Dangermond, J. 1991. Development and applications of GIS. Pages 101–103 *in* M. J. Heit and A. Shortreid, editors. GIS applications in natural resources. GIS World, Inc., Fort Collins, Colorado.

ESRI (Environmental Systems Research Institute). 1990. Understanding GIS, the ARC/INFO method, PC version. ESRI, Redlands, California.

Fowler, R. 1999. Digital orthophoto concepts and applications: a primer for effective use. Geo World 12(7):42–46.

Galtsoff, P. S. 1964. The American oyster *Crassostrea virginica* Gmelin. U.S. National Marine Fisheries Service Fishery Bulletin 61:1–480.

Gill, T. A., M. E. Monaco, S. K. Brown, and S. P. Orlando. 2001. Three GIS tools for assessing or predicting distributions of species, habitats, and impacts: Coastal Ocean Resource Assessment (CORA), Habitat Suitability Modeling (HSM), and Coastal Assessment and Data Synthesis (CA&DS). Pages 401–416 *in* Proceedings of the first international symposium on GIS fishery sciences. Fishery GIS Research Group, Saitama, Japan.

Griffith, D. A. 1996. Introduction: the need for spatial statistics. Pages 1–15 *in* S. L. Arlinghaus, editor. Practical handbook of spatial statistics. CRC Press, New York.

Haddad, K. D., G. M. MacAulay, and W. H. Teehan. 1996. GIS and fisheries management. Pages 28–38 *in* P. J. Rubec and J. O'Hop, editors. GIS applications for fisheries and coastal resources management. Gulf States Marine Fisheries Commission, Ocean Springs, Mississippi.

Haddad, K. D., and W. K. Michener. 1991. Design and implementation of a coastal resource geographic information system: administrative consideration. Pages 1958–1967 *in* O. T. Magoon, H. Converse, V. Tipple, L. T. Tobin, and D. Clark, editors. Coastal zone '91, proceedings of 7th symposium on coastal and ocean management. American Society of Civil Engineers, New York.

Holden, H., and E. LeDrew. 1998. The scientific issues surrounding remote detection of submerged coral systems. Progress in Physical Geography 22:190–221.

Ibrahim, S., and I. Hashim. 1990. Classification of mangrove forests by using 1:40, 000 scale aerial photographs. Forest Ecology and Management 33/34:583–592.

Isaak, D. J., and W. A. Hubert. 1997. Integrating new technologies into fisheries science: the application of geographic information systems. Fisheries 22(1):6–10.

Isaaks, E. H. and R. M. Srivastava. 1989. Applied geostatistics. Oxford University Press, Oxford, UK.

Kendall, M., C. Kruer, M. E. Monaco, and J. D. Christensen. 2000. Benthic habitats of Puerto Rico and the U.S. Virgin Islands: habitat classification scheme. National Oceanic and Atmospheric Administration, National Ocean Service, Center for Coastal Monitoring and Assessments, Silver Spring, Maryland.

Laurini, R., and D. Thompson. 1992. Fundamentals of spatial information systems. Academic Press, San Diego, California.

Levin, S. 1992. The problem of pattern and scale in ecology. Ecology 73:1943–1967.

Livingston, R. J. 1997. Trophic response of estuarine fishes to long-term changes in river runoff. Bulletin of Marine Science 60:984–1004.

Livingston, R. J., F. G. Lewis III, G. C. Woodsum, X.-F. Niu, B. Galperin, W. Huang, J. D. Christensen, M. E. Monaco, T. A. Battista, C. J. Klein, R. L. Howell, IV, and G. L. Ray. 2000. Modeling oyster population response variation to variation in freshwater input. Estuarine Coastal and Shelf Science 50:655–672.

Logan, B. J. 1993. Digital orthophotography bolsters GIS base for wetlands project. GIS World 6(special issue):58–60.

MacKenzie, C. L., Jr. 1977. Development of an aquaculture program for rehabilitation of damaged oyster reefs in Mississippi. Marine Fisheries Review 39:1–13.

Maniere, R., C. Bouchon, and Y. Bouchon-Navaro. 1994. Mapping of seagrass beds in the Bay of Fort-De-France (Martinique, French West Indies) by digitized aerial photographies. Proceedings of the first international airborne remote sensing conference and exhibition 3:735–743.

Maniere, R., J. Courboules, C. Bouchon, J. Jaubert, and D. Mahasneh. 1986. Application of high-resolution remote sensing and geographic information systems to the Jordanian coast. Proceedings of the twentieth international symposium on remote sensing of environment 2:771–780.

Mayer, L., J. H. Clarke, and S. Dijkstra. 1999. Multibeam sonar: potential applications for fisheries research. Journal of Shellfish Research 17:1463–1467.

Minello, T. J. 1999. Nekton densities in shallow estuarine habitats of Texas and Louisiana and the identification of essential fish habitat. Pages 43–75 *in* L. R. Benaka, editor. Fish habitat: essential fish habitat and rehabilitation. American Fisheries Society, Symposium 22, Bethesda, Maryland.

Monaco, M. E., and J. D. Christensen. 1997. Biogeography program: coupling species distributions and habitat. NOAA Technical Memorandum NOAA-TM-NMFS-SWRSC-239:133–139.

Monaco, M. E., S. B. Weisberg, and T. A. Lowery. 1998. Summer habitat affinities of estuarine fish in U.S. mid-Atlantic coastal systems. Fisheries Management and Ecology 5:161–171.

Mumby, P. J., E. P. Green, A. J. Edwards, and C. D. Clark. 1999. The cost-effectiveness of remote sensing for tropical coastal resource assessment and management. Journal of Environmental Management 55:157–166.

NOAA (National Oceanic and Atmospheric Administration). 1996. Magnuson-Stevens Fishery Conservation and Management Act as amended through 11 October 1996. NOAA Technical Memorandum NMFS-F/SPO-23.

NOAA (National Oceanic and Atmospheric Administration). 1998. Product overview: products and services for the identification of essential fish habitat in Puerto Rico and the U.S. Virgin Islands. National Oceanic and Atmospheric Administration, National Ocean Service, Strategic Environmental Assessments Division, Silver Spring, Maryland.

NOAA (National Oceanic and Atmospheric Administration). 1999a. Habitat suitability modeling extension. Available: *http://biogeo.nos.noaa.gov/products/apps/hsm/* (September 2003).

NOAA (National Oceanic and Atmospheric Administration). 1999b. Analysis of alternative freshwater inflows on managed biological resources of Apalachicola Bay, Florida. Available: *http://biogeo.nos.noaa.gov/descriptions/apalach.shtml* (September 2003).

NOAA (National Oceanic and Atmospheric Administration). 2000a. Benthic habitat mapping of Puerto Rico and the U.S. Virgin Islands. Available: http://biogeo.nos.noaa.gov/products/benthic/ (September 2003).

NOAA (National Oceanic and Atmospheric Administration). 2000b. Habitat digitizer extension. Available: *http://biogeo.nos.noaa.gov/products/apps/digitizer/* (September 2003).

NOAA (National Oceanic and Atmospheric Administration) 2002. Benthic Habitat Mapping of the main Hawaiian Islands. Available: *biogeo.nos.noaa.gov/projects/mapping/pacific/main8/* (September 2003).

NOS (National Ocean Service). 1997. Work plan: products and services for the identification of essential fish habitat in the Gulf of Mexico. National Oceanic and Atmospheric Administration, National Ocean Service, Strategic Environmental Assessments Division, Silver Spring, Maryland.

Oracle. 1999. Using Oracle8i *inter*Media for location queries and data warehousing. Oracle Corporation, Redwood Shores, California.

Parker, H. D. 1988. The unique qualities of geographic information system: a commentary. Photogrammetric Engineering and Remote Sensing 54:1547–1549.

Pasqualini, V., C. Pergent-Martini, C. Fernandez, and G. Pergent. 1997. The use of airborne remote sensing for benthic cartography: advantages and reliability. International Journal of Remote Sensing 18:1167–1177.

Rubec, P. J., M. S. Coyne, R. H. McMichael, Jr., and M. E. Monaco. 1998. Spatial methods being developed in Florida to determine essential fish habitat. Fisheries 23(7):21–25.

Rubec, P. J., S. G. Smith, M. S. Coyne, M. White, D. Wilder, A. Sullivan, R. Ruiz-Carus, T. MacDonald, R. H. McMichael, Jr., M. E. Monaco, and J. S. Ault. 2001. Spatial modeling of fish habitat suitability in Florida. *In* Spatial processes fisheries symposia. Alaska Sea Grant College Program, Fairbanks, Alaska.

Rulli, J., and N. Shah. 1998. Keep it simple: orthorectification alternatives save time and money. GIS World 11(6):58–62.

Sabins, F. F. 1996. Remote sensing: principles and interpretation, 3rd edition. Freeman, New York.

Scott, J. M., F. Davis, B. Csuti, R. Noss, B. Butterfield, C. Groves, H. Anderson, S. Caicco, F. D'Erchia, T. C. Edwards, Jr., J. Ulliman, and R. G. Wright. 1993. Gap analysis: a geographic approach to protection of biological diversity. Wildlife Monographs 123:1–41.

Shepard, C. R. C., K. Matheson, J. C. Bythell, P. Murphy, C. B. Myers, and B. Blake. 1995. Habitat mapping in the Caribbean for management and conservation: use and assessment of aerial photography. Aquatic Conservation: Marine and Freshwater Ecosystems 5:277–298.

Sotheran, I. S., R. L. Foster-Smith, and J. Davies. 1997. Mapping of marine benthic habitats using image processing techniques within a raster-based geographic information system. Estuarine Coastal and Shelf Science 44:25–31.

Thorpe, J. 1993. Aerial photogrammetry: state of the industry in the U.S. Photogrammetric Engineering and Remote Sensing 59:1599–1604.

Wiens, J. A. 1989. Spatial scaling in ecology. Functional Ecology 3:385–397.

Wilber, D. H. 1992. Associations between freshwater inflows and oyster productivity in Apalachicola Bay, Florida. Estuarine Coastal and Shelf Science 35:179–190.

Zeiler, M. 1999. Modeling our world: the ESRI guide to Geodatabase design. Environmental Systems Research Institute, Inc., Redlands, California.

Chapter 8

Geographic Information Systems Applications in Offshore Marine Fisheries

ANTHONY J. BOOTH AND BRENT WOOD

8.1 INTRODUCTION

Marine ecosystems are under direct threat from growing anthropological influences such as habitat degradation, overfishing, and global warming. This places pressure on both scientists and managers to make informed decisions that help mitigate a growing suite of problems. Scientists, therefore, are required to make inferences based on knowledge about the marine system being studied (from samples collected or the literature consulted) in order to provide qualified recommendations. Scientific recommendations usually are transferred to the managers who, assuming a lack of political intervention, then draft the necessary policies to be implemented. In many cases, critical information is not available or there is a lack of fundamental knowledge of the dynamics of the system. In these situations, the "precautionary principle" (FAO 1995) would be invoked, that is, when data or information pertaining to the status of a resource are limited, then the most conservative harvesting strategy should be implemented.

Geographers and fisheries scientists have, over the past two decades, fostered a shared interest in the spatial analysis of physical and biological resources. Geographers are skilled in understanding the concept of space and relationships among geographic entities. Fisheries scientists, on the other hand, are skilled in the separation of stocks, life histories, biological interaction, resource abundance, recruitment dynamics, and stock assessment. Synergy between the two disciplines has resulted in the adoption of geographic information systems (GIS) to collect, store and retrieve, and analyze fisheries related data that are spatially referenced (Caddy and Garcia 1986; Giles and Nielsen 1992; Isaak and Hubert 1997; Meaden 1999; Nishida and Booth 2001).

The marine environment can be divided into coastal and offshore zones, with the scope of this chapter being restricted to the latter. The marine limits of the coastal zone are difficult to define, but typically this zone includes areas visible from land such as the intertidal zone, bays, and estuaries (see Chapter 7). Offshore areas are defined, for the purposes of this chapter, as those marine areas not visible from land. These encompass pelagic and demersal environments, including the continental shelf and slope, abyssal plain, and pinnacles. This domain is arguably the most complex aquatic environment, encompassing 71% of the earth's surface area, and it also is arguably the area on earth, and within fishery science, that is the least understood.

Geographic information systems can assist in solving certain classes of fisheries problems by providing data management tools (storage, retrieval, and manipulation), visualization capabilities, and a platform to investigate spatial relationships within the data. This chapter reviews the challenges presented in using GIS in an offshore marine context, summarizes the techniques available for analysis of fisheries problems, and provides an overview of research and management applications in offshore marine environments.

8.2 CHALLENGES OF USING GIS IN OFFSHORE ENVIRONMENTS

Despite being an established tool in terrestrial disciplines, only recently has GIS been widely used for offshore marine applications. For instance, the first American doctoral dissertation involving multidisciplinary marine science and GIS was submitted in the mid-1990s (Wright 1994). Many reasons for the slow adoption of this technology have been proposed (Caddy and Garcia 1986; Meaden and Kapetsky 1991; Simpson 1992; Meaden 1996); these include spatiotemporal variability of marine environments, data acquisition and analysis costs, and technological constraints.

8.2.1 Spatiotemporal Variability

The offshore environment is a vast aquatic domain that forms structures that vary their relative positions and values over time and space. Data collected, therefore, are voluminous, dynamic, and spatiotemporally variable (Goodchild 1992; Li and Saxena 1993; Lockwood and Li 1995).

Sources of variation, composed of both process and measurement error components, need to be isolated and quantified so that trends and patterns can be recognized. Process error occurs because biological and physical processes lack stability (entropy principal) and often exhibit a high degree of interaction between each other. Measurement error, on the other hand, arises because sampling equipment will never completely be precise and because samples are discrete units, while the biological and physical phenomena that are being investigated are continuous. Sample data, therefore, only represent approximations of the phenomena being observed. Accounting for, and reducing sources of, this variability has led to the development of advanced quantitative modeling, particularly in the fields of fisheries stock assessment, oceanography, and data collection techniques.

Variability in fisheries data, while being recognized, is not formally considered in most GIS applications (Lucas 1999). Outputs appear imprecise and difficult to visualize. To date, little progress has been made to visualize variability in a satisfactory manner by using traditional cartographic methods, such as thematic maps or charts. In terrestrial environments, geographic position is (relatively) fixed, such as distance from roads, mountain peaks, and topographic beacons, and can be measured by cost-effective techniques involving foot, motor vehicle, or low altitude aircraft. By contrast, the marine environment is a dynamic fluid domain, and offshore boundaries often are indeterminate (Burrough 1996). Consequently, data structures can be perceived only as relative in both time and space. Despite having global positioning system (GPS) technology, the collection of data are still reduced to using acoustic transducers *underneath*; nets *behind;* conductivity, temperature, and depth (CTD) probes *next to*; submersibles *away* from vessels (Wright and Goodchild 1997); or satellites *above* Earth. Collecting and collating data within varying tidal patterns, swell heights, and ocean currents pose similar problems.

Spatial scales vary within and between systems, ranging from isolated volcanic atolls, patchy plankton blooms, winding thermal front boundaries, contiguous deep reef systems, expansive boundary currents, and even entire ocean masses. Biological activity also is patchy, with large geographical distances often separating populations or communities. For example, reef resident species are found in patches (<10 km^2), with each patch separated from each other by large sand or mud interfaces. Pelagic fish often are patchily distributed as discrete schools at frontal interfaces (Nishida and Miyashita 2001). The life history patterns of many fish species cover a variety of habitats and typically exhibit the trend of inshore spawning with ontogenetic migrations of adult fish to deeper waters (e.g., dusky rockfish *Sebastes ciliatus*) (Reuter 1999).

Offshore marine environments vary temporally. Biological cycles and life spans differ from a few minutes or hours (bacteria), to days (plankton), years (fish), or even centuries (cetaceans). Marine organisms move, and there can be displacements and localized depletions of fish over the decades, commensurate with global warming and fishing practices (Francis and Sibley 1991; Ray et al. 1992; Francis and Hare 1994). Many offshore marine fish migrate with movements varying from diel vertical migrations (plankton and pelagic fish) of a couple hundred meters a day (Luo et al. 2000), to long distance feeding or reproductive migrations (whales and tunas) covering thousands of kilometers over a period of months (Nishida and Miyashita 2001).

Combining these sources of variation (Figure 8.1) leads to important issues regarding accuracy and precision. For example, what happens at a fine spatial scale cannot necessarily be inferred to occur at a coarser resolution. A suitable example would be in the differences between high-resolution and low-resolution images. Resolution can be reduced but never increased. Geographic information system output summarizing any spatiotemporal trend, such as the distribution and abundance of a species, oceanographic conditions at a

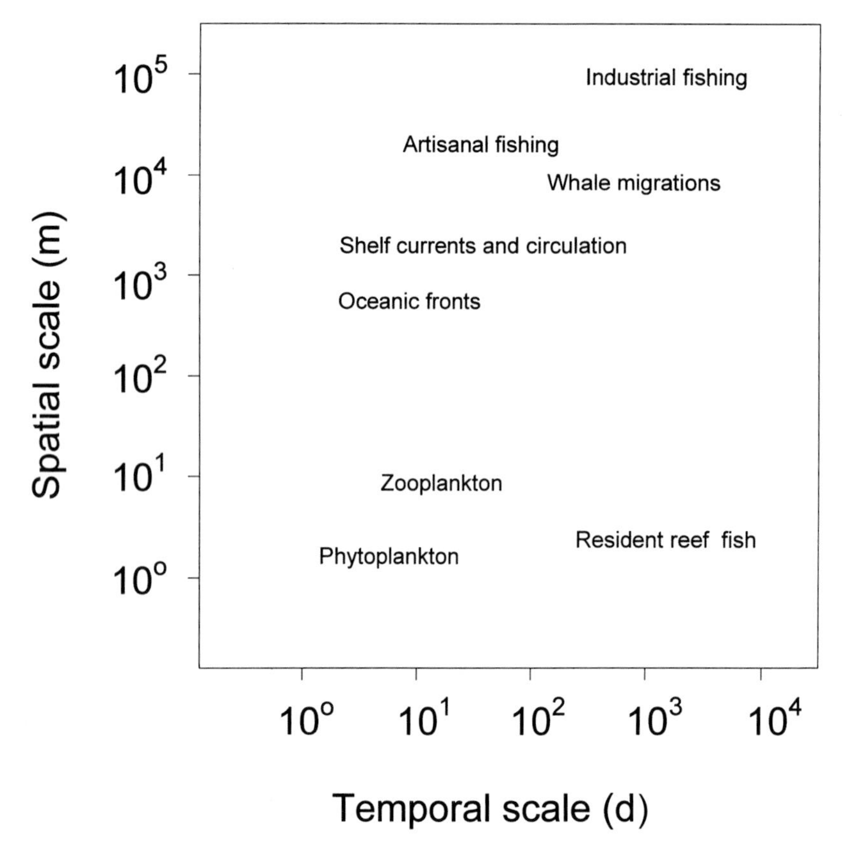

Figure 8.1 Spatiotemporal variability in marine organism life histories, physical processes, and anthropological activities that occur within the offshore marine environment.

specific area, or the locations of fishing vessels, will be outdated even before the data can be analyzed. Variation due to process and measurement error is difficult to analyze, and it is not surprising that this variation has been "conveniently" ignored. Outputs without corresponding estimates of variation (which unfortunately includes most published studies) should, therefore, be viewed with caution.

Appropriate decision making relies on a reasonable understanding of the physical, biological, and anthropological processes under investigation. The level of understanding of phenomena and the predictability and confidence in results obtained are similarly dependent on the quality of data available as input. Data quality has two facets: choosing those data that are appropriate to the investigation and the actual collection of samples by using suitable methods. Before these aspects are considered, a data model needs to be chosen that will dictate how the data will be stored, analyzed, and visualized. This is an ideal situation that rarely occurs in practice because data storage usually is perceived as an analysis-related issue. An example would be the estimation of pelagic fish biomass from acoustic surveys. An appropriate data model would be one that can cope with data in at least three dimensions (preferably four), as the objective of the analysis would be to calculate volume (that of fish biomass). Any GIS, like other mathematical constructs, is a representation of reality. If objects within the model are to represent those features of the "real world," then a conceptual model is imperative (Peuquet 1994)

Marine data, as with other temporally variant data, have a life span. Data only are valid if constantly changing physical and biological processes are taken into consideration. For example, within hours of sampling plankton, current patterns may have changed or the phytoplankton bloom dissipated. Ideally, data need to be synoptic, that is, they represent both physical and biological phenomena simultaneously. Resolving temporal conflicts in highly mobile data that are valid in space for only a brief time, such as pelagic or migratory stocks and fishing fleet dynamics—particularly those where there is a discrepancy in life spans (whale production versus plankton productivity)—can be problematic.

One of the benefits of GIS is the ability to collate diverse data sources. This strength also is one of its greatest weaknesses, as the lowest common denominator principal prevails with the overall quality of the collated data dependent on the lowest quality data set (Von Meyer et al. 1999). Data may have different temporal and spatial scales as a result of having been collected by using a variety of methods that include differing process and measurement errors. Dijkema (1991), for example, noted that data available for compiling a marine habitat map of the Netherlands varied considerably in spatial scale. Bathymetric map scales ranged between 1:10,000–1:25,000, while sediment map scales ranged between 1:25,000 and 1:200,000, with the poorest resolution being a coarse 1:700,000.

Much has been written on the incompatibility of marine data because of differences in scale, but standardization initiatives are in progress, and significant developments are anticipated in the near future. Alternative data models are being considered to circumvent these scale issues and to facilitate dynamic three-dimensional (3-D) and four-dimensional (4-D) modeling (Li 1999; Meaden 2001) (Figure 8.2). Currently used data models are an abstraction of the modular way humans tackle complicated problems. These data models then are implemented within an object-orientated GIS framework (Burrough and Frank 1996), where geographical and biological phenomena have been reduced to "crisply" (as opposed to "fuzzy") defined entities (Woodcock and Gopal 2000). Hierarchical data models, nesting within each other in a fractal fashion, are one solution. One derivative, a recursive tesselation of nested 3-D tiles, also known as the octree representation outlined by Li (1991), is a compact 3-D volume-based representation with predefined minimum-sized cubes or tiles to restrict database size. In using this method, storage is also relatively low in comparison with other volume-based representations. In addition, it is a suitable data structure for efficient spatial analysis, Boolean representation, and database management due to its hierarchical structure.

The measurement of the temporal domain also has been proposed, either by noting that time, being topologically similar to space,

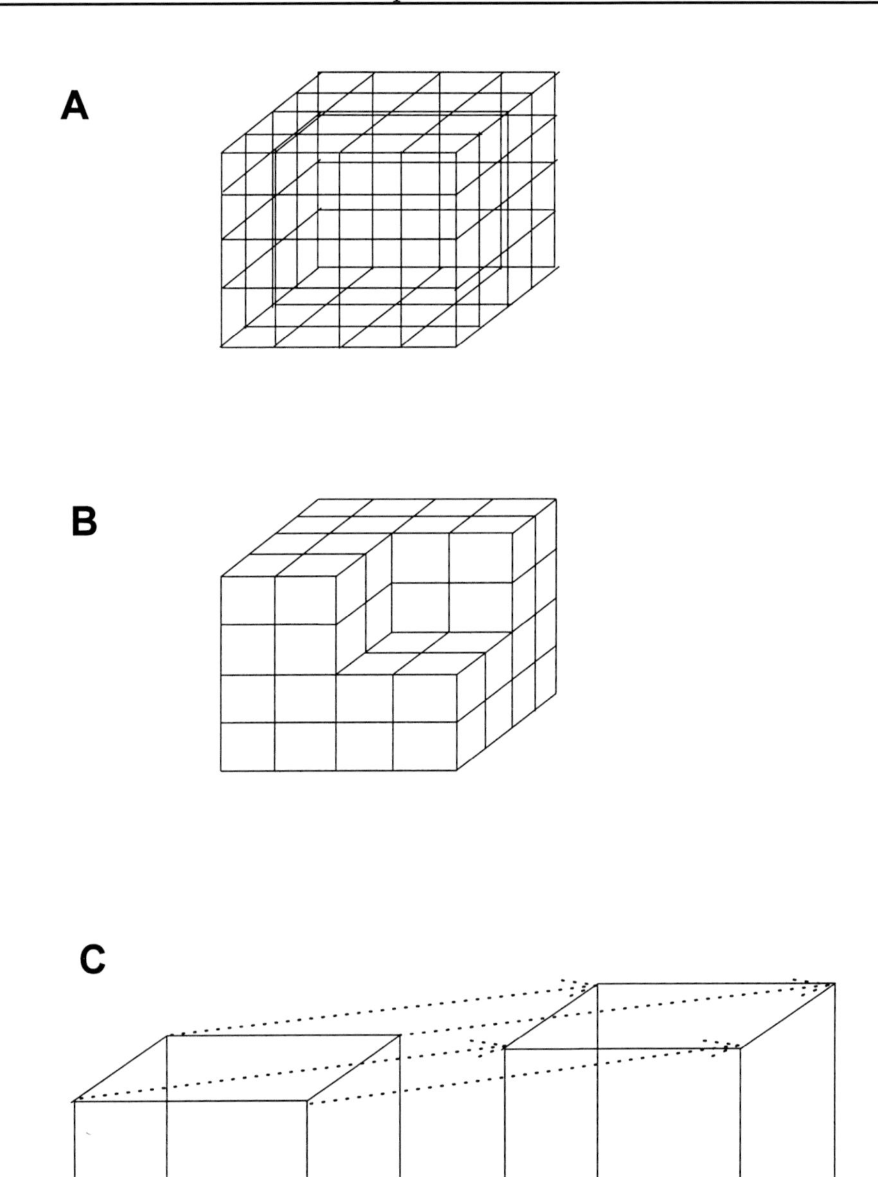

Figure 8.2 Three-dimensional and four-dimensional data models that could be used to guide data collection. A 3-D array (A) where each sampling unit would be within each small cube, an octree (B) illustrating recursive tessellated cubes where the sampling unit could be on a variety of scales being recursively tessellated within each other, and the 4-D hypercube (C) incorporating the temporal dimension where the sampling unit would be sampled at successive time intervals.

can allow for spatial data structures to be modified to include an additional dimension and establish a temporal topology (Langran and Chrisman 1988) or via the hypercube where the vertices of the cubes represent the temporal component within the data.

Finally, financial costs associated with the collection of marine biological, physicochemical and sedimentation data are high. Typical daily research vessel running costs range between approximately US$10,000 and $25,000 (Wright 1999). These costs, together with ancillary costs associated with the synthesis and storage of the data, in a compatible format, have hindered the development of aquatic and, in particular, marine GIS.

8.2.2 Technological Considerations

Most GIS platforms contain two-dimensional (2-D) database functionality with few progressing to 3-D data models. Spatial capabilities have been added to interface standard databases to the GIS, resulting in heterogeneous systems composed of separate, interfacing components. No generic commercial-specific or marine-specific packages are available that can efficiently handle fisheries data. In effect, this limits fisheries GIS to those users with advanced technical experience.

Papers presented during both the First International Symposium on GIS in the Fisheries Sciences (Nishida et al. 2001) and the recent Second Symposium on GIS/Spatial Analysis in Fishery and Aquatic Sciences (University of Sussex, Brighton, UK, September 2002) were based primarily on commercially available terrestrial 2-D or 2+D GIS software. Considering that traditional GIS were not designed to deal with temporal data, their geographic nature generally has restricted them to work with surfaces rather than volumes. This is popularly called 2+D visualization. Modern visualization and analysis software, such as seePower, Mapinfo, SQS/Sybase, S+, Matlab, IDL, and Vis5D are able to query relational database management systems (RDBMS) to retrieve data and then store and manage the data in internal multi-dimensional arrays to improve performance. These software platforms handle fisheries and oceanography data to a limited extent, specializing in only a few specific functions such as simple presentation, navigation systems (electrical charts), satellite data processing, contour estimation (2+D visualization), database functionality, vertical profiling for oceanographic information, and bathymetric mapping. Although these systems were functional in the strictest sense, they were not generic in being able to incorporate all of these specific functions into a single system. New software is needed for conducting spatial numerical analyses and modeling that are linked to stock assessments, simulations, and ecosystem management. Furthermore, such software must be user-friendly and able to run without requiring any programming because fishery scientists in many countries have limited funding to hire GIS specialists and do not have the time or the skills to do the programming themselves. Therefore, the development of easily available and understandable marine GIS software is required. Several systems are in the developmental stages (Itoh and Nishida

2001; D. A. Kiefer et al., University of Southern California, unpublished abstract from the First International Symposium on GIS in Fishery Science, 1999) and their release is anticipated. For example, Itoh and Nishida's (2001) menu-driven Marine Explorer GIS software is suited to the dedicated storage and manipulation of fisheries and oceanography data.

8.3 GIS TECHNIQUES IN OFFSHORE ENVIRONMENTS

8.3.1 Database Tools

From a database perspective, there is no inherent distinction to differentiate between data collected in offshore versus other marine environments. From a data perspective, however, there are some problems because the nature of the offshore water environment tends to make much of the data collected less accurate than similar data from other environments.

The relational model underlying RDBMS software has proved to be a robust foundation for storing and querying conventional attribute data. However, it fails to provide comparable functionality for spatial data. Relational database management systems were common and successful well before GIS began to appear as a mainstream technology. Geographic information system developers built on the success of relational database technology by adding value and functionality to existing databases. Hence, the usual approach to integrating spatial data with traditional RDBMS is to add spatial data capability to the existing RDBMS. This typically is achieved by inserting a spatial query server between the client (user) and database server. This spatial server addresses the spatial aspects of queries while returning nonspatial queries to the underlying database server. Implementations can use spatial servers supplied by the RDBMS vendor, such as Oracle's SDO, Sybase's SQS, Informix's Spatial Datablades, and the spatial datatypes available in PostgreSQL. An alternative approach is to use a spatial query engine from a different vendor; Environmental Systems Research Institute's SDE and Mapinfo's SpatialWare are examples.

The other approach is to implement a custom system that uses an RDBMS and a separate GIS. Virtually all of the available GIS packages provide links to RDBMS to enable attribute data management outside of the GIS package.

Users, therefore, have a choice between the security of a "one-stop shop," where the spatial and RDBMS components are developed and supplied by the same vendor, or the flexibility of selecting the GIS and DBMS software independently to best meet their requirements.

8.3.1.1 ***Metadatabase Considerations.*** Simplistically, metadata is "data about data"— much like an index or catalog. Many large databases are analogous to encyclopedias, in that they contain huge amounts of potentially useful information, yet have no equivalent of an index. The data are there but difficult to locate and extract, making the database relatively worthless. One example of a successful online maritime metadatabase is MarLIN (Rees and Ryba 1998), which consists of an Internet interface to an Oracle database, containing data describing the data sets held at the Commonwealth Scientific and Industrial Research Organization Marine Research Divisional Data Centre. Despite the obvious benefits of such systems, they have not proliferated, primarily because of the costs associated with populating them.

8.3.2 Analytical Tools

Various analytical tools are available on most proprietary systems to investigate the relationships between spatial elements and to improve inferential statistical capabilities. Inferential statistics are requisite if hypotheses are to be tested or risks associated with management recommendations quantified.

Statistical tools such as principal component analysis, linear and logistic regression, and generalized models are available in some but not in all GIS software programs. Analysis only recently has included spatial statistical methods—principally through third party statistical vendors—and these have the ability to analyze point-pattern, lattice, and geostatistical data. Examples include the commercially available statistical environment S+ (Insightful Corporation 2003), GAUSS-based package SpaceStat (TerraSeer, Inc. 2003), Matlab's mapping module (The MathWorks, Inc. 2003), or the open source statistical environment R (R Foundation 2003). S+, and, more recently, R, satisfy the definition of a GIS in having the requisite database tools such as a suite of graphical functions, a collection of nonspatial and spatial (through third-party add-ons) statistical routines, and the ability to incorporate custom-written third-party computer code by using standard programming languages like C, C++, and FORTRAN. The use of open source software is a growing trend because it facilitates the adoption of modern statistical methods, including contributed spatial statistical code, to a growing body of users.

Additional custom "add-ins" have been released into the public domain on condition that intellectual property is acknowledged. Spatial Tools and Animal Movement (Hooge and Eichenlaub 1997) provides users with a suite of more than 40 additional data manipulation tools to perform geometric transformations, mosaicking, resolution alteration, data clean-up, and spatial analysis.

8.3.3 Data

Data collection in the marine environment is not restricted to sampling with a bottle over the gunwhale of a vessel, haphazardly measuring catches, or interviewing fishers to log their fishing activities. Primary and secondary types of data are available (Meaden 1996). Primary data are original data collected manually or automatically during the course of an investigation. Secondary data, also known as derived data, involves information taken from preexisting sources and usually is housed in remote-sensing centers, government agencies, and research institutes in either hard copy or digital formats.

Hard copy formats include thematic maps, nautical charts, output from models, tables, photographs, maps, and charts. Digital sources include remote-sensed imagery, hydrographic and topographic maps, and digitized bathymetry and coastlines. For secondary data, the temporal aspect usually lost with processing, may be reincorporated through splicing together data sets in a temporal sequence. Meaden and Do Chi (1996) outlined the extensive Kingfisher maps for use by British fishers in the North and Irish seas and some French and Icelandic waters. Detailed depth information is available, together with the locations of wrecks, cables, pipelines, and wellheads. Secondary data, while being cheap and easily accessible compared with primary data, need to be assessed carefully before incorporation. There is a need for clearly defined scales and projections, together with an awareness of whether the information is outdated.

8.3.4 Sources and Collection Methods

Fisheries data from offshore marine environments are represented in at least 4-D space— geographic position in three dimensions, and one dimension representing the measured variable. The magnitude measure of the data refers to information that can be quantified, such as the volume or number of fish sampled, depth or type of substrate, or temperature reading. Adding a temporal component to data will increase the dimensionality to at least five dimensions.

Point-based fisheries data, unlike its mathematical construct that has zero dimensions, is located in 3-D space with coordinates along x, y, and z axes (Figure 8.3). Such data include Secchi depth measurements, fish catches recorded on log-sheets, and plankton net samples. The latter are frequently treated as point samples. Density can be estimated easily but the spatial arrangement within the volume sampled cannot be explained (Kracker 1999).

Profile data usually are collected along a near-vertical profile (such as with a CTD probe) or horizontally with trawls or line transects. Vertical profiles are those data that are recorded at a specific latitude-longitude location but contain locations on the vertical plane (one x, one y, and many z coordinates; Figure 8.3). There are, however, no data between data points, around data points, or in the time between sampling repeats. The CTD profilers provide such information on conductivity, temperature, and depth taken at intervals from a stationary ship, providing information through the water column. Similar spatiotemporal data also could be collected by mooring the instrument and logging many data points through time, but at a single x, y, and z. Rectangular midwater trawl plankton samplers also would fit into this category because they sample plankton at discrete intervals.

Plane data includes data collected on a plane in three dimensions (one x but many y and z coordinates; Figure 8.3). These data usually are collected by remote means, such as acoustic instruments onboard fishing or research vessels or from satellites. Such approaches provide near synoptic sampling if we assume that pelagic fish are slower than the survey vessel in the case of acoustic data. Data col-

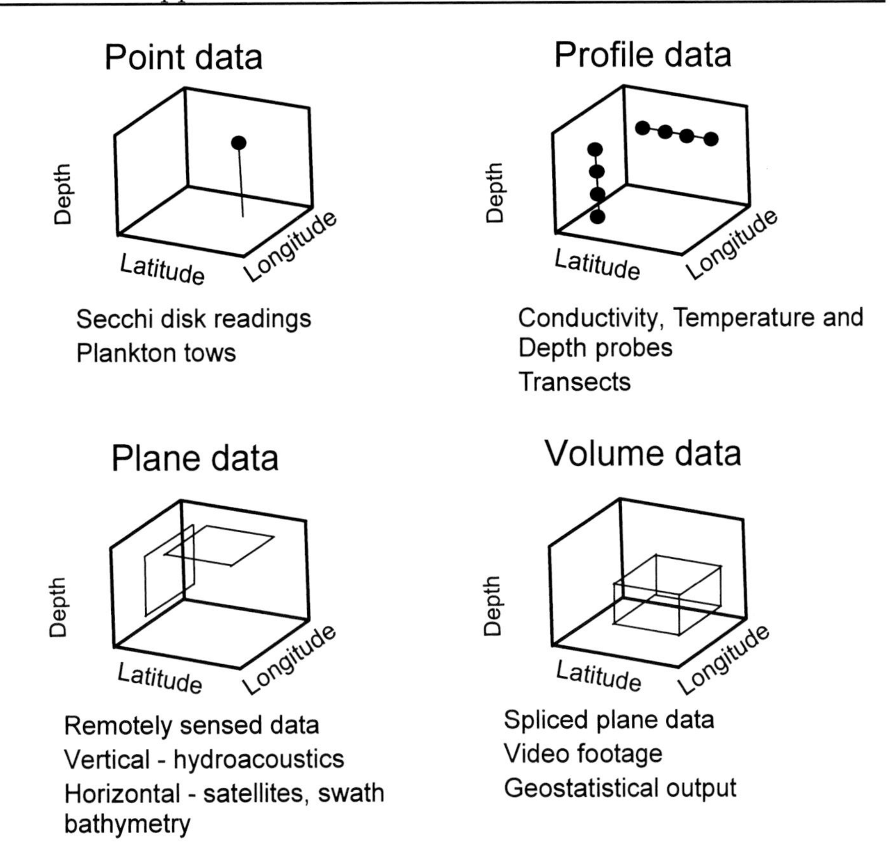

Figure 8.3 Data categories for offshore fisheries GIS applications. Note that the locations of the data are in three dimensions, with each sample including a fourth dimension (which is not illustrated) that refers to the measurement of the sample.

lected by using acoustic methods sample the aquatic environment vertically and horizontally when using swath bathymetry.

Volumetric data includes data collected in at least four dimensions (many x, y, and z coordinates plus time; Figure 8.3). These usually are derived data, such as plane data collated over time, to implicitly include a temporal dimension or in the form of underwater video. Hydroacoustics, in contrast to horizontal satellite derived data or point data from research or commercial vessels, gives a sense of volume. This method can sample a large volume of water in a relatively short period of time and, because of this advantage together with significant improvements in echo-sounder and echo-integration technology, is becoming favored over other sampling methods. Strictly speaking, acoustic and satellite data are a number of discrete samples collected with pings or reflected radiation. Petitgas (1993), Meaden (1996), and Kracker (1999) outline problems with summary statistics for these data and suitable statistical tools to parse the spatially referenced data sets that are generated through the echo-integration process. The data are complex and tend to be condensed during analysis to reduce their complexity, thereby losing resolution.

8.3.5 Data Logging Systems

There are two main categories of systems designed specifically for logging data. One uses fixed sampling devices to collect information. Often multiple instruments are used at selected sites to collect data over time, extrapolating data from the sites to derive a model covering a wide area. Perhaps the best maritime example of such a system is the TAO (tropical atmosphere ocean) project (McPhaden et al. 1998). This project involves an array of deep ocean moored buoys covering the equatorial Pacific Ocean. Instruments are attached to the buoys to record surface meteorological and subsurface oceanic data, which are transmitted to shore via satellite and are used for a variety of projects.

The second type of data logging system is mobile. Typical examples are swath bathymetry systems (both towed and hull mounted), as well as various instruments connected to recording devices on research and fishing vessels. Many offshore fishing vessels are equipped with personal computer-based navigation software, capable of capturing position and depth data. Temperature, salinity, and meteorological data logging capabilities are less common.

Simple data capture systems are effective, but there frequently are problems with erroneous data. Echo sounders may report misleading depths due to rapid changes in the depth or record thermoclines or schools of fish as the bottom depth. Any analysis of captured data must implement some method of filtering these errors. All such errors are captured along with "good" data. It is not until some examination of the data are undertaken that most such errors are discovered.

8.4 APPLICATIONS OF GIS IN OFFSHORE MARINE ENVIRONMENTS

Is GIS a "tool" or a "science"? This question over GIS within the "tool–science" continuum is important because the definition adopted will dictate how a problem is approached (Wright et al. 1997; Booth 2001).

To "do GIS" merely refers to using a computerized toolbox; an analogy would be using a spreadsheet to do calculations or a word processor to manipulate text. The choice of tool is context specific. The "toolbox" position sees GIS as the use of a particular piece of software, the associated hardware, and the spatially referenced data for some specific purpose. This tool is inherently useful, and its development and availability are independent to its application. The "toolbox" aspect is applicable where the problem that has been identified can be solved by using the technology available. It is not surprising then that this approach is successful with government, business, and management.

In contrast, the "science of GIS" infers an intimate connection between tool and science, involving research on a set of basic problems, each of which probably existed prior to the development of GIS but which now is solvable because of technology. Wright et al. (1997)

used the analogy of "computer science." The development of computing technology provided the opportunity to solve fundamental research problems that were mathematical in origin.

To assess where GIS fits into the continuum, the user merely has to ask whether they are using a sequence of spatial analytical commands within a software program to solve a problem or are they dealing with analysis of issues that surround the use of GIS. In the context of this chapter, we have adopted the definition of a "geographic information system" (as opposed to "geographic information science") and, therefore, GIS is considered a spatial tool.

8.4.1 The GIS Amoeba

Before considering specific current offshore marine GIS applications, it is necessary to understand how GIS is developing, thereby providing insight into potential problems that can be addressed in the future.

Problem solving within a GIS environment can be viewed as a constantly evolving and dynamic system. Figure 8.4 illustrates this

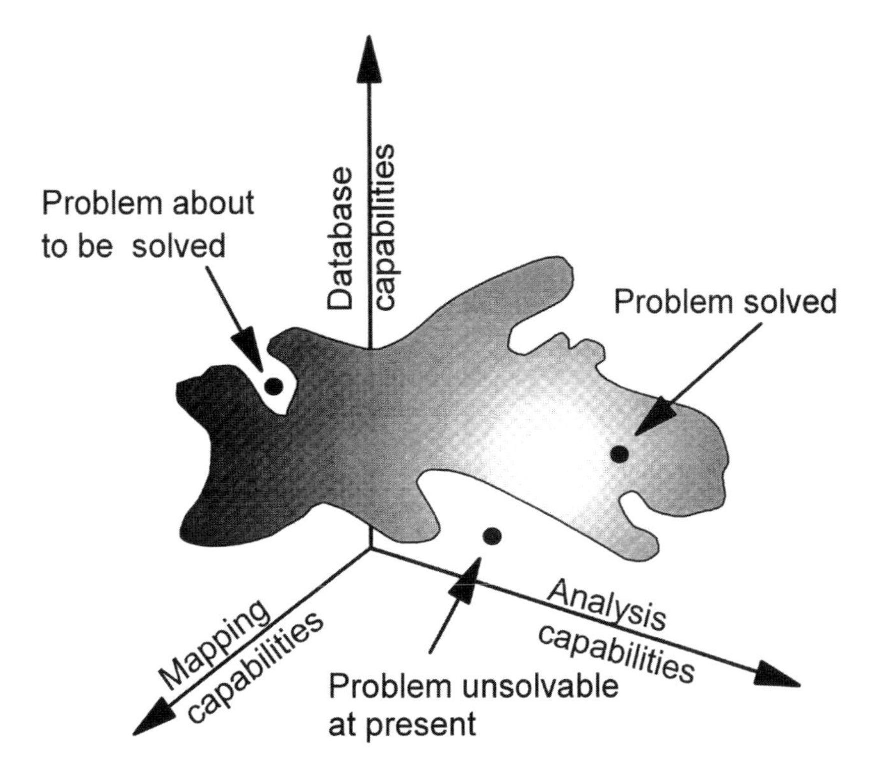

Figure 8.4 The GIS amoeba that illustrates those classes of problems that will be tractable with concurrent advances in three core GIS functions: database, analysis, and mapping capabilities. The amoeba's pseudopodia represent nonlinear advances in the three core areas, which together give the amoeba a 3-D reality. Classes of problems are represented as circles with their tractability determined by whether they have been ingested by the pseudopodia.

concept from a 3-D perspective. In reality, this figure would be hyper-dimensional, but it, nevertheless, illustrates the concept, given the human difficulty in visualizing highly dimensional diagrams. The three axes represent increasing capability (and complexity) in mapping, databases, and analysis. Three classes of spatial problems have been illustrated, one falling within the amoeba, one about to be ingested by the amoeba's pseudopodia, and a third out of reach of the pseudopodia. These three problems are positioned to represent their tractability. The first has been solved and represents a standard GIS task (such as 2-D mapping or buffer zone analysis). The second is a class of problem that has been unsolvable in the past but due to advances in any two or more core areas, will be solvable in the near future. An example in this category would be proper 3-D (not a 2+D representation) volume analysis that is currently restricted to specialized systems but will be made available on most commercially available platforms. The third class of problem is currently intractable because all three GIS areas are immature in development. Progress in GIS development, as with most technologies, has followed an amoebic locomotory development history, with progress in the core areas developing in a nonlinear direction as specialized research groups solve fundamental problems.

GIS can be applied toward management goals or the resolution of scientific questions. The examples of management-related GIS are numerous in terrestrially based disciplines such as engineering, town planning, and waste management (Marble et al. 1984; Smith et al. 1987; Star and Estes 1990; Maguire et al. 1991). In contrast, no fisheries systems are dedicated to management yet. Management agencies do, however, contract or outsource research to address management needs. The contracted agencies are often the GIS users, developing systems that are context specific. Scientifically related GIS include the development of smaller systems with a short life-span, designed for rapid exploratory data analysis and hypothesis testing, in an attempt to understand the spatial relationships among variables.

8.4.2 Scientific-Driven or Hypothesis-Driven Approaches

8.4.2.1 ***Visualization.*** There has been a steady move toward visualizing data, often prior to analysis, in order to examine possible trends, to identify outliers, and to gain overall insight into the problem at hand. While humans are able to perceive data within two or even three dimensions, there are difficulties visualizing higher dimensional data.

There have been several strategies for overcoming this limitation. A commonly used method is to provide a number of different views of the same data. For example, data could be decomposed into an array of relatively simple 2-D or 3-D graphs with different axes. This is the rationale behind the Trellis Graphics function in S+. Similarly, data are integrated over space or time and played back, such as videosequencing images, to the investigator. An excellent example would be the study by Ault et al. (1999) that presents oceanographic conditions, together with biological processes as a temporal sequence of frames.

Figures 8.5 and 8.6 illustrate two examples of scientific visualization using two open source software packages—OpenDX (2003) and Generic Mapping Tools (Wessel and Smith 2003). Figure 8.5 is a single frame from an animation of the 2001 fishing season for hoki *Macruronus novaezelandiae* off South Island, New Zealand. The animation allows users to view changes in catch, effort, or catch rate. This provides a better understanding of the dynamics and the effects of depletion of the fishery that targets spawning aggregations of hoki around the Hokitika Canyon. Changes in the densities of the New Zealand dredge oyster *Tiostrea chilensis* (also known as *T. lutaria*) in the Foveaux Strait, New Zealand, from three surveys are illustrated

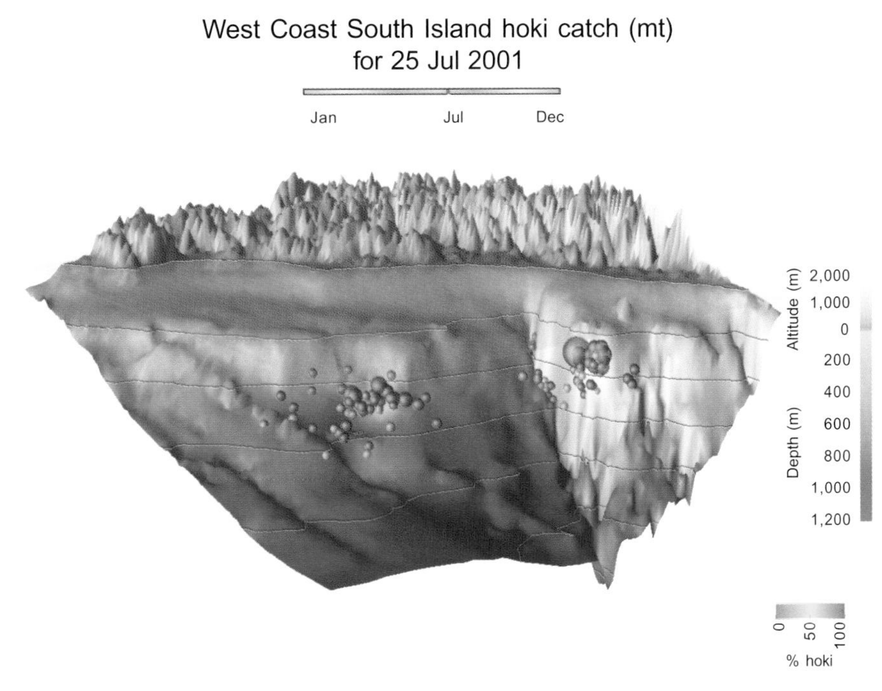

Figure 8.5 A three-dimensional (3-D) visualization of the northwestern coast of the South Island of New Zealand. Shown are a terrain model (with the land area having a vertical scale of one-fifth of the sea area), 200-m contour lines below sea level, and catches of hoki for 25 July 2001. The catch glyphs are plotted in the appropriate location and depth in the 3-D volume shown. The size reflects the size of the catch, and the color represents the percentage of hoki comprising the catch. This fishery is based on annual spawning aggregations around the Hokitika Canyon, a feature clearly shown in the figure. A transparent layer representing the sea surface also is shown to provide spatial context. This image is one frame from an animation showing the complete 2001 season. The number of tows was 130, the number of vessels was 38, hoki catch was 2,219 metric tons, and total catch was 2,350 metric tons.

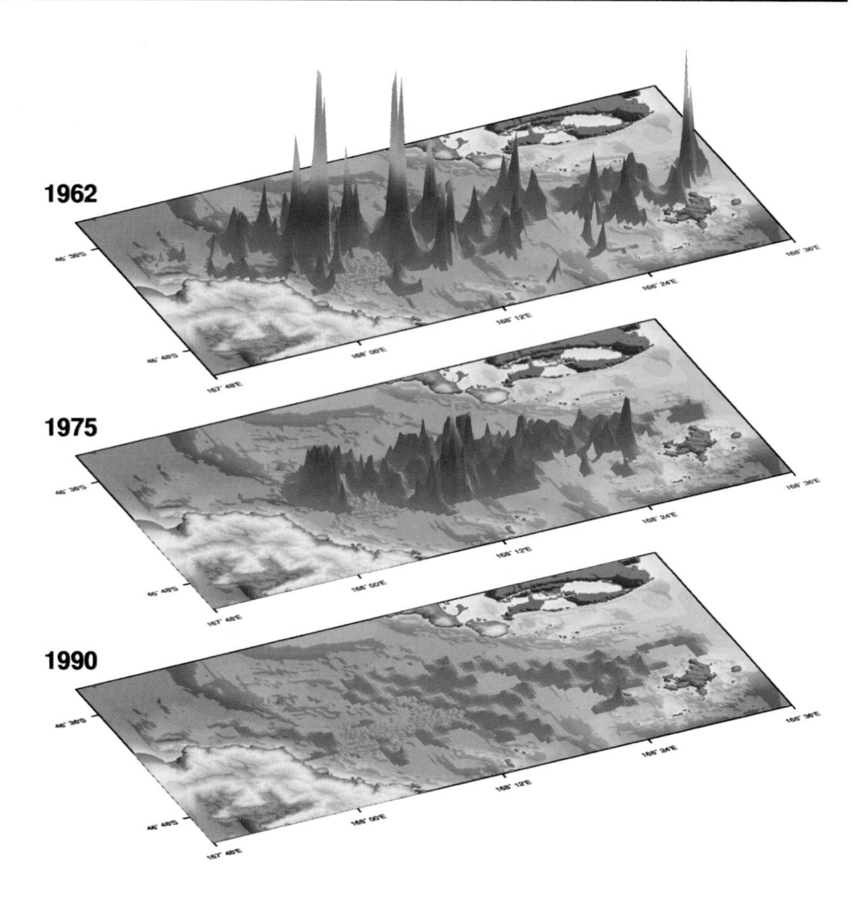

Figure 8.6 New Zealand dredge oyster densities from three surveys are shown as a 2+D surface draped over a terrain model of Foveaux Strait and the neighboring South and Stewart islands of New Zealand. Areas of high oyster density are represented as peaks on the map. The primary reason for the illustrated decline in oyster densities is believed to be due to the protozoan parasite *Bonamia* sp.

in Figure 8.6. The image, a condensed version of the full sequence, provides insight into the decline of the oyster resource over the past four decades due to the protozoan parasite *Bonamia* sp.

8.4.2.2 ***Habitat Mapping.*** Close to 99% of the planet Venus has been mapped to a resolution of 50 m. In stark contrast, only 10% of the oceans, which comprise 71% of the earth's surface, have been mapped to this resolution. Such maps provide baseline information for any research or management work to be conducted on the marine benthic environment.

In the past, marine surveyors have interpreted depth-sounder information when classifying the seabed. This was based on inferences from echo intensity and echo-to-echo coherence, based on the visual interpretation of analog records. These results were highly dependent on human interpretation and were tedious, inefficient, and lacked resolution and reproducibility.

Rapid advancements have been made in classifying the seabed. Technology now is available that can use high-speed digital processing of acoustic data from standard echo sounders, together with the identification of larger scale features with side-scan acoustic imagery. Sonographs provide more complete spatial coverage of the seabed but provide little information on aspects of the substrate, thereby permitting only qualitative interpretations of sediment texture. Digital processing of echo-sounder returns by using systems such as Simrad's EP60, Stenmar's Roxann, and Quester Tangent's QTCView, integrate a number of sedimentary and topographical factors into dimensionless parameters that can be used to identify differences in sediment composition and structure (Figure 8.7).

8.4.2.3 ***Fish Distribution and Abundance.*** Crucial to any study investigating the distribution and abundance of a fish species is the construction of some type of a map. This may take the form of a simple 2-D graphical representation of presence or absence, or density, or even a thematic map illustrating, in a schematic way, temporal aspects of the species' life history or migration patterns (Caddy and Garcia 1986). A GIS is well suited to this task because mapping functionality is one of its strongest and most widely used core functions. It, therefore, comes as no surprise that GIS was initially used for this task. Understanding habitat, distribution, and abundance is an important issue in fisheries management. In the United States, the 1996 reauthorization of the

Figure 8.7 A 2+D composite image illustrating seabed classification within four sediment classes. Data were generated by QTC-MULTIVIEW (Quester Tangent 2003) and then draped onto the underlying bathymetry. Single-beam seabed classification overlaid on multibeam bathymetry of Race Rocks, a marine protected area in the Juan de Fuca Strait, British Columbia. Data were collected by the Canadian Hydrographic Service in Sidney, British Columbia. Seabed classification data were collected with QTC VIEW and processed by using QTC IMPACT. Brown = exposed bedrock, gray = bedrock with boulder and cobble veneer, orange = coarse clastics, and red = fine clastics.

Magnuson-Stevens Fishery Conservation and Management Act requires amendments of all U.S. federal fisheries management plans to describe, identify, conserve, and enhance essential fish habitat. This will necessitate characterizing habitat for life stages of each species and identifying current and potential threats to this habitat.

A GIS is much more than just a glorified mapping tool (Booth 2001). With the incorporation of spatial tools, GIS is becoming a versatile platform that also explores data and performs various analyses. Linear and nonlinear models now are being developed to quantify correlations between the distribution and abundance of fish and their biotic or abiotic environment (Booth 1998; Waluda and Pierce 1998; Rubec et al. 2001). In the United States, GIS are being used for the initial characterization of habitat, the spatial correlation of potential threats to habitat, the evaluation of cumulative impacts, and the monitoring of habitat quality and quantity. Habitat mapping, modeling, and the determination of essential habitat are now commonly addressed within a GIS framework (Moses and Finn 1997; Reuter 1999; Ross and Ott 2001; M. Parke, Joint Institute of Marine and Atmospheric Research, University of Hawaii, unpublished abstract from the First International Symposium on GIS in Fishery Science, 1999).

8.4.2.4 ***Fisheries Oceanographic Modeling.*** Research in fisheries oceanography involves explaining the relationships among oceanography, fish, and harvesting fisheries. A source of information well used by terrestrial and inshore researchers has been remote-sensing data, largely consisting of satellite imagery. This trend has moved offshore. Meteorological satellites are adept at picking out characteristics of the oceans such as temperature, silt or sediment patterns, wave trains, and current circulation. Satellites used to sense these and other properties of the ocean surface include Seasat, Radarsat, the Coastal Zone Color Scanner on Nimbus 7, and SeaWiFS. Sandwell and Smith (1996) presented a global topography data set that was created in part from gravity models from satellite-derived data. The National Institute of Water and Atmospheric Research, New Zealand, provides forecasts for commercial tuna fishers in the New Zealand region, based on satellite-determined sea surface temperature data and known preferences and behavioral patterns of the species involved.

8.4.2.5 ***Spatiotemporal Analysis of Resources and Harvesting Fisheries.*** Edwards et al. (2001) have addressed ecosystem-based management of fishery resources in the Northeast shelf ecosystem by using a GIS. The objective was to determine whether management of marine fisheries resources in the northeast region of the United States was consistent with ecosystem-based management for an aggregated sustainable yield of commercially valuable species. Distributions of species, fishing effort, and landing revenues, based on 10-min squares over Georges Bank during a 3-year period, were spatially analyzed by using a GIS. Maps of fishing for both fish trawls and scallop dredges suggested the scope for likely bycatch. An indication of the economic importance of the closed areas to other fisheries, especially to the

Atlantic fishery for sea scallops *Placopecten magellanicus*, was suggested by revenue coverages. As a result, a GIS could handle the spatial analysis of ecological, technological, and economic relationships, and could facilitate reviews of management plans for their consistency with ecosystem requirements. An essential component of this study was a clear understanding of the spatial distribution of interactions among species, fishing effort, and technologies and markets for fisheries products. Edwards et al. (2001) concluded that GIS was the only tool for such complex spatial analyses, and their research is now progressing with this particular objective.

Geographic information systems have been developed to understand the catch per effort patterns in industrialized fisheries. A system developed by Fairweather (2001) was able to detect patterns in fishing effort directed at the cape hake *Merluccius capensis* off the western coast of South Africa. Figure 8.8 illustrates a simple report from the queryable fish atlas on billfish and tuna catches that is maintained by the Food and Agriculture Organization of the United Nations.

Remotely sensed data are becoming useful in detecting pelagic fish and fishing patterns. For example, by using imagery from the Defense Meteorological Satellite Program (DMSP) and the Operational Linescan System (OLS), Kiyofuji et al. (2001) investigated the spatial and temporal distribution of squid fishing boats fishing during the day and at night. The relationship between sea surface temperature obtained from the National Oceanic and Atmospheric Administration Advanced Very High Resolution Radiometer and fishing boat distribution also was investigated. The preliminary conclusion was that visible images from DMSP/OLS could provide both the position where fishing boats gather and the relationship between fishing boat location and sea surface temperature.

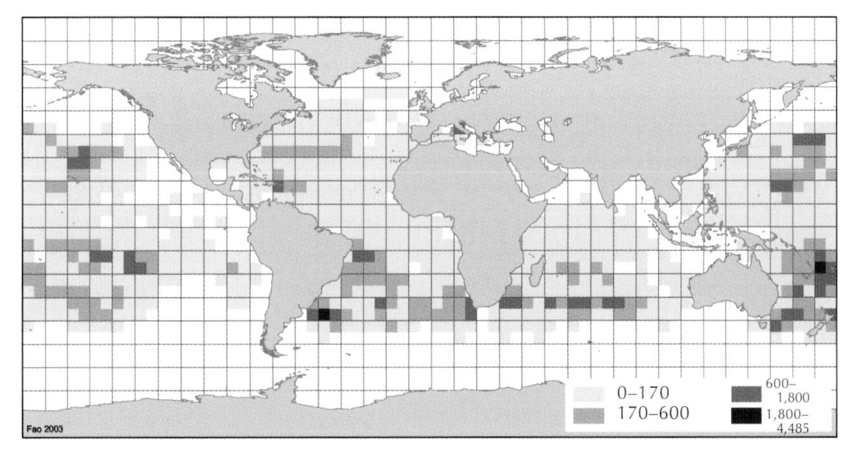

Figure 8.8 World average annual catches of albacore *Thunnus alalunga* by using longlines for the period 1996–1998. Data from the Food and Agriculture Organization's online atlas of billfish and tuna catches (FAO 2003).

Studies of migration dynamics often include habitat profiles for pelagic and highly migratory species. S. Saitoh et al. (Hokkaido University, unpublished abstract from the First International Symposium on GIS in Fishery Science, 1999) investigated the relationships between oceanic conditions and the migration patterns of Japanese saury *Engraulis japonicus* (also known as Japanese anchovy) by observing the movement of pursuing fishing vessels obtained from satellite imagery. The results of overlaying of sea surface temperature data with fishing boat movement along the Oyashio front clearly indicated Japanese saury migration dynamics. Habitats of the beaked whale (family: Ziphiidae) and the sperm whale *Physeter macrocephalus* in shelf-edge and deeper waters off the northeastern United States have also been investigated (G. T. Waring et al., National Marine Fisheries Service, unpublished abstract from the First International Symposium on GIS in Fishery Science, 1999). By using sighting data and corresponding information on bathymetry, slope, oceanic fronts, and sea surface temperature, a logistic regression analysis was conducted to show that the distribution of sperm whales was more related to depth and slope, while that of beaked whales more dependent on sea surface temperature.

8.4.2.6 ***Stock Assessment.*** The need to manage fisheries from a spatial perspective is clear (Hinds 1992). Few attempts, however, have been made to incorporate the spatial variability of stock age structure, maturity, and growth patterns, together with catch and effort data into an assessment framework. Commercial catches are georeferenced, with fish being harvested at specific geographic locations as a function of the fishing effort and stock abundance at that location. By neglecting this spatial component, existing stock assessment models evaluate the status and productivity of the stock based on pooled data for catch at age, fishing effort, and key population parameters. Currently, there is a growing interest in the development of marine GIS, both to visualize these spatial data sets and to provide a platform for further stock assessments and forecasting.

Booth's (2001; unpublished abstract from the First International Symposium on GIS in Fishery Science, 1999) studies correlated fishing effort with observed age-structured fishing mortality to present a spatial perspective of localized resource status. Per-recruit modeling was expanded in Maury and Gascuel's (1999) study, providing insight into spatial problems inherent to the delineation of marine protected areas and how these might affect fishing operations. These models encapsulate three fundamental spatiotemporal aspects of fisheries modeling: the environment, the fish stock, and the fishing fleet.

The equilibrium biomass production modeling approach has been used by Corsi et al. (2001) to assess the abundance of the Italian demersal resources as a function of spatially distributed fishing effort off the coast of Italy. Peña et al. (2001) further simplified the stock dynamics model and used a GIS to estimate the nominal yield of jack mackerel *Trachurus symmetricus* by using fishing ground information, observed yield, and sea surface temperature gradients.

By using real-time fishing catch and location data, together with near-real-time satellite imagery, a transition probability matrix was used to calculate nominal yields at various thermal gradients. Cruz-Trinidad et al. (1997) conducted a cost returns analysis of the trawl fishery of Brunei Darussalam where optimal fishing patterns were determined by using profitability indicators under various economic and operational scenarios. Walden et al. (2001) also developed a simple, yet real-time, GIS for a demersal fishery in New England. This GIS was used to evaluate various fine-scale time-area closures to assess the projected mortality reductions and losses in revenue of three principal demersal fish species, namely Atlantic cod *Gadus morhua*, haddock *Melanogrammus aeglefinus*, and yellowtail flounder *Pleuronectes ferrugineus*.

Several studies have addressed the estimation of population size from fisheries independent surveys by using GIS. Nishida and Miyashita (2001) estimated age-1 southern bluefin tuna *Thunnus maccoyii* (also known as *T. maccoyi*) recruitment, by using information obtained by omni-scan sonar. In their study, a linear relationship was estimated between the strength of the currents into the survey area and the average school size as recorded by the sonar. Nishida and Miyashita (2001) noted that young southern bluefin tuna schools were transported to the survey area depending on the strength of the currents. Recruitment abundance was estimated by standardization with respect to current strength. Similarly, R. Ali et al. (Southeast Asian Fisheries Development Center, unpublished abstract from the First International Symposium on GIS in Fishery Science, 1999) used echo sounders to investigate the fish abundance off the coast of Malaysia. Line transects were deployed to estimate population size and quantify other ecological parameters. In-built kriging procedures facilitated the quantification of biomass.

8.4.3 Management-Driven Approaches

This area is, regrettably, the most poorly represented in the GIS literature. This is principally due to the "in-house" nature of many GIS studies and their unsuitability, in many instances, for publication in the peer-reviewed literature (Booth 2001).

8.4.3.1 ***Fishing Fleet Behavior.*** Because the restriction of fishing effort is a direct way to mitigate fishing impacts on marine resources, fisheries managers have been giving some priority to monitoring the locations of fishing vessels. The GIS software now allows for GPS integration for onboard data capture. Some GPS capability is used by fishermen for relocation to good fishing grounds by analyzing historical data using GIS (P. K. Simpson and D. J. Anderson, Scientific Fishery Systems, Inc., unpublished abstract from the First International Symposium on GIS in Fishery Science, 1999).

Data sources include vessel monitoring systems, catch-reporting and logging (Meaden and Kemp 1996; Kemp and Meaden 1998; Long et al. 1994; Fox and Bobbitt 1999; Ward et al. 1999), and remote-sensed imagery of fishing areas (Kiyofuji et al. 2001; Peña et al. 2001). Foucher et al. (1998) describe a prototype GIS that uses simple overlaying tools to quantify areas of conflict between competing fisheries for the octopus and groundfish stocks off Senegal. Unfortunately, the data used by many of these systems are often entered from handwritten or hard copy catch return reports. This implies a lag from the event to the time at which it could be used as information. There is a definite move toward collecting the data in a digital format and transmitting it from vessels still at sea, increasing the adaptability of the GIS (Meaden 1993; Pollitt 1994; Meaden and Kemp 1996).

8.4.3.2 ***Bycatch and Conflict Management.*** The use of GIS as a bycatch mitigation tool is in its formative stages. The lack of commercial software programs has been cited as a stumbling block to progress. Hopefully, when this issue is resolved, a large number of applications should be expected because bycatch management is arguably one of the most urgent and serious issues in world fisheries. Published results illustrate that GIS can specify (even pinpoint) the habitat areas of bycatch species on a fine spatiotemporal resolution. R. M. Mikol (University of Alaksa, Fairbanks, unpublished abstract from the First International Symposium on GIS in Fishery Science,1999) has investigated vessels targeting Pacific hake *Merluccius productus* off the coast of Washington and Oregon. D. Ackley (Alaska Department of Fish and Game, unpublished abstract from the First International Symposium on GIS in Fishery Science, 1999) assessed Alaska's groundfish bycatch problem. Ackley's GIS investigated time-area closures necessary to minimize bycatch of red king crab *Paralithodes camtschaticus,* blue king crab *P. platypus,* chinook salmon *Oncorhynchus tshawytscha,* and chum salmon *O. keta* in the Bering Sea as part of a larger fisheries management procedure for the North Pacific Fisheries Management Council.

Caddy and Carocci (1999) described a GIS for aiding fishery managers and coastal area planners in analyzing the likely interactions of ports, offshore stocks, and local nonmigratory inshore stocks. This tool provides a flexible modeling framework for making decisions on fishery development and zoning issues, and has been applied to the scallop fishery in the Bay of Fundy, Canada, and the demersal fishery off the northern Tyrrhenian coastline, Italy.

8.5 CONCLUSIONS

The marine environment is difficult to quantify. Aquatic organisms complete different stages of their life histories in different habitats, with each habitat often characterized by unique spatiotemporal, physical, and chemical characteristics. To exacerbate these complexities, almost all biotic and abiotic processes occur outside the scope of our observations. Rapid technological advancements have facilitated a synergistic relationship between the geographic, oceanographic, and biologic disciplines in the form of GIS. Terrestrial advances in GIS have sparked interest in possible fisheries applications but progress has been hampered by data quality and availability issues. Marine data have more dimensions and are more variable than terrestrial equivalents; the costs associated with data collection are higher and technical experience less mature. Despite the limitations of GIS in offshore environments, current applications are diverse and span most disciplines within what is considered fisheries science. These applications include essential habitat classification, life history elucidation, catch and effort estimation, stock assessment, and monitoring.

A glimpse into the crystal ball reveals the following developments during the next 5 years. It is our opinion that GIS will be used routinely to store, to manipulate, and to analyze marine data, so that improved management advice can be provided. Geographic information systems will, in effect, become like word processors and spreadsheets—increasingly user-friendly and readily accessible. Despite the adoption of this technology, marine-specific GIS has specific hurdles to clear, such as the highly dimensional spatiotemporal nature of the data. Society's dependence on fisheries resources and its concern for global warming will, however, ensure that adequate funding will be obtained for fundamental and applied research, resulting in breakthroughs in temporal topologies, multidimensional data management, and the development of new spatial visualization and statistical tools.

8.6 REFERENCES

Ault, J. S., J. Luo, S. G. Smith, J. E. Serafy, J. D. Wang, G. Diaz, and R. Humston. 1999. A dynamic spatial multistock production model. Canadian Journal of Fisheries and Aquatic Sciences 56(Supplement 1):4–25.

Booth, A. J. 1998. Spatial analysis of fish distribution and abundance patterns: a GIS approach. Alaska Sea Grant College Program Report AK-SG-98-01:719–740.

Booth, A. J. 2001. Are fisheries geographical information systems merely glorified mapping tools? Pages 366–378 *in* Nishida et al. (2001).

Burrough, P. A. 1996. Natural objects with indeterminate boundaries. Pages 3–28 *in* P. A. Burrough and A. U. Frank, editors. Geographic objects with indeterminate boundaries. Taylor and Francis, London.

Burrough, P. A., and A. Frank. 1996. Geographic objects with indeterminate boundaries. Taylor and Francis, London.

Caddy, J. F., and F. Carocci. 1999. The spatial allocation of fishing intensity by port-based inshore fleets: a GIS application. ICES Journal of Marine Science 56:388–409.

Caddy, J. F., and S. Garcia. 1986. Fisheries thematic mapping: a prerequisite for intelligent mapping and development of fisheries. Océanographie Tropicale 21:31–52.

Corsi, F., S. Agnesi, and G. Ardizonne. 2001. Integrating GIS and surplus production models: a new approach for spatial assessment of demersal resources? Pages 143–156 *in* Nishida et al. (2001).

Cruz-Trinidad, A., G. Silvestre, and D. Pauly. 1997. A low-level geographic information system for coastal zone management, with application to Brunei Darusalam. Naga 20:31–36.

Dijkema, K. S. 1991. Towards a habitat map of The Netherlands, German and Danish Wadden Sea. Ocean and Shoreline Management 16:1–21.

Edwards, S. F., B. P. Rountree, D. D. Sheehan, and J. W. Walden. 2001. An inquiry into ecosystem-based management of fishery resources on Georges Bank. Pages 202–214 *in* Nishida et al. (2001).

Fairweather, T. P. 2001. An analysis of the trawl and longline fisheries for *Merluccius capensis* off the west coast of South Africa. Master's thesis. Rhodes University, Grahamstown, South Africa.

FAO (Food and Agriculture Organization of the United Nations). 1995. Code of conduct for responsible fisheries. FAO, Rome.

FAO (Food and Agriculture Organization of the United Nations). 2003. Atlas of tuna and billfish catches. FAO. Available: *www.fao.org/fi/atlas/tunabill/english/mapset.htm* (September 2003).

Foucher, E., M. Thiam, and M. Barry. 1998. A GIS for the management of fisheries in West Africa: preliminary application to the octopus stock in Senegal. South African Journal of Marine Science 20:337–346.

Fox, C. G., and A. M. Bobbitt. 1999. The National Oceanic and Atmospheric Administration's Vents program GIS: integration, analysis, and distribution of multidisciplinary oceanographic data. Pages 163–176 *in* Wright and Bartlett (1999).

Francis, R. C., and S. R. Hare. 1994. Decadal-scale regime shifts in the large marine ecosystems of the Northeast Pacific: a case for historical science. Fisheries Oceanography 3:1–13.

Francis, R. C., and T. H. Sibley. 1991. Climate change and fisheries: what are the real issues? Northwest Environmental Journal 7:295–307.

Giles, R. H., Jr., and L. A. Nielsen. 1992. The uses of geographical information systems in fisheries. Pages 81–94 *in* R. H. Stroud, editor. Fisheries management and watershed development. American Fisheries Society, Symposium 13, Bethesda, Maryland.

Goodchild, M. F. 1992. Geographic data modeling. Computers and GeoSciences 18:401–408.

Hinds, L. 1992. World marine fisheries: management and development problems. Marine Policy 16:394–403.

Hooge, P. N., and B. Eichenlaub. 1997. Animal movement extension to ArcView, version 1.1. U.S. Geological Survey, Alaska Biological Science Center, Anchorage.

Insightful Corporation. 2003. Data mining software and statistical software. Available: *www.insightful.com* (September 2003).

Isaak, D. J., and W. A. Hubert. 1997. Integrating new technologies into fisheries science: the application of geographical information systems. Fisheries 22(1):6–10.

Itoh, K., and T. Nishida. 2001. Marine Explorer: marine GIS software for fisheries and oceanographic information. Pages 427–437 *in* Nishida et al. (2001).

Kemp, Z., and G. J. Meaden. 1998. Towards a comprehensive fisheries management information system. Pages 522–531 *in* A. Eide and T. Vassal, editors. IIFET '98 proceedings, volume 2. University of Tromso, Norwegian College of Fisheries Science, Tromso, Norway.

Kiyofuji, H., S. Saitoh, Y. Sakuri, T. Hokimoto, and K. Yoneta. 2001. Spatial and temporal analysis of fishing fleets distribution in the southern Japan Sea using DMS/OLS visible data in October 1996. Pages 178–185 *in* Nishida et al. (2001).

Kracker, L. M. 1999. The geography of fish: the use of remote sensing and spatial analysis tools in fisheries research. Professional Geographer 51:440–450.

Langran, G., and N. Chrisman. 1988. A framework for temporal geographic information. Cartographica 25:1–14.

Li, R. 1991. An algorithm for building octree from boundary representation. Pages 13–23 in A. Bagchi and J. J. Beaman, editors. American Society of Mechanical Engineers, Production Engineering Division, New York.

Li, R. 1999. Data models for marine and coastal geographic information systems. Pages 25–36 *in* Wright and Bartlett (1999).

Li, R., and N. K. Saxena. 1993. Development of an integrated marine geographic information system. Marine Geodesy 16:294–307.

Lockwood, M., and R. Li. 1995. Marine GIS—what sets them apart? Marine Geodesy 18: 157–159.

Long, B., T. Skewes, M. Bishop, and I. Poiner. 1994. Geographic information system helps manage Torres Strait fisheries. Australian Fisheries 53(2):14–15.

Lucas, A. 1999. Representation of reality in marine environmental data. Pages 53–74 *in* Wright and Bartlett (1999).

Luo, J., P. B. Ortner, D. Forcucci, and S. R. Cummings. 2000. Diel vertical migration of zooplankton and mesopelagic fish in the Arabian Sea. Deep-Sea Research 47:1451–1474.

Maguire, D. J., M. F. Goodchild, and D. W. Rhind, editors. 1991. Geographic information systems—principles and applications. Wiley, London.

Marble, D. E., H. W. Calkins, and D. J. Peuquet. 1984. Basic readings in geographic information systems. SPAD Systems, Williamsville, New York.

Maury, O., and D. Gascuel. 1999. SHADYS, a GIS-based numerical model of fisheries. Example application of a marine protected area. Aquatic Living Resources 12:77–88.

McPhaden, M. J., A. J. Busalacchi, R. Cheney, J. R. Donguy, K. S. Gage, D. Halpern, M. Ji, P. Julian, G. Meyers, G. T. Mitchum, P. P. Niiler, J. Picaut, R. W. Reynolds, N. Smith, and K. Takeuchi. 1998. The tropical ocean–global atmosphere (TOGA) observing system: a decade of progress. Journal of Geophysical Research 103:14169–14240.

Meaden, G. J. 1993. Instigation of the world's first marine fisheries GIS. ICES Statutory Meeting, C.M. 1993/D:64, Dublin.

Meaden, G. J. 1996. Potential for geographical information systems (GIS) in fisheries management. Pages 41–79 *in* B. A. Megrey and E. Moksness, editors. Computers in fisheries research. Chapman and Hall, London.

Meaden, G. J. 1999. Applications of GIS to fisheries management. Pages 205–226 *in* Wright and Bartlett (1999).

Meaden, G. J. 2001. GIS in fisheries science: foundations for the new millennium. Pages 3–29 *in* Nishida et al. (2001).

Meaden, G. J., and T. Do Chi. 1996. Geographical information systems: applications to marine fisheries. FAO Fisheries Technical Paper 356.

Meaden, G. J., and J. M. Kapetsky. 1991. Geographical information systems and remote sensing in inland fisheries and aquaculture. FAO Fisheries Technical Paper 318.

Meaden, G. J., and Z. Kemp. 1996. Monitoring fisheries effort and catch using a geographical information system and a global positioning system. Pages 238–248 *in* D. A. Hancock, D. C. Smith, A. Grant, and J. P. Beumer, editors. Developing and sustaining world fisheries resources. Commonwealth Scientific and Industrial Research Organization, Australia.

Moses, E., and J. T. Finn. 1997. Using geographic information systems to predict North Atlantic right whale (*Eubalena glacialis*) habitat. Journal of Northwest Atlantic Fishery Science 58:393–409.

Nishida, T., and A. J. Booth. 2001. Recent methods and approaches using GIS in the spatial analysis of fish populations. University of Alaska Sea Grant Report AK-SG-01-02:19–36.

Nishida, T., P. J. Kailola, and C. E. Hollingworth, editors. 2001. Proceedings of the first symposium on GIS in fishery science. Fishery GIS Research Group, Saitama, Japan.

Nishida, T., and K. Miyashita. 2001. Spatial dynamics of southern bluefin tuna (*Thunnus maccoyii*) recruitment. Pages 89–106 *in* Nishida et al. (2001).

OpenDX. 2003. Open visualization data explorer. Available: *www.opendx.org* (September 2003).

Peña, H., C. González, and F. Véjar. 2001. Spatial dynamics of jack mackerel fishing grounds and environmental conditions using a GIS. Pages 107–113 *in* Nishida et al. (2001).

Petitgas, P. 1993. Geostatistics for fish stock assessments: a review and an acoustic application. ICES Journal of Marine Science 50:285–298.

Peuquet, D. J. 1994. It's about time: a conceptual framework for the representation of temporal dynamics in geographic information systems. Annals of the Association of American Geographers 84:441–461.

Pollitt, M. 1994. Protecting Irish interests: GIS on patrol. GIS Europe 3:18–20.

Quester Tangent. 2003. Quester Tangent—marine–transit. Available: *www.questertangent.com* (September 2003).

Ray, G. C., B. P. Hayden, A. J. Bulger, Jr., and M. G. McCormick-Ray. 1992. Effects of global warming on the biodiversity of coastal-marine zones. Pages 91–104 *in* R. L. Peters and T. E. Lovejoy, editors. Global warming and biological diversity. Yale University Press, New Haven, Connecticut.

Rees, A. J. J., and M. M. Ryba. 1998. MarLIN—a metadatabase for research data holdings at CSIRO Marine Research. CSIRO Marine Research, Hobart, Tasmania.

Reuter, R. F. 1999. Describing dusky rockfish (*Sebastes ciliatus*) habitat in the Gulf of Alaska using historical data. Master's thesis. California State University, Hayward.

R Foundation. 2003. The R project for statistical computing. Available: *www.r-project.org* (September 2003).

Ross, S. W., and J. Ott. 2001. Development of a desktop GIS for estuarine resource evaluation with an example application for fishery habitat management. Pages 229–241 *in* Nishida et al. (2001).

Rubec, P. J., S. G. Smith, M. S. Coyne, M. White, D. Wilder, A. Sullivan, R. Ruiz-Cruz, T. MacDonald, R. H. McMichael, G. E. Henderson, M. E. Monaco, and J. S. Ault. 2001. Spatial modeling of fish habitat in Florida. University of Alaska Sea Grant Report AK-SG-01-02:1–18.

Sandwell, D. T., and W. H. F. Smith. 1996. Global bathymetric prediction for ocean modelling and marine geophysics. Available: *http://topex.ucsd.edu/marine_topo/text/topo.html*. (July 2003)

Simpson, J. J. 1992. Remote sensing and geographical information systems: their past, present and future use in global marine fisheries. Fisheries Oceanography 1:238–280.

Smith, T. R., S. Menon, S. Start, and J. L. Estes. 1987. Requirements and principles for the implementation and construction of large-scale geographic information systems. International Journal of Geographical Information Systems 1:13–31.

Star, J. and Estes, J. 1990. Geographic information systems—an introduction. Prentice Hall, Upper Saddle River, New Jersey.

TerraSeer, Inc. 2003. SpaceStat, software for spatial regression modeling. TerraSeer, Inc. Available: *www.terraseer.com/Spacestat.html* (September 2003).

The MathWorks, Inc. 2003. The Mathworks: developers of MATLAB and Simulink for technical computing. The MathWorks, Inc. Available: *www.mathworks.com* (September 2003).

Von Meyer, N., K. E. Foote, and D. J. Huebner. 1999. Information quality considerations for coastal data. Pages 295–308 *in* Wright and Bartlett (1999).

Walden, J. B., D. Sheenan, B. Roundtree, and S. Edwards. 2001. Integrating GIS with mathematical programming models to assist in fishery management decisions. Pages 167–177 *in* Nishida et al. (2001).

Waluda, C. M., and G. J. Pierce. 1998. Temporal and spatial patterns in the distribution of squid *Loligo* spp. in United Kingdom waters. South African Journal of Marine Science 20:323–336.

Ward, R, C. Roberts, and R. Furness. 1999. Electronic chart display and information systems (ECDIS): state-of-the-art in nautical charting. Pages 149–161 *in* Wright and Bartlett (1999).

Wessel, P., and W. H. F. Smith. 2003. GMT—the generic mapping tools. Available: *http://gmt.soest.hawaii.edu/* (September 2003).

Woodcock, C. E., and S. Gopal. 2000. Fuzzy set theory and thematic maps: accuracy assessment and area estimation. International Journal of Geographic Information Systems 14:153–172.

Wright, D. J. 1994. From pattern to process on the deep ocean floor: a geographic information system approach. Doctoral dissertation. University of California, Santa Barbara.

Wright, D. J. 1999. Down to the sea in ships: the emergence of marine GIS. Pages 1–10 *in* Wright and Bartlett (1999).

Wright, D. J., and D. Bartlett, editors. 1999. Marine and coastal geographical information systems. Taylor and Francis, London.

Wright, D. J., and M. F. Goodchild. 1997. Data from the deep: implications for the GIS community. International Journal of Geographical Information Science 11:523–528.

Wright, D. J., M. F. Goodchild, and J. D. Proctor. 1997. Demystifying the persistent ambiguity of GIS as "tool" versus "science." Annals of the Association of American Geographers 87:346–362.

Chapter 9

Geographic Information Systems in Marine Fisheries Science and Decision Making

KEVIN ST. MARTIN

I shall suggest that in the future, fisheries management and its associated science will have to deal with "places" far more than they have in the recent past. Indeed, I shall suggest that they will have to return, in many cases, to ancient modes of allocating fisheries resources to local communities, rooted in physical places. (Pauly 1997)

9.1 INTRODUCTION

The use of geographic information systems (GIS) in decision making and policy development is growing rapidly in many fields of resource management. While these applications are often limited to inventories and basic GIS techniques (e.g., database query), the use of GIS, nevertheless, is making resource management more explicitly spatial. In fisheries, however, the use of GIS has been much more limited, and its impact has yet to be felt to any great degree (Isaak and Hubert 1997; Fisher and Toepfer 1998). This is true particularly for marine fisheries science and management, where GIS is used only occasionally to support or illustrate fisheries assessments or as a supplement to general environmental analysis.[1] The use of GIS by managers themselves as an active aid for decision making, scenario testing, site suitability analysis, or socioeconomic analysis has yet to be established in marine fisheries.

While GIS seldom appears within the institutionalized systems of science and management, there are many examples of fisheries scientists and community organizations who are integrating GIS into innovative research and management initiatives. These alternatives

[1] The word "management" will be used throughout this chapter to mean marine fisheries decision making and policy development. And, unless otherwise stated, the management context will be the federal waters of the United States. In addition, because of the close relationship between government-funded fisheries science (e.g., that performed by the National Marine Fisheries Service) and regional fisheries management councils, "management" is understood here to include the management-driven science, both biological and social science, that informs fisheries decision making.

demonstrate the potential for GIS to contribute to a "spatial turn" in marine fisheries science and management. This chapter will suggest that specific institutionalized aspects of the current regime present significant barriers to change; current practices often stifle both a shift toward geographic methods of analysis and management, and the use of GIS. Therefore, if GIS is to be adopted widely for marine fisheries decision making, change must occur within a variety of scientific and management institutions. Such a "paradigm shift" in fisheries likely will have an impact on a broad range of issues from fisheries data collection to stock allocation and property rights.

This chapter will explore the current and potential uses of GIS in marine fisheries decision making. It will focus first on the current regime of science and management and how it has limited the use of GIS and spatial approaches generally. What traditionally constitutes fisheries data, how it is used analytically, and how it is integrated into management will be discussed in turn. The last section of the chapter will focus on some of the implications of GIS for fisheries decision making.

9.2 THE CURRENT REGIME IN FISHERIES AND GIS

The slow emergence of GIS in marine fisheries is due in part to the unique and inherent characteristics of fisheries resources that make their representation and analysis difficult when using GIS (e.g., the mobility of fish stocks and the three dimensionality of the ocean environment; Meaden and Chi 1996). It also is due to socioeconomic constraints, such as the level of education or financial resources needed to implement the technology (Meaden 1999). These aspects of fisheries represent formidable barriers to the adoption of GIS for both biological assessment and management. In addition, institutionally determined barriers also need to be considered not only to explain but also to address the absence of GIS in fisheries decision making and policy.

The current regime of marine fisheries management in the United States was institutionalized in the 1970s with passage of the Magnuson-Stevens Fishery Conservation and Management Act. This act formalized a system of national control over marine fisheries where management of the resource was intimately linked to the goals of fisheries bioeconomics.[2] The basic elements of this approach are now standard not only within the United States but also internationally via their promotion by organizations (e.g., the United Nations) (Gulland

[2] Although the primary goal of the act was ostensibly noneconomic (i.e., the maximum sustainable yield of fish), the act implemented a new regime of management set on maximizing production within the newly enclosed 322-km (200-mi) exclusive economic zone by controlling the fishing industry. Other, explicitly bioeconomic, goals such as maximum economic yield and optimum yield made clear to management that the rationalization and privatization of the industry would increase net economic benefits to the nation and avert an inevitable "tragedy of the commons" (Hardin 1968) in U.S. fisheries.

1984; Garcia and Hayashi 2000). The goal is to limit fishing effort and, therefore, sustain levels of harvest and maintain a viable fishing industry. Methods for limiting effort range from gear restrictions to harvesting quotas but all emanate from common assumptions about fishers' behavior and the space of fishing: fishers are individuals set on maximizing individual profits, while space is the container of those fishers as well as the fish they catch enumerated by single species. These assumptions, institutionalized at several levels, determine what is possible or imaginable for management. They reify a particular story about individual "fishermen"[3] and their aggregate effort in an equilibrium relationship with single species of fish within a spatial domain that generally defaults to very large regions of national or international interest divided only for numeric or statistical purposes (St. Martin 2001). For more on the history of statistical regions used for fisheries management, see Clay (1996), and, for a detailed history of fisheries science and management, see Smith (1994).

Assessing aggregate numbers of fish and adjusting fishing effort accordingly does not, however, require any spatial analysis per se. This approach relies upon and reproduces a space where fish and fishing effort are sampled statistically; it does not incorporate the environmentally heterogeneous spaces of ecosystems that fisheries are a part of nor does it incorporate the intimate and socially produced spaces of fishing communities. In short, the dominant form of fisheries science and management uses a homogenous notion of space that is well suited to the calculation of aggregate numbers of fish and fishing effort for a large management area. While GIS has a role to play at such scales and within this numeric science (e.g., Fogarty and Murawski 1998), its strength is in its ability to make visible and to analyze heterogeneous processes across a diverse landscape.

In contrast to the initial assumptions of the current regime of management (i.e., fishers as individuals on an homogeneous commons), recent fisheries research, as well as legislation suggest two alternative categories with which to conceptualize fisheries: communities and ecosystems. These categories are central to a growing body of literature focused on a "paradigm shift" in fisheries science and management that involves a shift toward communities (rather than individuals) in the social sciences (e.g., Jentoft 1999) and a shift toward ecosystems (rather than single species) in the biological sciences (e.g., Botsford et al. 1997; Costanza et al. 1998; Sainsbury 1998). In this and other literature advocating alternative systems of fisher-

[3] Here I use the term "fisherman" to signify the imagined individual, independent, and competitive fisher of bioeconomic theory. The term "fisher," used throughout this paper, refers to all people who work as harvesters of fish within commercial enterprises regardless of their position, and it assumes nothing about their relationships with each other (e.g., cooperative or competitive). A fisher may be any one of several crew members (e.g., captain, mate, engineer, deckhand) and may or may not own a fishing boat.

ies science and management, it is often clear that alternatives to the current regime will likely give a higher priority to the spatial aspects of fisheries (cf. Booth 2000).

In addition to recent research directions, new federal legislation has included explicitly the categories of community and ecosystem as mandated objects of fisheries science and management. Impacts on "fishing communities" must now be assessed in every fishery management plan applying for federally managed species.[4] While there remains considerable debate as to what constitutes a community, the category is gaining acceptance within science and management institutions. Similarly, recent legislation clearly points to the consideration of an ecosystems-based management regime (EPAP 1999). While the management of fisheries is still based on individual species rather than ecosystems, the introduction of new requirements such as the delineation (definition and mapping) of essential fish habitat (EFH) for each species suggests the growing importance of ecosystem interactions and processes (cf. Witherell et al. 2000).[5]

Both communities and ecosystems suggest a heterogeneous marine environment at multiple scales. Although often associated with the local and proximate, the spatial extent of one community may vary widely from that of another community. Whereas some communities have local and discrete domains (e.g., those harboring only a small-boat lobster fishery), others overlap and extend across the entire Northeast Atlantic continental shelf (e.g., those associated with the industrial scallop fishery). Similarly, ecosystems are associated with a particular spatial scale despite being constituted by multiple processes at multiple scales; they typically are thought of as very large biological or physical systems (e.g., Sherman and Skjoldal 2002). While the full extent of an ecosystem is important to consider and define, it is the complex of processes and interactions within the system (e.g., between species, between species and habitats, and be-

---

[4] The Magnuson-Stevens Fishery Conservation and Management Act (reauthorized and amended by the Sustainable Fisheries Act [SFA], 1996) established the current system of fisheries management in the United States. Eight regional councils have the authority (with the oversight of the National Marine Fisheries Service) to produce fishery management plans that govern access to and use of fisheries resources. The 1996 amendment (SFA), for the first time, made it mandatory that social and economic impacts upon communities be assessed for each fishery management plan.

[5] The Magnuson Act requires the regional councils to consider ecosystems and suggests an expanded use of ecosystem principles in fisheries management. The SFA commissioned a review and report on ecosystem approaches (i.e., EPAP 1999). In addition, the National Marine Fisheries Service Strategic Plan for Fisheries Research (also a product of the SFA) emphasizes research on species interactions, habitat, and other ecosystemic processes. Finally, the SFA requires regional councils to protect and enhance "essential fish habitat," including preservation of "habitat areas of particular concern" where necessary.

tween fishers and the environment) that are the focus of most ecosystem analysis and ideally the basis for policy development (Langton et al. 1995). These interactions are typically local events demanding local data collection and analyses that can then be integrated with analyses at other scales. A shift toward communities and ecosystems reduces neither science nor management to a particular scale; it does, however, necessitate a spatial understanding of the processes that constitute communities and ecosystems.

A shift toward these alternative categories is compatible with GIS insofar as both communities and ecosystems can be said to exist in discrete places and that they are elements of a heterogeneous fishing landscape. Geographic information systems give us the tools to visualize this landscape, to make these alternative categories and processes as analytically important as individual fishers and populations of a single species of fish. Either directly or indirectly, the categories of ecosystem and community have influenced the development of several initiatives that will undoubtedly change the current regime of fisheries management. Examples include the federal mandate to define an EFH; the increasingly popular idea of designating marine protected areas (MPA) as biological reserves safe from commercial fishing (Hall 1998; Committee on Evaluation, Design, and Monitoring of Marine Reserves and Protected Areas in the United States 2001); and fishery management plans that increasingly use "area-based" management strategies such as subregional areas that are seasonally closed to commercial fishing. Such initiatives constitute important elements in an ecosystems approach (e.g., Witherell et al. 2000).

Increased pressure (e.g., governmental mandates to consider communities and ecosystems) facilitates change by challenging the standard, nonspatial methods of fisheries science and management. Scientists and managers are reevaluating methods and practices that have long been barriers to GIS. While this situation is creating opportunities for GIS in fisheries, there remains little research that illustrates what forms of spatial analysis are available and how they might be implemented or merged with current forms of analysis and management. The following section looks at two general aspects of the current regime of fisheries management in the United States. The first is data acquisition, and the second is scientific analysis and management practices. These general areas constitute the foundations of the decision making and policy development process in the United States and in many other countries. While there are clear limitations on the incorporation of GIS within each, there are important examples to the contrary that point to the potential for GIS.

9.3 GIS AND FISHERIES DATA ACQUISITION

Fisheries science and management relies on two general categories of data collection: biological surveys to measure quantities of fish in the sea and landings or catch tallies to measure fishing effort and fishing mortality (Pierce and Hugl 1979). Because the methods of data collection are designed to produce aggregate measures of these key variables for the management region as a whole, the data are of limited use for spatial analysis. This approach is widespread and is truly foundational for fisheries management. In the New England region of the United States, for example, these data sets have been collected for several decades and have revealed conclusively several crises of overfishing. They are the basis for virtually every fishery management plan in terms of demonstrating depleted fish stocks and setting targets for numerically rebuilding them.

Biological surveys of fish stocks are performed by using sampling techniques that rely upon specific mappings of the ocean (e.g., a regular grid or zonation modified by bathymetry) designed to estimate accurately fish population levels at the scale of the management region; for smaller areas, the usefulness of data collected at these scales becomes doubtful (Wilson et al. 1999). Similarly, the geography and scale at which landing information is coded (typically 30-min or 10-min grids) also is designed for analysis over large areas. In the first case, the geography used for biological surveys only loosely corresponds to ecosystem or environmental boundaries, and, in the second case, the geography, although concerned with harvest locations, is distinctly different from that used by fishers themselves (Clay 1997). Neither the environmental nor the social "landscape" (that is, geographies constituted by variability over space) of fishing are used as a spatial frame for data acquisition.

The intention of data collection under the current regime of fisheries management is to accurately assess fish stocks and to adjust levels of fishing effort accordingly. It is a numeric science focused on the *aggregation* of data across space, while GIS is typically focused on *variation* across space (MacCall 1990). While standard data sets do have a spatial component that can form the basis of use with GIS (i.e., survey locations or blocks in which fish are harvested), this level of data does not take full advantage of the capabilities of GIS and must be used with caution. Indeed, using GIS with current data sets usually reduces them to a simple map representation of survey points and their attributes. While it is useful to visualize this data, the resolution of the data makes spatial analysis of subregions, the strength of GIS, nearly impossible.

The recent use of EFH demonstrates both the limitations of current fisheries data and the potential for GIS. An EFH, as mandated by federal law, must be mapped for all managed species in each of the management regions (Figure 9.1). These designations, while concerned with individual managed species, have the potential to disrupt and transform the traditional approach to fisheries management insofar as EFH shifts the focus of management from solely numeric to include spatial considerations. In addition, once specified, the spaces of EFH will have some relationship to ecosystem and community spaces; they might overlap, coincide, be proximate, or be distant from either. Essential fish habitats and other essentially spatial initiatives (e.g., marine protected areas) contribute to a new spatialization of management re-

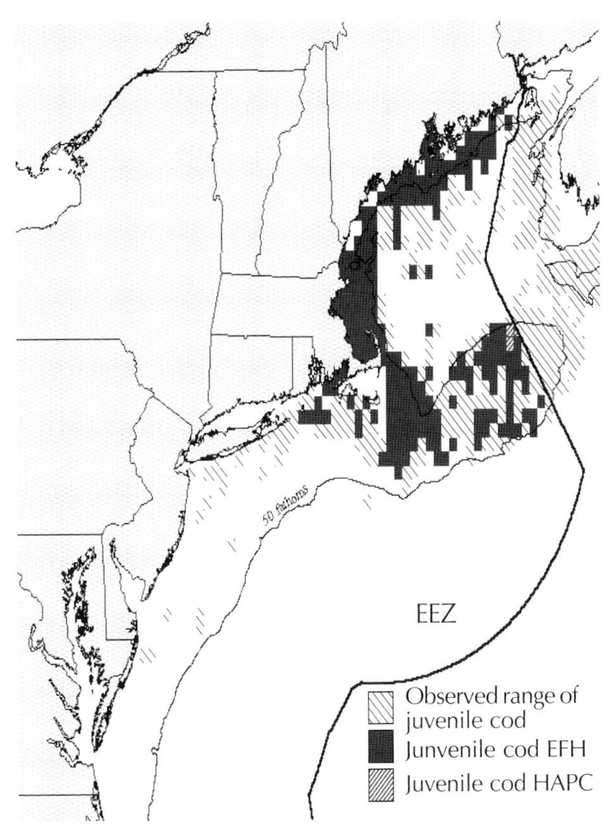

Figure 9.1 Essential fish habitat (EFH) and habitat areas of particular concern (HAPC) for juvenile cod within the exclusive economic zone (EEZ) of the Northeast United States. Data expressed in terms of 30′ squares (NEFMC 2003).

gions where the image of the ocean as a container for fish stocks is replaced by a heterogeneous "landscape" of habitats, species domains, zones of species' interactions, and community resource areas.

The EFH designation, however, is primarily dependent upon existing fisheries surveys where data on fish presence or abundance has been collected by using a simple, habitat-based spatial sampling scheme. Survey locations, initially designed to assess numbers of fish, are used to map fish habitat based on the assumption that fish are found within habitats essential to them. In addition to these spatial limitations, there also are temporal problems to consider. Associating EFH with presence will not capture the habitat that is no longer inhabited to the degree it may have been in the past. For example, local fishers have identified several historic spawning grounds for Atlantic cod *Gadus morhua* and haddock *Melanogrammus aeglefinus* in the Gulf of Maine, areas that are not now significant spawning grounds but may prove important to the reestablishment of cod stocks (Ames 1997; see also Graham et al. 2002). While other data also have been incorporated into current EFH designations for specific fishery management plans, they often suffer from similar spatial and temporal problems, and there are clearly "data gaps" relative to EFH (NEFMC 1998). The resultant maps are rough approximations of a species' EFH despite the use of GIS techniques for compilation and visualization.

The limitations of standard data sets for EFH designation, however, make clear the need to develop and incorporate other, intentionally spatial, fisheries data. Data sources such as targeted biological surveys that sample smaller areas at a finer resolution, detailed locational data from boat tracking devices, vessel trip logbook information, and even satellite imagery already have proven compatible with GIS and are potentially useful to management. Geographic information systems, however, also can be used to derive new layers of information from existing sources. For example, maps that show distance and travel time from features such as ports, measures of effort by location, or the territories of groups of boats can be calculated easily (Caddy and Carocci 1999).

Perhaps the most promising source of local information for GIS analysis is that which is available from fishers themselves. Many have noted the potential of fishers to supply local environmental information or social and economic information (e.g., the historic study by Ames 1997; Pederson and Hall-Arber 1999; Maurstad 2000). While there are convincing examples of initiatives based on fishers' knowledge (e.g., Hutchings and Ferguson 2000), few have detailed the types of data available or the methods necessary to collect it relative to standard fisheries variables and scientific methods. An exception is the work by Neis et al. (1999), who point to three types of fishers' knowledge that are obtainable by using interview methods: information on seasonal and directional fish movements, information pertaining to stock structure, and catch rate data that indicates local abundance.

Curiously, most examples of fishers' knowledge for fisheries management, whether community or science driven, cite the use of maps combined with qualitative interviews as an important (if not primary) tool for data collection.

> During the interviews, I used a map of the area with a mylar overlay, on which place names, areas associated with certain behaviors, paths and directions associated with migratory and other movements, bathymetry, and other information were recorded. The map was an invaluable aid to the interview, since the depiction of the physical world provided a common reference point. (Huntington 1998:239)

Fishers' local knowledge is place-bound and specific, and mapping makes this information tangible, as well as appropriate for use in GIS. Once mapped and digitized, data from multiple fishers can be merged, corroborated, and analyzed in a variety of ways (e.g., Graham et al. 2002). Unfortunately, it is unlikely that fishers' knowledge will be used while regional scale surveys remain institutionally dominant even for EFH. At the regional level, where the focus remains on aggregate numbers of fish, fishers' local knowledge is largely superfluous (Wilson et al. 1994). The EFH initiative, as a subregional and qualitative assessment, might benefit from the incorporation of fishers' knowledge concerning the habits and habitats of fish (cf. Manson and Die 2001 on MPAs).

The use of GIS, even with poor resolution data, has contributed to the analysis of EFH. For example, in many EFH designations, survey data were overlaid with layers of information such as bathymetric data, proximity to coastal features, and other expert knowledge. To these quantitative layers could be added other vital qualitative information such as fishers' local knowledge. In addition, GIS database development might combine not only layers of environmental information but social and economic information as well (e.g., revenues, boat characteristics, travel distances, and crew sizes by location). The potential of GIS for EFH designation or other initiatives, however, will be difficult to realize without a fundamental shift in current methods of data collection.

9.4 ANALYTICAL TECHNIQUES AND THEIR APPLICATION TO MANAGEMENT

Fisheries analysis, built upon standard data collections, involves numeric equilibrium models devoid of spatial variation (Caddy 1996). These models assume a single homogenous space within which the variables of overall stock and fishing effort exist (Wilson et al. 1999). While fisheries science is much broader than this characterization, most fisheries management is based upon these relatively simple and nonspatial models (Booth 2000). In addition, the poor spatial resolution of standard data sets makes it difficult to rework fisheries models to include a spatial component. That is, modeling for multiple areas (subregions) as opposed to one large area requires finer data sets (e.g., more biological survey locations over smaller areas) than are available. These models are continuous with the initial assumptions that drive data collection (i.e., individuals on a homogenous commons) and leave out other ontological categories such as communities and ecosystems. These other categories are not represented within the equilibrium equations where variables for overall fishing effort and fish populations are balanced.

While traditional models are not designed to incorporate multiple areas and their differences, GIS (and spatial analysis generally) is a tool for exploring differences from one location to the next. A GIS focuses on the relationships between locations as opposed to averaging or smoothing values from many locations as is practiced in standard fisheries models. These institutionalized forms of analysis will be difficult to change. In the United States, for example, laws exist that mandate single species assessments of overfishing for entire management regions (e.g., Sustainable Fisheries Act [SFA], Public Law 104-297, 104th Congress, 11 October 1996). Approaches that are subregional or consider other factors and species coincident in space are a lower priority than modeling the stock overall. Classic equilibrium models, excellent for aggregate numbers, are difficult to adapt to area-based needs given the data traditionally provided (Caddy 1996).

While standard data and forms of analysis make the use of GIS unlikely, these standards are changing and more spatial considerations are emerging. The already implemented spatial approaches to management such as EFH and MPAs are using GIS for basic visualization functions but could be greatly enhanced by the use of spatial analytical techniques also available through GIS. In addition, possible future approaches focused explicitly on communities or ecosystems will likely depend upon GIS as essential to their incorporation into fisheries analysis and decision making. A GIS will make it easier to ask questions such as: What resource areas are most important to this community of fishers? Where is environmental change most rapidly occurring? These and other spatial questions might be answered by using standard forms of spatial analysis such as database query, distance calculations, context operations, spatial statistics, and multicriteria evaluation for decision making.

Database query is a fundamental yet very powerful technique in GIS that allows the user to query by either location (what is happening in this location) or by attribute (show me all the locations that have this characteristic). Social impact analyses of management initiatives is an example where such spatial queries already have informed decision making and may become standard practice (McCay et al. 2002a, 2002b). This technique particularly is applicable when the management initiative in question is "area-based" (e.g., the "rolling closures" in the Gulf of Maine or the rotational closures in the fisheries for scallops *Placopecten magellanicus* of the Northeast) and the social (and economic) impacts will likely vary from one community to the next (Edwards et al. 2001; Walden et al. 2001).

Database query, of course, depends upon a quality database that contains information pertinent to the application. Analyses that attempt to assess the social impact of an area closure, for example, need regularly collected data that link offshore activities or locations to onshore communities. To assess the impact of an area closure, reliable data that can be used to find the communities that have been historically dependent upon access to the area is essential. Vessel trip report data collected by the National Marine Fisheries Service (NMFS) provides information about individual fishing trips by boats carrying federal fishing permits. The information includes species and quantity caught, number of crew members, and, importantly, the trip location. While these data are self-reported by fishers and contain substantial errors, in the aggregate, it is a good estimate of the relative use of particular resource areas by particular communities of fishers, and it has been incorporated into at least two social impact analyses in the mid-Atlantic region (McCay et al. 2002a, 2002b). These analyses query the trip report data for all previous fishing trips in the area to be closed and then characterize the attributes of those trips (e.g., species caught, quantity, length of trip, etc.) and the boats involved (e.g., gear type, average size, and, importantly, port association, which provides the link between offshore trip location and onshore community). The result is a series of tables and maps characterizing the fishing activity in the potentially closed area and the dependence of ports on that area (Figure 9.2).

Vessel tracking data obtained from tracking devices mounted on boats and other spatially coded information collected by NMFS also could contribute to spatial database query related to decision making. Community-based data (e.g., information from fishers concerning fishing grounds), once digitized and incorporated into a GIS database, also could be used. The latter could verify and add to information collected by the government, as well as provide unique informa-

Figure 9.2 Communities and ports (pie charts) potentially impacted by a proposed closure (option 3; gray shaded polygon) of fishing grounds to *Loligo* squid and other fishing. Analysis based on vessel trip report data (McCay et al. 2002a). The size of each pie chart corresponds to the port's percentage contribution to the total quantity of *Loligo* squid reported in 2000. Dark shaded areas of pie charts = percent value from option 3; gray shaded areas of pie charts = percent value outside option 3. Pie charts only approximate actual port locations. Black dots are trips within option 3 area; gray dots are trips outside option 3 area. For the lines passing through the area, the top line is the 50 fathoms mark, the middle line is the 100 fathoms mark, and the bottom line is the 500 fathoms mark. Data from state, county, and depth files; Northeast Fisheries Science Center; and Environmental Systems Research Institute.

tion available nowhere else (e.g., environmental knowledge about specific locations). A GIS lends itself to the combination of disparate layers of information making possible, for example, queries that regularly use collected vessel trip data and spatial data from interviews with fishers.

In addition to database query, other standard forms of GIS analysis could be extremely useful for fisheries decision making. Distance and cost–distance surface calculation can provide information about the accessibility of fisheries resources to particular communities and ports. This is a type of derivative mapping where new layers of information are derived from existing layers of information. Distance surfaces simply record the distance to some feature (e.g., a port) from all other locations (e.g., fishing grounds). A cost–distance surface is similar but incorporates frictions to travel, making the modeling of fishers' spatial search or travel behavior more realistic. For example, physical barriers to movement such as currents or landforms, as well as boat- or port-based constraints, such as boat size or fuel prices, can be integrated into an analysis that estimates the relative cost of reaching particular fishing grounds from a specific port.

Distance and cost surfaces can be used to predict fishing effort and its distribution offshore for particular ports, to indicate which ports may have more or less capacity relative to available (i.e., spatially accessible) resources, and to reveal areas of potential overlap or conflict between ports (see Caddy and Carocci 1999 for a detailed example). As fisheries management becomes more area based (e.g., the "rolling closures" implemented in the Gulf of Maine), there is a growing need to assess the differential impacts of management on particular communities and ports due to resource access and constraints on mobility. This type of GIS-based location/allocation modeling is a growing and important field of inquiry (e.g., Densham and Rushton 1996) but with few applications to fisheries.

Other forms of spatial analysis also might prove useful to fisheries management. Vessel trip report data, essential to impact analysis of closed areas (see above), also provide trip locations, which can be analyzed for their patterns. Point pattern analysis includes functions that can recognize significant clusters of fishing trips, produce maps of trip or catch density by location (Figure 9.3), and calculate the "home range" or probability that a port or other group of boats will use an area. These functions appear as generic tools within many GIS (e.g., the "Spatial Analyst" tool for ESRI's ArcView software or modules for distance and context operations found in Clark University's IDRISI software). They also are found in more specialized software packages such as CrimeStat for analysis of crime patterns (U.S. Department of Justice 2003), Animal Movement for the spatial analysis of wildlife movements and home range (U.S. Geological Survey 2003), or any of several similar stand-alone packages.

Point pattern functions can be used to derive maps of discrete areas representing resource locations associated with a set of trips by a particular set of boats (e.g., areas used by the inshore trawler fleet of

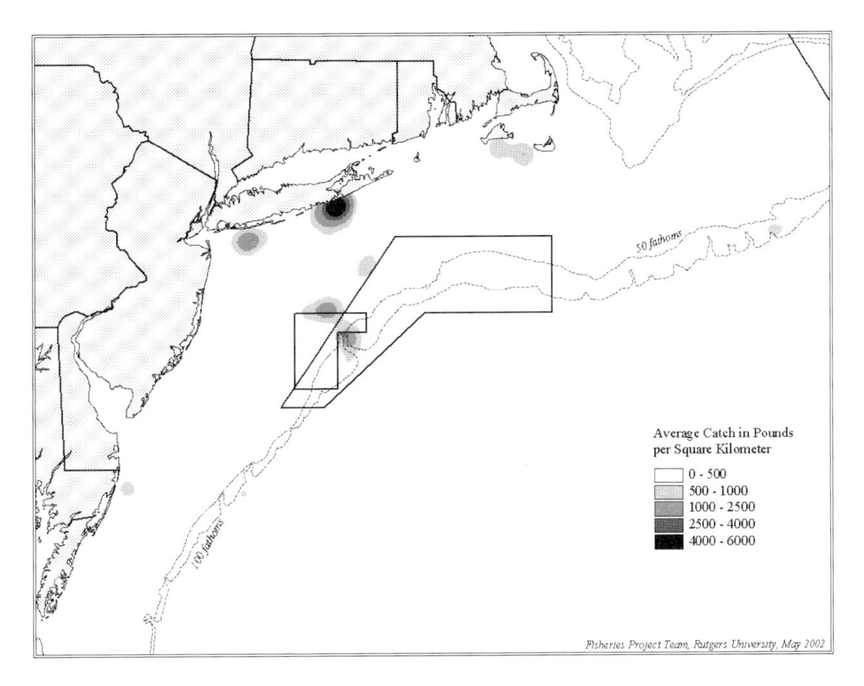

Figure 9.3 Density surface produced by using a density analysis of vessel trip report data from 2000 for all boats catching *Loligo* squid (Rutgers University Fisheries Project). Solid lines define two potential area closures. Density was calculated using a 20-km (radius) kernal and a Gaussian density function.

Massachusetts in 2001, see Figure 9.4). These areas can be calculated in terms of not just frequency of use but quantity caught, value, crew size, for example, and can be interpreted as, for example, the resource areas upon which a community or port is most dependent. The mapping of such areas, however, must not be considered complete given just a single (and disputed) data source such as vessel trip reports; other information, such as from fisher interviews, is essential to either corroborate or correct patterns derived from vessel trip data alone.

Once discrete areas representing the areas regularly frequented by various groups of fishers are derived, they can be analyzed further by using pattern analysis techniques that are typical in terrestrial wildlife studies and landscape architecture. These techniques can be used to measure the diversity of a particular resource area in terms of gear type, boat type, or port participation. Indexes of overlap, potential conflict, port dominance, and even fishing effort could be developed for specific areas, information vital to area-based decision making and fisheries management.

So far, GIS techniques have been discussed in terms of the production of "social" layers of information (i.e., layers representing a heterogeneous landscape of fishing and fishing practices) and their incorporation into fisheries decision making via GIS. Clearly, information on the locations of fishing by particular communities or ports and the characterization of those locations is essential to impact analysis

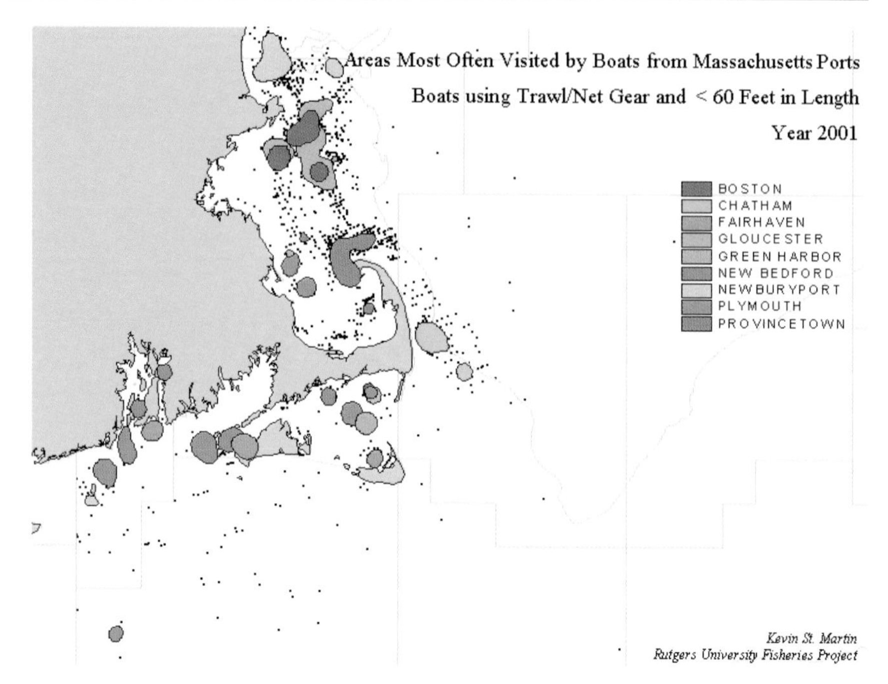

Figure 9.4 Home range analysis of inshore trawl boats whose principal port is located in Massachusetts. Ports with fewer than four boats excluded. Map produced for the Rutgers University Fisheries Project.

(see above). These layers, however, also might become important to analyses that combine social with biological considerations. That is, rather than just considering the social landscape in terms of the social and economic impact of particular management plans, the social landscape could become integral to the analyses upon which management plans are built.

By using GIS analytical techniques, layers of information representing various aspects of the social landscape can be produced and used in combination with biological and physical data layers. Doing so, however, leads to using the GIS for more than just database query or derivative mapping but as a tool for modeling complex interactions of social and biological or physical processes over space. Standard stock assessment models could be adapted to a GIS environment where they would operate on a more localized basis. This would allow the integration of layers of information pertaining to the spatial differentiation of management areas due to "social factors" such as proximity to ports, vessel range, port capacity, differential fishing practices, fishers' local environmental knowledge, and other localized variables. Within the GIS, data layers (biological or social) can then be combined as variables in equations by using map algebra techniques.

More interactive modeling techniques that are built upon map algebra concepts and tools are becoming available within GIS. For example, methods that combine data layers within an explicit decision-making or modeling environment increasingly are integrated into

GIS software (e.g., the IDRISI software). Data layers are thought of as criteria for decision making and can be developed, standardized, weighted relative to each other, and, finally, aggregated in a number of ways so as to manage the level of trade-off between criteria and overall level of risk (Malczewski 1996; Malczewski et al. 1997; Jiang and Eastman 2000). These developments in GIS software are finding application in resource and wildlife management and are proving useful for conflict resolution related to resource management (e.g., Brown et al. 1994, 2000; Kyem 2001).

In marine fisheries, there are few examples of the application of these techniques for fisheries management; there are, however, related examples. Wright et al. (1998) applied multicriteria evaluation techniques to locate the most suitable sites for an artificial reef in the Moray Firth of Scotland. Their model considers a range of physical, biological, and social variables (e.g., bathymetry, seabed composition, fisheries, other species, recreation, and commercial fishing locations). Other fisheries-related references to the use of multicriteria tools include Kitsiou et al. (2002) on ranking coastal areas, Brown et al. (2000) on habitat suitability for particular species, Garibaldi and Caddy (1998) on the characterization of faunal provinces in the Mediterranean, and Rubec et al. (1998) on determining EFH in Florida. For fisheries management per se, multicriteria analysis could be applied to any decision-making problem that involves site suitability analysis such as determining EFH, MPAs, or closure areas. Multiple criteria (i.e., layers of biophysical as well as social and economic data) can be weighed against each other and aggregated to create new maps representing the level of suitability for all locations (cf. Eastman et al. 1995). This result may then be input into further analysis focused on assessing the impacts of closing the most suitable areas (e.g., what boats will be displaced).

A GIS also incorporates techniques and heuristics that allow decision makers to interactively alter the criteria considered, the weights given to individual criteria, and the formulas used for combinations of criteria. The presumption is that of an iterative decision-making process, where the GIS is used to examine initial outcomes, assess their impacts, and readjust criteria weights and aggregation formulas (in terms of the desired levels of trade-off and risk) until the results are acceptable to all participants (Eastman et al. 1998). In addition to techniques that can handle multiple criteria, GIS also can accommodate multiple objectives, as well as "fuzzy" thresholds of suitability (Jiang and Eastman 2000).

Multicriteria evaluation methods could be combined with existing models that predict the effects of area closures. Walden et al. (2001) have produced a model for the Gulf of Maine that uses existing NMFS data and a GIS interface. It allows users to select areas for potential closure and returns projected mortality reductions as well as revenue losses by a fishing fleet (where "fleet" is associated with a regional port grouping). The strength of the model is its easy-to-use interface and emphasis on iterative scenario testing; multiple areas can be closed and results quickly returned in an iterative process accessible to managers.

This model is a progressive step toward the greater use of GIS for analysis that goes beyond simple mapping and database query. Using GIS to assess both biological and social and economic impacts of different management scenarios speaks to the potential of this technology within (and not necessarily before) the decision-making process. Further adjustment of the model would add to this potential. For example, finer resolution data on catch per effort by particular fishing communities (perhaps only available from fishers themselves), weighting criteria that are social, economic, and biological, or adjusting the model for varying levels of trade-off and risk could be incorporated.

Fisheries management has been institutionalized such that it focuses primarily on large areas and sustaining quantities of fish by single species. It is typically numeric (e.g., annual total allowable catches or quota systems) and, as a result, does not require GIS; spatial considerations insofar as they are part of management are broad (e.g., mesh regulation areas) and do not involve the spaces of either ecosystems or communities per se. While the main concern of management is to maximize the sustainable return available from fisheries for the management area as a whole, it also must balance the various competing demands of the industry itself. These demands can be sectoral (e.g., gear type), but they also can be location or community specific (e.g., port-based communities of fishers), implying a role for GIS at the level of management (e.g., incorporating iterative decision-making tools and heuristics), as well as fisheries analysis.

9.5 THE IMPLICATIONS OF A GIS APPROACH TO FISHERIES MANAGEMENT

The regular use of GIS for fisheries science and management will not occur without institutional change at several levels. This chapter has suggested that practices at all levels (e.g., data acquisition, scientific analysis, and management and decision making) must be modified to accommodate spatial concerns. Such changes would, indeed, constitute a paradigm shift whose effect will likely go beyond new forms of data acquisition, analysis, and an increased sensitivity to spatial considerations. A spatial approach suggests new ways of thinking about fisheries (e.g., communities and ecosystems) that are unimaginable given the current focus on individual fishermen and single species of fish. In particular, debates concerning property rights to fisheries resources (a central issue for all who are engaged in imagining a sustainable fisheries management system) might be enriched by a spatial focus and a concomitant consideration of communities and ecosystems.

The current regime of science and management constructs fisheries as essentially problematic; there are always "too many fishermen chasing too few fish." Within this conceptualization, individuals and their behavior relative to single species are the cause of resource depletion. The solutions to this problem, informed by bioeconomic theory, are obvious and limited: management can continue its unsustainable attempt to reduce effort by restricting technology and limiting aggregate harvests, or management essentially can privatize the resource by implementing quotas allocated to individuals (i.e., individually transferable quotas [ITQ], which are currently unpopular within fishing communities). If the problem of fishing is, however, recast in spatial rather than numeric terms (for example, in terms of communities and ecosystems), alternative solutions to the "problem of fishing" based on communities and their territories rather than individual private property may emerge (St. Martin 2001).

Community-based or comanagement systems that would make ITQs unnecessary (Pinkerton 1989; Leal 1996; McCay and Jentoft 1996) are unimaginable within federally managed fisheries, precisely because these fisheries are constituted only in terms of individual fishermen competing on a homogenous commons. The spatial heterogeneity that results from localized and particular community practices is obfuscated given current systems of data collection and scientific analysis that are the foundations for management. While revealing the heterogeneity of fisheries (social, biological, and physical) via GIS does not necessitate a movement toward community-based management strategies, it does provide a basis for such strategies that did not exist previously.

The growing need for more localized data (in terms of local habitat and environment, as well as local community analyses) suggests a participatory role for communities in both science and management. Fishers can act as sources of valuable localized information within a regime of trust and mutual benefit. Similarly, community members can provide information necessary for informed and reliable impact analyses at the community level, information that is simply not otherwise available within current federal and state databases. Importantly, GIS increasingly is associated with community participation in both science and assessment of policy impacts (e.g., Cornett 1994; Sheppard 1995; Kyem 2002). Public participatory GIS case studies and experiences (e.g., Craig et al. 2002) exist in a number of urban settings and rural-resource settings and document a growing potential for GIS to act as a bridge between a spatially knowledgeable public development and science-policy development. Several papers on public participatory GIS were compiled and made available through the Varenius Project of the National Center for Geographic Information and Analysis (NCGIA 2003). These examples will likely prove invaluable to both fishing communities and fisheries scientists and managers as the use of GIS in fisheries grows and eventually is institutionalized.

9.6 CONCLUSION

Fisheries science and management in the United States and elsewhere has been institutionalized to develop sustainable fish resources involving individual species located within a homogeneous and region-wide commons. Under these conditions, it is not surprising that the adoption of GIS has been slow. Significant institutional changes must take place before GIS will be widely used for marine fisheries science and management. The direction of change is, perhaps, indicated by current calls for a "paradigm shift" and by recent initiatives to accommodate a new administrative focus on communities and ecosystems. These pressures have in common a distinct spatialization of fisheries and a growing imperative to focus on more localized forms of science and management; they indicate not only a potential for change but for the effective use of GIS.

The emergence of new geographies of ecosystems and communities in fisheries science and management suggests a growing need to compile data from several sources, to focus analyses on local areas, and to provide mechanisms for multiobjective and multicriteria decision making; in short, it suggests a need for GIS. This need, however, can only be met if local data and knowledge of the environment is collected and integrated with standard data, if analyses are broadened to include spatial processes that are both environmental and social, and if management can be adjusted to work more closely with local fishing communities (e.g., in terms of localized data collection and enforcement).

As indicated by Meaden (1999), GIS is a promising technology for fisheries science and management. However, the promise of GIS goes beyond supplementing current numeric methods with a new technology, it implies performing fisheries science and management in new ways at several institutional levels. The promise of GIS is that its incorporation into science and management might, at the same time, create new opportunities to combine social data with biological data, to enhance cooperation between fisher communities and fisheries science, and to make management more participatory and multiobjective.

9.7 REFERENCES

Ames, E. P. 1997. Cod and haddock spawning grounds in the Gulf of Maine. Island Institute, Rockland, Maine.

Booth, A. J. 2000. Incorporating the spatial component of fisheries data into stock assessment models. ICES Journal of Marine Science 57:858–865.

Botsford, L., J. C. Castilla, and C. H. Peterson. 1997. The management of fisheries and marine ecosystems. Science 277:509–515.

Brown, S., H. Schreier, W. A. Thompson, and I. Vertinsky. 1994. Linking multiple accounts with GIS as decision support system to resolve forestry/wildlife conflicts. Journal of Environmental Management 42:349–364.

Brown, S. K., K. R. Buja, S. H. Jury, M. E. Monaco, and A. Banner. 2000. Habitat suitability index models for eight fish and invertebrate species in Casco and Sheepscot bays, Maine. North American Journal of Fisheries Management 20:408–435.

Caddy, J. F. 1996. Regime shifts and paradigm changes: is there still a place for equilibrium thinking? Fisheries Research 25:219–230.

Caddy, J. F., and F. Carocci. 1999. The spatial allocation of fishing intensity by port-based inshore fleets: a GIS application. ICES Journal of Marine Science 56:388–403.

Clay, P. M. 1996. Management regions, statistical areas and fishing grounds: criteria for dividing up the sea. Journal of Northwest Atlantic Fisheries Science 19:103–125.

Clay, P. M. 1997. Fishermen views on property in the northeast fishery: private versus common, quotas versus territory. National Marine Fisheries Service, Woods Hole, Massachusetts.

Committee on Evaluation, Design, and Monitoring of Marine Reserves and Protected Areas in the United States. 2001. Marine protected areas: tools for sustaining ocean ecosystems. National Academy Press, Washington, D.C.

Cornett, Z. J. 1994. GIS as a catalyst for effective public involvement in ecosystem management decision-making. *In* V. A. Sample, editor. Remote sensing and GIS in ecosystem management. Island Press, Washington D.C.

Costanza, R., F. Andrade, P. Antunes, M. van den Belt, D. Boersma, D. F. Boesch, F. Catarino, S. Hanna, K. Limburg, B. Low, M. Molitor, J. G. Pereira, S. Rayner, R. Santos, J. Wilson, and M. Young. 1998. Principles for sustainable governance of the oceans. Science 281:198–199.

Craig, W., T. Harris, and D. Weiner, editors. 2002. Community participation and geographic information systems. Taylor and Francis, London.

Densham, P. J., and G. Rushton. 1996. Providing spatial decision support for rural public service facilities that require a minimum workload. Environment and Planning B: Planning and Design 23:553–574.

Eastman, J. R., H. Jiang, and J. Toledano. 1998. Multi-criteria and multi-objective decision making for land allocation using GIS. *In* E. Beinat and P. Nijkamp, editors. Multicriteria analysis for land-use management. Kluwer Academic Publishers, Dordrecht, The Netherlands.

Eastman, J. R., W. Jin, P. A. K. Kyem, and J. Toledano. 1995. Raster procedures for multi-criteria/multi-objective decisions. Photogrammetric Engineering and Remote Sensing 61:539–547.

Edwards, S. F., B. Rountree, J. W. Walden, and D. D. Sheehan. 2001. An inquiry into ecosystem-based management of fishery resources on Georges Bank. Pages 202–214 *in* T. Nishida, P. J. Kailola, and C. E. Hollingworth, editors. Proceedings of the first international symposium on geographic information systems (GIS) in fishery science. Fishery GIS Research Group, Saitama, Japan.

EPAP (Ecosystem Principles Advisory Panel). 1999. Ecosystems-based fisheries management: a report to Congress. National Oceanic and Atmospheric Administration, National Marine Fisheries Service, Silver Spring, Maryland.

Fisher, W. L., and C. S. Toepfer. 1998. Recent trends in geographic information systems education and fisheries research applications at U.S. universities. Fisheries 23(5):10–13.

Fogarty, M. J., and S. A. Murawski. 1998. Large-scale disturbance and the structure of marine systems: fishery impacts on Georges Bank. Ecological Applications 8(1):S6–S22.

Garcia, S. M., and M. Hayashi. 2000. Division of the oceans and ecosystem management: a contrastive spatial evolution of marine fisheries governance. Ocean and Coastal Management 43:445–474.

Garibaldi, L., and J. F. Caddy. 1998. Biogeographic characterization of Mediterranean and Black seas faunal provinces using GIS procedures. Ocean and Coastal Management 39:211–227.

Graham, J., S. Engle, and M. Recchia. 2002. Local knowledge and local stocks: an atlas of groundfish spawning in the Bay of Fundy. The Centre for Community-Based Management, Extension Department, St. Francis Xavier University, Antigonish, Nova Scotia.

Gulland, J. A. 1984. Fisheries: looking beyond the golden age. Marine Policy 8:137–150.

Hall, S. 1998. Closed areas for fisheries management—the case consolidates. Tree 13:297–298.

Hardin, G. 1968. The tragedy of the commons. Science 162:1243–1248.

Huntington, H. P. 1998. Observations on the utility of the semi-directive interview for documenting traditional ecological knowledge. Arctic 51:237–242.

Hutchings, J. A., and M. Ferguson. 2000. Links between fishers' knowledge, fisheries science, and management: Newfoundland's inshore fishery for northern Atlantic cod, *Gadus morhua*. *In* B. Neis and L. Felt, editors. Finding our sea legs: linking fishery people and their knowledge with science and management. ISER Press, St. John's, Newfoundland.

Isaak, D. J., and W. A. Hubert. 1997. Integrating new technologies into fisheries science: the application of geographic information systems. Fisheries 22(1):6–10.

Jentoft, S. 1999. Healthy fishing communities: an important component of healthy fish stocks. Fisheries 24(5):28–29.

Jiang, H., and J. R. Eastman. 2000. Application of fuzzy measures in multicriteria evaluation in GIS. International Journal of Geographical Information Systems 14:173–184.

Kitsiou, D., H. Coccossis, and M. Karydis. 2002. Multi-dimensional evaluation and ranking of coastal areas using GIS and multiple criteria choice methods. Science of the Total Environment 284:1–17.

Kyem, P. A. K. 2001. Embedding GIS applications into resource management and planning activities of local communities: a desirable innovation or a destabilizing enterprise? Journal of Planning 20:176–186.

Kyem, P. A. K. 2002. Promoting local community participation in forest management through a PPGIS application in Southern Ghana. Pages 218–231 *in* W. Craig, T. Harris, and D. Weiner, editors. Community participation and geographic information systems. Taylor and Francis, London.

Langton, R. W., P. J. Auster, and D. C. Schneider. 1995. A spatial and temporal perspective on research and management of groundfish in the Northwest Atlantic. Reviews in Fisheries Science 3:201–229.

Leal, D. R. 1996. Community-run fisheries: avoiding the 'Tragedy of the Commons.' *In* J. S. Shaw, editor. PERC Policy Series, Issue Number PS-7, Bozeman, Montana.

MacCall, A. D. 1990. Dynamic geography of marine fish populations. University of Washington Press, Seattle.

Malczewski, J. 1996. A GIS-based approach to multiple criteria group decision-making. International Journal of Geographical Information Systems 10:955–971.

Malczewski, J., M. Pazner, and M. Zaliwska. 1997. Visualization of multicriteria location analysis using raster GIS: a case study. Cartography and Geographic Information Systems 24:80–90.

Manson, F. J., and D. J. Die. 2001. Incorporating commercial fishery information into the design of marine protected areas. Ocean and Coastal Management 44:517–530.

Maurstad, A. 2000. Trapped in biology: an interdisciplinary attempt to integrate fish harvesters' knowledge into Norwegian fisheries management. *In* B. Neis and L. Felt, editors. Finding our sea legs: linking fishery people and their knowledge with science and management. ISER Press, St. John's Newfoundland.

McCay, B. J., and S. Jentoft. 1996. From the bottom up: participatory issues in fisheries management. Society and Natural Resources 9:37–250.

McCay, B. J., B. Oles, B. Stoffle, E. Bochenek, K. St. Martin, G. Graziosi, T. Johnson, and J. Lamarque. 2002a. Port and community profiles, amendment 9, squid, Atlantic mackerel, and butterfish fishery management plan. Report to the Mid-Atlantic Fishery Management Council, Fisheries Project, Rutgers University, New Brunswick, New Jersey.

McCay, B. J., D. C. Wilson, J. Lamarque, K. St. Martin, E. Bochenek, B. Stoffle, B. Oles, and T. Johnson. 2002b. Port and community profiles and social impact assessment, amendment 13 of the surfclam and ocean quahog fishery management plan. Report to the Mid-Atlantic Fishery Management Council, Fisheries Project, Rutgers University, New Brunswick, New Jersey.

Meaden, G. J. 1999. Applications of GIS to fisheries management. *In* D. Wright and D. Bartlett, editors. Marine and coastal geographical information systems. Taylor and Francis, London.

Meaden, G. J., and T. D. Chi. 1996. Geographical information systems: applications to marine fisheries. Food and Agriculture Organization, Rome.

NCGIA (National Center for Geographic Information and Analysis). 2003. Papers submitted for the NCGIA specialist meeting on "Empowerment, marginalization, and public participation GIS." NCGIA. Available: *www.ncgia.ucsb.edu/varenius/ppgis/papers* (September 2003).

NEFMC (New England Fishery Management Council). 1998. New England Fisheries Management Council essential fish habitat amendment, volume 1. NEFMC, Newburyport, Massachusetts.

NEFMC (New England Fishery Management Council). 2003. New England fishery management council. NEFMC. Available: *www.nefmc.org* (September 2003).

Neis, B., D. C. Schneider, L. Felt, R. L. Haedrich, J. Fischer, and J. A. Hutchings. 1999. Fisheries assessment: what can be learned from interviewing resource users. Canadian Journal of Fisheries and Aquatic Sciences 56:1949–1963.

Pauly, D. 1997. Putting fisheries management back in places. Reviews in Fish Biology and Fisheries 7:125–127.

Pederson, J., and M. Hall-Arber. 1999. Fish habitat: a focus on New England fishermen's perspectives. Pages 188–211 *in* L. R. Benaka, editor. Fish habitat: essential fish habitat and rehabilitation. American Fisheries Society, Symposium 22, Bethesda, Maryland.

Pierce, D. E., and P. E. Hugl. 1979. Insight into the methodology and logic behind National Marine Fisheries Service fish stock assessments. Massachusetts Division of Marine Fisheries, Coastal Zone Management Office, Boston.

Pinkerton, E., editor. 1989. Co-operative management of local fisheries. University of British Columbia Press, Vancouver.

Rubec, P. J., M. S. Coyne, R. H. McMichael Jr., and M. E. Monaco. 1998. Spatial methods being developed in Florida to determine essential fish habitat. Fisheries 23(7):21–25.

Sainsbury, K. 1998. Living marine resource assessment for the 21st century: what will be needed and how will it be provided? Pages1–40 *in* F. Funk, T. J. Quinn II, J. Heifetz, J. N. Ianelli, T. F. Powers, J. F. Schweigert, P. J. Sullivan, and C. I. Zhang, editors. Proceedings of the international symposium on fishing stock assessment models for the 21st century. Lowell Wakefield Fisheries Symposium, Anchorage, Alaska.

Sheppard, E. 1995. GIS and society: towards a research agenda. Cartography and Geographical Information Systems 22:5–16.

Sherman, K., and H. R. Skjoldal, editors. 2002. Large marine ecosystems of the North Atlantic: changing states and sustainability. Elsevier, Amsterdam.

Smith, T. D. 1994. Scaling fisheries: the science of measuring the effects of fishing 1855–1955. Cambridge University Press, Cambridge, UK.

St. Martin, K. 2001. Making space for community resource management in fisheries. Annals of the Association of American Geographers 91:122–142.

U.S. Department of Justice. 2003. National Institute of Justice: the research, development, and evaluation agency of the U.S. Department of Justice. U.S. Department of Justice. Available: *www.ojp.usdoj.gov/nij/* (September 2003).

U.S. Geological Survey. 2003. GIS tools. U.S. Geological Survey, Glacier Bay Field Station. Available: *www.absc.usgs.gov/glba/* (September 2003).

Walden, J. B., D. Sheehan, B. Pollard Rountree, and S. F. Edwards. 2001. Integrating GIS with mathematical programming models to assist in fishery management decisions. Pages 167–177 *in* T. Nishida, P. J. Kailola and C. E. Hollingworth, editors. Proceedings of the first international symposium on geographic information systems (GIS) in fishery science. Fishery GIS Research Group, Saitama, Japan.

Wilson, J., B. Low, R. Costanza, and E. Ostrom. 1999. Scale misperceptions and the spatial dynamics of a social-ecological system. Ecological Economics 31:243–257.

Wilson, J. A., J. M. Acheson, M. Metcalfe, and P. Kleban. 1994. Chaos, complexity and community management of fisheries. Marine Policy 18:291–305.

Witherell, D., C. Pautzke, and D. Fluharty. 2000. An ecosystem-based approach for Alaska groundfish fisheries. ICES Journal of Marine Science 57:771–777.

Wright, R., S. Ray, D. R. Green, and M. Wood. 1998. Development of a GIS of the Moray Firth (Scotland, UK) and its application in environmental management (Site Selection for an 'Artificial Reef'). Science of the Total Environment 223:65–76.

Chapter 10

Future of Geographic Information Systems in Fisheries

William L. Fisher

10.1 INTRODUCTION

Nielsen (1999) developed a conceptual model of fisheries management that included biological, engineering, sociological, economic, and political disciplines encompassed by the three overlapping components of a fishery: organisms, habitats, and people. Fundamental to this model is the overlap among the components, which provide opportunities for enhancing the value of fisheries resources (Nielsen 1999). By using Nielsen's concept, a model of fisheries geographic information systems (GIS) might be viewed as the overlay of maps and information on species distributions, aquatic habitats, and the effort by people (recreational and commercial fishermen) using fisheries resources (Figure 10.1). It is at the intersection of these three components that GIS can be extremely useful, for example, in relating fishing effort to habitat conditions and, thus, to fish species' distributions. The overlay of multiple maps is a fundamental GIS process that is increasingly improving our ability to manage fisheries resources.

The purpose of this chapter is to portend the future of fisheries GIS. To place this chapter in the proper context, I begin by briefly examining the history of GIS and, in particular, fisheries GIS. I then summarize what the chapter authors see as the future of fisheries GIS. Finally, I look at current trends in the GIS industry to discern how they might influence the future of fisheries GIS. It is my hope that this chapter will provide a springboard for the next generation of fisheries professionals who plan to use GIS as an aid in solving fisheries research and management problems.

10.2 BRIEF HISTORY OF GIS IN FISHERIES

Modern GIS was influenced by early 20th century landscape architects and planners in Europe and North America who performed rudimentary spatial analyses with multiple map overlays (Foresman 1998). The actual origin of GIS can be traced to the early 1960s when Dr. Roger Tomlinson and a group of Canadian geographers and com-

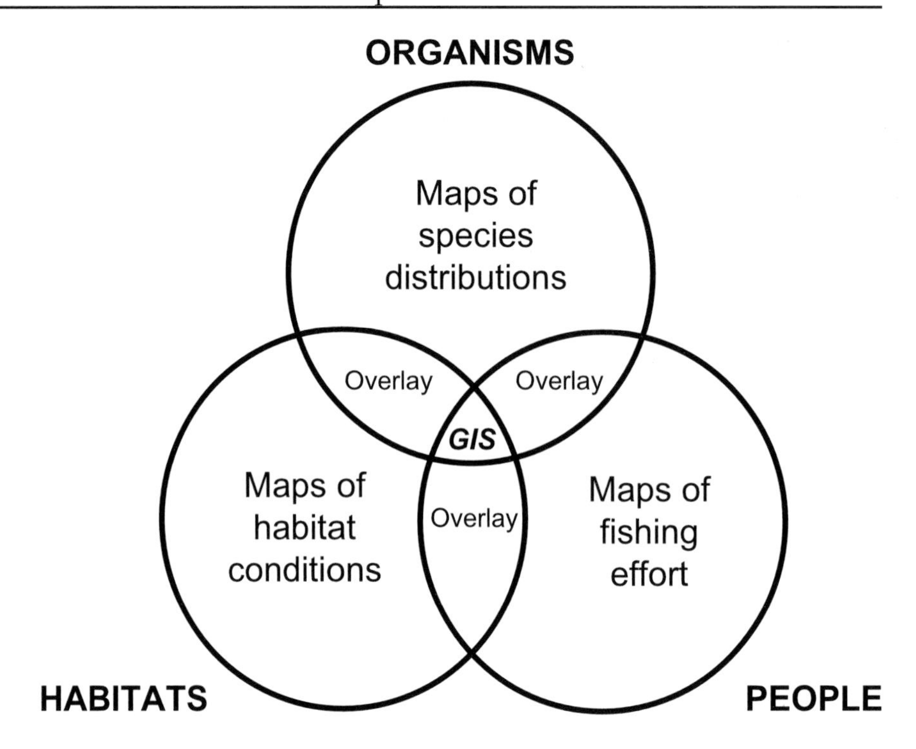

Figure 10.1 A conceptual model of fisheries GIS depicted as the overlay of maps and spatial information about the three components of a fishery: organisms, habitats, and people.

puter scientists developed the first GIS, the Canada Geographic Information System. And it was Tomlinson who coined the term "geographic information system." Fuel for development of GIS came from state, provincial, and federal agencies in North America that sought automated solutions for landscape-level resource inventories and assessments, primarily in response to greater involvement by the Canadian government in land-use management in the 1950s and early 1960s, and the environmental movement of the late 1960s and the early 1970s in the United States. Since then, GIS has evolved into a complex architecture of computer hardware, software, and data that is used in a variety of applications from land-use planning to fish conservation and fisheries management.

Meaden (2001) dates the first fisheries GIS publications to the mid-1980s. Mooneyhan (1985) developed a GIS and remote sensing exercise for a training course that estimated inland aquaculture potential in Southern Florida, and Caddy and Garcia (1986) promoted the use of GIS through visual-based mapping techniques for fisheries development and management. These early publications were followed by a boom in GIS fisheries publications in the 1990s (Meaden 2001), which has continued into the 21st century. Comprehensive compilations of GIS applications in fisheries include reports by Meaden and Kapetsky (1991) and Meaden and Chi (1996), proceedings from an international GIS symposium by Nishida et al. (2001), and a book on GIS in oceanography and fisheries by Valavanis (2002). Although the field of fisheries GIS has begun to gel, Meaden (2001), in his review of the literature, found a lack of uniformity or cohesion of the methods and standards in GIS applications and suggested that fisheries GIS exists more as an adaptation of technology toward solving problems in fisheries science rather than a recognized field of study. Nonetheless, the number and variety of GIS applications in both marine and freshwater fisheries and aquaculture have provided a foundation for the future of fisheries GIS.

10.3 FUTURE OF GIS IN FISHERIES

Each chapter in this book concludes with the authors' speculation about the emerging techniques, technologies, and issues in fisheries GIS. The chapter authors have provided their unique perspective on what they see as the future of GIS in fisheries research and management. What follows is a summary of their perspectives, grouped by the challenges and fishery activities in aquatic environments.

The growth of fisheries GIS entails overcoming challenges. Meaden identified four interrelated categories of challenges facing those working with GIS in aquatic environments (see Chapter 2). These were (1) intellectual and theoretical challenges of optimizing the mapping and visualization of variables that are dynamic at multiple spatial and temporal scales in three-dimensional (3-D) aquatic environments; (2) practical and organizational challenges of dealing with obtaining, organizing, and producing GIS data; (3) economic challenges of obtaining and sustaining funding to support fisheries GIS operations; and (4) social and cultural challenges to overcome organizational inertia, political boundaries, and to develop geographic cognition within the unique culture of organizations. Meaden concluded that the growth and expansion of fisheries GIS will occur as the cost of data is reduced through increased data-gathering technologies and data dissemination, better cooperation and organization among practitioners at institutions around the world, increased dissemination of GIS applications in fisheries in recognized and accessible publications and at a broader range of conferences, demonstrations of well-designed and well-engineered GIS projects, development of international standards, and increased progress toward data storage and modeling of 3-D and four-dimensional (4-D) GIS. Some of these challenges are being met through the increased availability of inexpensive yet powerful computers, user-friendly software, data distribution through the Internet, and integration of GIS into organizations for decision making.

The future of fisheries GIS in freshwater environments, such as streams, rivers, reservoirs, and lakes, will depend on obtaining environmental data more economically and integrating those data with spatially explicit population dynamic models. Creating accurate two-dimensional (2-D) and 3-D maps of freshwater habitats is labor intensive and time consuming. Increasing availability and use of remote-sensing instruments, such as low-altitude airborne scanners (to map water temperature, water depths, geomorphic channel units, and woody debris in streams and rivers; see Chapter 3) and satellite imagery, light detection and ranging (LIDAR) airborne scanners, multibeam, and side-scan sonar sensors (to map water depth and temperature, and the distribution of phytoplankton and emergent vegetation in lakes and reservoirs; see Chapters 4, 5) is improving our ability to efficiently gather large amounts of spatial data over long distances (e.g., streams and rivers) and large areas (e.g., reservoirs and lakes). Furthermore, relating temporal changes in freshwater fish population characteristics to watershed land-use activities will require periodic classification and analysis of changes in landscape features based on satellite imagery. However, remote-sensing data are expensive to acquire and process, and agencies and institutions will need to provide sufficient funds to support data-gathering missions. Future applications of GIS to freshwater fisheries management will be further enhanced by fisheries data collected in a spatially explicit manner by using locational receivers such as global positioning systems (GPS). Geographic locations of fish collections in specific habitats (stream channel units, lake and reservoir embayments) or of individual fish with telemetry transmitters are needed to determine spatial relationships among fish collections and angler locations, and to expand fish abundance estimates over larger areas (see Chapters 3 and 4). Coupling spatially explicit, dynamic models of fish population abundance with multidimensional (2-D, 3-D, 4-D) maps of freshwater environments is still in its infancy (see Chapter 5), but is critical for the advancement of fisheries GIS and our ability to forecast changes in populations and communities in response to environmental changes such as global warming.

Aquaculture operations occur in both freshwater and marine environments; however, most GIS applications in aquaculture have occurred in coastal areas (see Chapter 6). Geographic information systems have been used to estimate two criteria for aquaculture: (1) suitability of a site for a specific culture method, such as the production of fish in ponds, raceways, or floating cages; and (2) suitability of a site for the organism that is to be cultured (Kapetsky 2001). These applications are relatively well developed (see Chapter 6). Despite that, Kapetsky sees a largely undeveloped role for GIS in the development and management of aquaculture, particularly in geographic areas like the Far East where the value and production of aquaculture are high. Among the unrealized opportunities of GIS applications in aquaculture are the following: monitoring and predicting the environmental impacts of culture systems; evaluating sea-ranching aquaculture in coastal marine environments; socioeconomic analysis of culture systems, including planning, management, distribution, and marketing of aquaculture products; and creating Internet-based aquaculture information systems (see Chapter 6). However, before GIS in aquaculture can expand, administrators must be informed about the value of GIS and support technical staff and the GIS enterprise.

Fisheries in marine environments are the most spatially extensive economic activity in the world (Meaden and Chi 1996) and one of the most imperiled natural resources. Rapid technological advancements have not only improved the efficiency of fishing fleets but also the ability of scientists and managers to map and model marine environments and fisheries. These advancements, including GIS, have facilitated synergist relationships among geographers, oceanographers, and biologists (see Chapter 8). Marine scientists worldwide are using remote sensors on board satellites, airplanes, and vessels, and attached to moored and drifting buoys, and they are using rapid assessment techniques to survey and map marine environments (Valavanis 2002). Battista and Monaco (Chapter 7) suggest that the continued advancement of GIS applications for nearshore, coastal fisheries will depend on the development of more cost-effective and available remote-sensing technologies. New sensors (Ikonos; Sea-viewing wide field of view sensor, SeaWiFS; and Moderate-resolution imaging spectroradiometer, MODIS) and better algorithms for rapid data processing will enable both oceanographic and fisheries researchers to visualize complex relationships among fish populations and assemblages and their habitats in oceanic environments. Booth and Wood (see Chapter 8) believe that, in the future, the routine use of GIS to store, manipulate, and analyze marine data will improve management advise and decisions. Geographic information systems, coupled with intelligent databases, 3-D visualization, and various models (e.g., bioenergetic, statistical, mathematical) will improve the ability of scientists to forecast changes in marine fish populations and assemblages (see Chapter 7). However, none of this will be possible if there is not a commitment by the marine fisheries community to invest in data, technology, and training of students and professionals.

Physical and biological scientists focus on the organismal and habitat components of a fishery. Over the past decade, there has been increased emphasis by social and economic scientists on the human component of fisheries, particularly in freshwater environments. The human component, however, has long been an integral element of marine fisheries because of the strong dependence of fishing communities on the sustainability of the resource. St. Martin (see Chapter 9) envisions GIS playing a key role in the ongoing paradigm shift in marine fisheries from the current focus on individual fishermen competing for single species located in homogeneous common fishing grounds to a new focus on communities of fishers and fishes occupying spatially heterogeneous ecosystems. Geographic information systems increasingly are being used as a bridge between the public and policy makers by providing a medium for communicating science and assessing policy scenarios. St. Martin (see Chapter 9) suggests that the emergence of an ecosystem- and community-based approach in marine fisheries science and management will require compilation of data from multiple sources, a focus on analyses of local areas, and mechanisms for multiobjective and multicriteria decision making that can all be facilitated with the use of GIS. Incorporating GIS into science and management will likely provide opportunities to combine social and biological data, enhance cooperation between communities of fishers and fisheries professionals, and make fisheries management more participatory (see Chapter 9).

10.4 OPPORTUNITIES FOR FISHERIES GIS

The geospatial industry is rapidly evolving. A quick glance through the recent industry trade magazines, such as *Geoworld, Geospatial Solutions,* or *GPSworld,* finds natural resource-related articles on topics such as mapping wildlife species distributions by using interactive white boards and real-time digitizing; the use of LIDAR data to generate digital terrain models; the use of knowledge-based systems with landscape simulation models for ecological assessments and resource planning; the use of airborne scanners with high-resolution multispectoral sensors to map the Colorado River in the Grand Canyon, Arizona; and the use of GIS to model water demand on the lower Colorado River in Texas. The emphasis of many of these articles is the use of remote-sensing technologies to aid in natural resource management, which many authors in this book described as a need for future fisheries research and management. And although advances in technological tools and computer portability (e.g., mobile GPS and GIS systems, tablet personal computers, personal digital assistants, cellular phones) will continue to improve our ability to gather data, the emphasis is shifting toward making geospatial data accessible and usable by the public through the Internet.

Many of the GIS industry leaders see the distribution of GIS data and applications on the Internet and the creation of global standards as the biggest trend in the geospatial industry (Barnes 2003). Development of interoperability among disparate computer systems is the biggest challenge to distributing geospatial information on the Internet. However, there has been considerable progress on this front with the creation of OpenGIS through the Open GIS Consortium (OGC), an international industry consortium of 258 companies, government agencies, and universities (OGC 2003). The OGC uses a consensus process to develop publicly available geoprocessing specifications that support open interfaces and protocols among computer systems through OpenGIS. These specifications support interoperable solutions that enable the distribution of geospatial information through the Internet, wireless and location-based services, and mainstream information technology. The goal of the OGC is to empower technology developers to make complex spatial information and services accessible and useful with all kinds of applications on all computer platforms. This type of accessibility requires globally recognized data-exchange protocols for transferring information from server and client computers, including open-source software (e.g., Linux operating system, PostGIS database management system, Mapserver Internet map server) and interoperability standards (e.g., extensible markup language, XML; simple object access protocol, SOAP; universal description, discovery, and integration, UDDI; Web services description lanaguage, WSDL) standards for orchestrating Web services (WebServices 2002). The end result of this open environment will be the democratization of GIS, allowing access of geospatial information to people who are not just GIS professionals.

An open GIS environment for sharing geospatial information through the Internet will undoubtedly have an impact on the future of fisheries GIS. Internet-based information systems on fishes, such as FishBase (2003), a global information system on fishes; aquaculture, such as AquaGIS (2003; see Chapter 6), an information system on aquaculture for the Canadian provinces of Newfoundland and Labrador; and habitats, such as ReefBase (WorldFish Center and International Coral Reef Action Network 2003), an information system on coral reefs, all provide GIS-based mapping capabilities that are readily accessible.

Goodchild (1998) postulated that GIS is about to intrude into many aspects of the everyday life of people. The demand for accurate and accessible geospatial information will only increase as users become more aware of its utility. The challenge will be to provide reliable, up-to-date information. It is clear, however, that future management of freshwater and marine fisheries resources will require geospatial information and GIS technology that is not only accessible to the fisheries community, including fishers, fisheries scientists, and fisheries policy makers, but also to the public. Such access will hopefully increase awareness about the status of fisheries resources throughout the world and influence decisions about conserving them.

10.5 REFERENCES

AquaGIS. 2003. Newfoundland and Labrador aquaculture information and mapping. Government of Newfoundland and Labrador Fisheries and Aquaculture. Available: *http://gis.gov.nf.ca/aquagis/* (September 2003).

Barnes, S. 2003. Market map 2003: standards, open systems drive spatial marketplace. Geospatial Solutions 13(1):28–32.

Caddy, J. F., and S. Garcia. 1986. Fisheries thematic mapping—a prerequisite for intelligent management and development of fisheries. Oceanographie Tropicale 21:31–52.

FishBase Consortium. 2003. FishBase: a geographic information system on fishes. FishBase Consortium. Available: *www.fishbase.org* (September 2003).

Foresman, T. W. 1998. GIS early years and threads of evolution. Pages 3–17 *in* T. W. Foresman, editor. The history of geographic information systems: perspectives from the pioneers. Prentice Hall, Upper Saddle River, New Jersey.

Goodchild, M. F. 1998. What next? Reflections form the middle growth curve. Pages 369–381 *in* T. W. Foresman, editor. The history of geographic information systems: perspectives from the pioneers. Prentice Hall, Upper Saddle River, New Jersey.

Kapetsky, J. M. 2001. Recent applications of GIS in inland fisheries. Pages 339–359 *in* T. Nishida, P. J. Kailola, and C. E. Hollingworth, editors. Proceedings of the first international symposium on GIS in fishery science. Fishery GIS Research Group, Saitama, Japan.

Meaden, G. J. 2001. GIS in fisheries science: foundations for the new millennium. Pages 3–29 *in* T. Nishida, P. J. Kailola, and C. E. Hollingworth, editors. Proceedings of the first international symposium on GIS in fishery science. Fishery GIS Research Group, Saitama, Japan.

Meaden, G. J., and T. D. Chi. 1996. Geographical information systems: applications to marine fisheries. Food and Agriculture Organization of the United Nations, FAO Fisheries Technical Paper 356.

Meaden, G. J., and J. M. Kapetsky. 1991. Geographical information systems and remote sensing inland fisheries and aquaculture. Food and Agriculture Organization of the United Nations, FAO Fisheries Technical Paper 318.

Mooneyhan, W. 1985. Determining aquaculture development potential via remote sensing and spatial modeling. Pages 217–247 in Applications of remote sensing to aquaculture and inland fisheries. Report of the ninth UN/FAO international training course in Italy. FAO Remote Sensing Center, Series 27, Rome.

Nielsen, L. A. 1999. History of inland fisheries management in North America. Pages 3–30 *in* C. C. Kohler and W. A. Hubert, editor. Inland fisheries management in North America, 2nd edition. American Fisheries Society, Bethesda, Maryland.

Nishida, T., P. J. Kailola, and C. E. Hollingworth, editors. 2001. Proceedings of the first international symposium on GIS in fishery science. Fishery GIS Research Group, Saitama, Japan.

OGC (Open GIS Consortium). 2003. Open GIS Consortium, Inc. Available: *www.opengis.org* (September 2003).

Valavanis, V. D. 2002. Geographic information systems in oceanography and fisheries. Taylor and Francis, London.

WebServices. 2002. WebServices.Org—the web services industry portal. WebServices.Org. Available: *www.webservices.org* (September 2003).

WorldFish Center and International Coral Reef Action Network (ICRAN). 2003. ReefBase: a geographic information system on coral reefs. WordFish Center and ICRAN. Avaialble: *www.reefbase.org* (September 2003).

Index

F

G

H

I

K

L

O

P

R

W

Z